Polymer-Based Biomaterials and Tissue Engineering

Polymer-Based Biomaterials and Tissue Engineering

Editors

Roser Sabater i Serra
Ángel Serrano-Aroca

MDPI • Basel • Beijing • Wuhan • Barcelona • Belgrade • Manchester • Tokyo • Cluj • Tianjin

Editors

Roser Sabater i Serra
Centre for Biomaterials
and Tissue Engineering
Universitat Politècnica
de València
Valencia
Spain

Ángel Serrano-Aroca
Centro de Investigación
Traslacional San Alberto Magno
Universidad Católica de
Valencia San Vicente Mártir
Valencia
Spain

Editorial Office
MDPI
St. Alban-Anlage 66
4052 Basel, Switzerland

This is a reprint of articles from the Special Issue published online in the open access journal *Materials* (ISSN 1996-1944) (available at: www.mdpi.com/journal/materials/special_issues/polymer_biomaterials_tissue_engineering).

For citation purposes, cite each article independently as indicated on the article page online and as indicated below:

LastName, A.A.; LastName, B.B.; LastName, C.C. Article Title. *Journal Name* **Year**, *Volume Number*, Page Range.

ISBN 978-3-0365-8237-5 (Hbk)
ISBN 978-3-0365-8236-8 (PDF)

Contents

About the Editors

Roser Sabater i Serra

Roser Sabater i Serra is a Full Professor at the Universitat Politècnica de València (Spain) and a member of the Centre for Biomaterials and Tissue Engineering (CBIT-UPV). She is also an associate researcher at the Centro de Investigación Biomédica en Red in the subject area of Bioengineering, Biomaterials and Nanomedicine (CIBER-BBN, Spain). Her current research involves the design, synthesis, and physico-chemical and biological characterization of novel biomaterials for tissue engineering and regenerative medicine. Her other research interests include the development of biomaterials for the controlled release of bioactive factors and biomolecules. Her publications focus on the synthesis and characterization of the physico-chemical properties of polymeric biomaterials and hydrogel-based hybrid composites and scaffolds, as well as cell–biomaterial interaction in both 2D and 3D environments.

Ángel Serrano-Aroca

Ángel Serrano-Aroca is a Professor of Biotechnology and a Principal Investigator at the Biomaterials and Bioengineering Lab at the San Alberto Magno Translational Research Centre, Catholic University of Valencia Saint Vincent Martyr (CITSAM-UCV). The main line of research of his group consists of developing new antimicrobial materials for biomedical applications such as tissue engineering and drug delivery. His research focuses on bioengineering, biomaterials, nanobiotechnology, and biomedical engineering. Prof. Serrano-Aroca was enlisted in the "World Ranking of Top 2% Scientists" released by Stanford University (USA) in 2021 and 2022.

Preface to "Polymer-Based Biomaterials and Tissue Engineering"

Polymeric biomaterials are one of the key elements in tissue engineering and controlled drug release. In recent years, a wide range of biomaterials have been developed using different and more sophisticated approaches. This Special Issue aims to present novel developments of biomaterials based on polymeric matrices for biomedical applications, especially regenerative medicine, drug delivery, and surgical solutions, to address pathologies such as hernias. It includes several research studies performed in this area using different approaches, which could be of special interest for researchers working in the field of biomaterials.

We would like to thank all the authors and reviewers who have contributed to this Special Issue, as well as the MDPI staffs and editorial team of *Materials*. In particular, we would like to thank the assistant editor, Ms. Taylor Liang, for her invaluable help throughout the process leading up to the publication of this Special Issue.

Roser Sabater i Serra and Ángel Serrano-Aroca
Editors

materials

Editorial

Special Issue: "Polymer-Based Biomaterials and Tissue Engineering"

Roser Sabater i Serra [1,2,*] and Ángel Serrano-Aroca [3,*]

1 Centre for Biomaterials and Tissue Engineering, Universitat Politècnica de València, 46022 Valencia, Spain
2 Biomedical Research Networking Centre in Bioengineering, Biomaterials and Nanomedicine (CIBER-BBN), 46022 Valencia, Spain
3 Biomaterials and Bioengineering Lab, Centro de Investigación Traslacional San Alberto Magno, Universidad Católica de Valencia San Vicente Mártir, 46001 Valencia, Spain
* Correspondence: rsabater@die.upv.es (R.S.i.S.); angel.serrano@ucv.es (Á.S.-A.)

Polymers in the form of films, fibers, nano- and microspheres, composites, and porous supports are promising biomaterials for a wide range of advanced biomedical applications: wound healing, controlling drug delivery, anti-cancer therapy, biosensors, stem cell therapy, and tissue engineering. In this regard, polymer-based materials in the form of hydrogels; interpenetrated and semi-interpenetrated polymer networks; composites; or nanocomposites in their pure form, which can be functionalized or used in combination with other materials, nanomaterials, particles, or nanoparticles, can be exploited to produce a broad range of advanced nano- and macro-biomaterials. Specific features, such as mechanical performance, wettability, water diffusion, electroactivity, thermal properties, and even antimicrobial activity, can be tailored to engineer biomimetic microenvironments that are able to promote cellular interactions and tissue development for tissue engineering and regenerative medicine applications while preventing microbial infections. Furthermore, polymers are becoming increasingly important as a primary tool for controlling the release profile of drugs. New approaches have been developed to improve the efficacy of drug therapy, improving healing effectiveness.

This Special Issue presents new approaches in the areas of novel biomaterials for tissue engineering and drug delivery. It includes biomaterials based on polymeric matrices with various structures, such as porous composites, nanofibrous scaffolds, hydrogels, or meshes, which can include micro- or nanoparticles (graphene nanosheets, bioactive glass, and hydroxyapatite and halloysite nanotubes).

Stefaniak and Masek [1] summarized in a short review the copolymerization of the well-known poly(lactic acid) (PLA) with other polymers to improve PLA properties. PLA can be considered an ecological material because it can be produced using renewable resources. The review focuses on PLA copolymerization accomplishments for different applications, especially in pharmaceutical and biomedical fields.

Studies [2–5] reported novel strategies for the development of biomaterials in the form of composites, hydrogels, or electrospun fibers for tissue engineering applications. Głąb et al. [2] investigated the influence of collagen types on the physico-chemical properties of polyvinylpyrrolidone and poly(vinyl alcohol) composites that also included a ceramic phase (hydroxyapatite). The study demonstrated the great potential of collagen-modified composites for biomedical use, particularly for bone tissue engineering. Aparicio-Collado et al. [3] proposed a novel graphene-based poly(3-hydroxybutyrate-co-3-hydroxyvalerate)/polyvinyl alcohol semi-interpenetrated networks with low amounts of graphene (G) nanosheets within the polymer matrix to produce nanohybrid hydrogels with electroactive properties. The mechanical and electrical properties significantly increased in nanohybrid hydrogels with only 0.2% of G nanosheets, which showed good biocompatibility with muscle cells. The conductive hydrogels, with electrical conductivity

Citation: Sabater i Serra, R.; Serrano-Aroca, Á. Special Issue: "Polymer-Based Biomaterials and Tissue Engineering". *Materials* **2023**, *16*, 4923. https://doi.org/10.3390/ma16144923

Received: 3 July 2023
Accepted: 6 July 2023
Published: 10 July 2023

in the range of human skeletal muscle tissue, were able to induce myoblast proliferation, indicating its great potential for musculoskeletal tissue engineering. Schuhladen et al. [4] prepared polyhydroxyalkanoate (PHA)/bioactive glass (BG) composites. Bioglass 45S5 (in wt.%: 45.0 SiO_2, 24.5 Na_2O, 24.5 CaO, and 6.0 P_2O_5) and copper-doped 45S5 BG (in wt.%: 45.0 SiO_2, 24.5 Na_2O, 22.0 CaO, 6.0 P_2O_5, and 2.5 CuO) were used to engineer PHA/BG porous scaffolds (using the salt-leaching method). Murine stromal ST2 cells cultured in media with different dissolution products released from the scaffolds show a minor reduction in cell viability with an increase in VEFG release, which indicates that the composites are interesting for tissue engineering applications. In addition, Matschegewski et al. [5] prepared electrospun scaffolds from FDA-approved commercial medical-grade polymers (PLA, polycaprolactone (PCL), and polyamide) to assess the physicochemical and biological evaluation of cardiac implants. Untreated and plasma-activated polymeric nonwovens were analyzed to evaluate their influence on endothelial cell response. The study demonstrated the potential of plasma-activated electrospun scaffolds for advanced cardiac implant development.

Studies [6–8] focused on different approaches related to drug delivery. Haroosh et at. [6] developed PLA/PCL blends using halloysite nanotubes to obtain a sustained release of hydrophilic drugs. Using tetracycline hydrochloride (TCH) as a drug model, they found that when TCH was loaded into hydrophobic PLA/PCL blends with halloysite nanotubes, drug release decreased, overcoming the weak interaction between TCH and PLA/PCL blends. In another study, Bhanderi et al. [7] reported a novel delivery approach to deliver rivastigmine, a reversible cholinesterase inhibitor, for intranasal applications relative to brain delivery to treat neurodegenerative diseases such as Alzheimer's disease. The study reports the development of mucoadhesive rivastigmine loaded in chitosan and coated with Eudragit EPO, a cationic terpolymer from the poly(methacrylate) family, for intranasal delivery. This system could circumvent the first-pass metabolism of drugs and help achieve a sustained drug release. The development of microparticles that are able to encapsulate the nanoparticles of compounds for pulmonary drug-delivery applications has been addressed by Sato and Murakami [8]. Temperature-responsive polysaccharide microparticles containing nanoparticles were engineered, which can release two differently charged compounds in a two-step release. The delivery system has great potential to be used as a temperature-responsive drug carrier for various administration routes, such as pulmonary, transpulmonary, intramuscular, and transdermal administration.

Finally, Turlakiewicz et al. [9] focused on the problems related to parastomal hernia and the major properties of surgical meshes (mainly based on polymers but also biological meshes) available on the market. The review also includes the surgical techniques currently used to treat parastomal hernia and post-surgery complications.

We hope that the findings presented in this Special Issue will be useful in ongoing efforts to develop new biomaterials and novel approaches based on polymeric biomaterials for advanced biomedical applications.

Author Contributions: Conceptualization, R.S.i.S.; writing—review and editing, R.S.i.S. and Á.S.-A. All authors have read and agreed to the published version of the manuscript.

Funding: This research received no external funding.

Institutional Review Board Statement: Not applicable.

Informed Consent Statement: Not applicable.

Data Availability Statement: Data sharing is not applicable.

Conflicts of Interest: The authors declare no conflict of interest.

References

1. Stefaniak, K.; Masek, A. Green Copolymers Based on Poly(Lactic Acid)—Short Review. *Materials* **2021**, *14*, 5254. [CrossRef] [PubMed]
2. Głąb, M.; Drabczyk, A.; Kudłacik-Kramarczyk, S.; Kędzierska, M.; Tomala, A.; Sobczak-Kupiec, A.; Mierzwiński, D.; Tyliszczak, B. Investigations on the Influence of Collagen Type on Physicochemical Properties of PVP/PVA Composites Enriched with Hydroxyapatite Developed for Biomedical Applications. *Materials* **2022**, *15*, 37. [CrossRef] [PubMed]
3. Aparicio-Collado, J.L.; Zheng, Q.; Molina-Mateo, J.; Torregrosa Cabanilles, C.; Vidaurre, A.; Serrano-Aroca, Á.; Sabater Serra, R. Engineered Highly Porous Polyvinyl Alcohol Hydrogels with Poly(3-Hydroxybutyrate-Co-3-Hydroxyvalerate) and Graphene Nanosheets for Musculoskeletal Tissue Engineering: Morphology, Water Sorption, Thermal, Mechanical, Electrical Properties, and Biocompatibility. *Materials* **2023**, *16*, 3114. [CrossRef] [PubMed]
4. Schuhladen, K.; Lukasiewicz, B.; Basnett, P.; Roy, I.; Boccaccini, A.R. Comparison of the Influence of 45s5 and Cu-Containing 45s5 Bioactive Glass (BG) on the Biological Properties of Novel Polyhydroxyalkanoate (PHA)/BG Composites. *Materials* **2020**, *13*, 2607. [CrossRef] [PubMed]
5. Matschegewski, C.; Kohse, S.; Markhoff, J.; Teske, M.; Wulf, K.; Grabow, N.; Schmitz, K.P.; Illner, S. Accelerated Endothelialization of Nanofibrous Scaffolds for Biomimetic Cardiovascular Implants. *Materials* **2022**, *15*, 2014. [CrossRef] [PubMed]
6. Haroosh, H.J.; Dong, Y.; Jasim, S.; Ramakrishna, S. Improvement of Drug Release and Compatibility between Hydrophilic Drugs and Hydrophobic Nanofibrous Composites. *Materials* **2021**, *14*, 5344. [CrossRef] [PubMed]
7. Bhanderi, M.; Shah, J.; Gorain, B.; Nair, A.B.; Jacob, S.; Mohammed Basheeruddin Asdaq, S.; Fattepur, S.; Alamri, A.S.; Alsanie, W.F.; Alhomrani, M.; et al. Materials Optimized Rivastigmine Nanoparticles Coated with Eudragit for Intranasal Application to Brain Delivery: Evaluation and Nasal Ciliotoxicity Studies. *Materials* **2021**, *14*, 6291. [CrossRef] [PubMed]
8. Sato, T.; Murakami, Y. Temperature-Responsive Polysaccharide Microparticles Containing Nanoparticles: Release of Multiple Cationic/Anionic Compounds. *Materials* **2022**, *15*, 4717. [CrossRef] [PubMed]
9. Turlakiewicz, K.; Puchalski, M.; Krucińska, I.; Sujka, W. The Role of Mesh Implants in Surgical Treatment of Parastomal Hernia. *Materials* **2021**, *14*, 1062, Erratum in *Materials* **2021**, *14*, 3511. [CrossRef] [PubMed]

materials

Review

Green Copolymers Based on Poly(Lactic Acid)—Short Review

Konrad Stefaniak and Anna Masek *

Institute of Polymer and Dye Technology, Faculty of Chemistry, Lodz University of Technology, 90-924 Lodz, Poland; 237771@edu.p.lodz.pl
* Correspondence: anna.masek@p.lodz.pl

Abstract: Polylactic acid (PLA) is a biodegradable and biocompatible polymer that can be applied in the field of packaging and medicine. Its starting substrate is lactic acid and, on this account, PLA can also be considered an ecological material produced from renewable resources. Apart from several advantages, polylactic acid has drawbacks such as brittleness and relatively high glass transition and melting temperatures. However, copolymerization of PLA with other polymers improves PLA features, and a desirable material marked by preferable physical properties can be obtained. Presenting a detailed overview of the accounts on the PLA copolymerization accomplishments is the innovation of this paper. Scientific findings, examples of copolymers (including branched, star, grafted or block macromolecules), and its applications are discussed. As PLA copolymers can be potentially used in pharmaceutical and biomedical areas, the attention of this article is also placed on the advances present in this field of study. Moreover, the subject of PLA synthesis is described. Three methods are given: azeotropic dehydrative condensation, direct poly-condensation, and ring-opening polymerization (ROP), along with its mechanisms. The applied catalyst also has an impact on the end product and should be adequately selected depending on the intended use of the synthesized PLA. Different ways of using stannous octoate ($Sn(Oct)_2$) and examples of the other inorganic and organic catalysts used in PLA synthesis are presented.

Keywords: polylactic acid; copolymers; catalysts; polymer synthesis; ring-opening polymerization; medical application

Citation: Stefaniak, K.; Masek, A. Green Copolymers Based on Poly(Lactic Acid)—Short Review. *Materials* **2021**, *14*, 5254. https://doi.org/10.3390/ma14185254

Academic Editor: Roser Sabater i Serra

Received: 5 August 2021
Accepted: 9 September 2021
Published: 13 September 2021

1. Introduction

Approximately 140 million tons of petroleum-based synthetic polymers are manufactured globally per year, and a considerable amount of them perform in the ecosystem as industrial waste products [1]. Additionally, more and more countries are banning plastic grocery bags, which are said to be responsible for so-called "white pollution" around the world [2]. Even European Union law issues the requirements that concern balanced natural resources management and waste handling [3]. These issues encourage studies on bioplastics that, as biodegradable and ecological replacements, are expected to limit influence on the natural environment [2].

The above-mentioned issues regarding the need for environment protection, decreasing resources of crude oil in the world, and global warming have led to a search for new biodegradable materials. Bio-based plastics have been introduced in order to reduce carbon emissions (because the bio-based raw materials absorb CO_2 from the atmosphere), and biodegradable plastics have been developed to reduce plastic pollution (because they degrade faster than traditional plastics) [4]. One such material is polylactic acid (PLA). It is made from lactic acid (LA), which is a fermentation product of organic substances such as sweetcorn, rice, soya, potatoes, or whey that is a by-product of the dairy industry. This demonstrates that PLA is produced from renewable resources.

Ecological material has to be environmentally safe in every stage of its "life-cycle"—raw materials sourcing, production, and waste management. Ideal eco-material should be produced from renewable resources, have appropriate mechanical properties, and be easily degradable after exploitation [3]. PLA meets these requirements and, in comparison to other biodegradable plastics, has better mechanical strength, durability, and transparency [4]. PLA is widely applied in food packaging and the medical sector [5]. It is worth noting that several biomedical applications for PLA have been described, e.g., for orthopedic regenerative engineering, as surgical applications in human tissues, or as controlled delivery carriers [6].

In recent years, several studies have been made on the subject of polylactic acid, and new methods of synthesis have been proposed. Several methods of PLA structural modifications have been described [2,4,7–9].

This article presents an overview of the studies concerning improving PLA properties and its synthesis. At first, the overall characteristics of PLA is given. The different mechanisms of PLA synthesis and the catalysts used are then presented. Finally, this manuscript concentrates on the variety of PLA copolymers that enable them to change the physical properties of neat PLA. The main goal of this paper is to highlight the fact that, despite PLA being intensely studied, there are still many research aspects that have to be developed, e.g., to improve PLA's brittleness and gas barrier, by investigating new PLA copolymers. Moreover, attention to the topic of PLA should be drawn because it is a prospective polymer due to its biodegradability and biocompatibility. This review discusses and evaluates recent developments in PLA research with particular reference to its copolymers [6,10,11].

2. Lactic Acid (LA)—PLA Monomer

Lactic acid (2-hydroxy propionic acid) (LA) is a starting substrate for obtaining polylactic acid. It is an organic acid that contains asymmetric carbon atoms in its molecules and, as a result, it exists in two different isomers: the levorotatory form called R or D(−) lactic acid and the dextrorotatory form called S or L(+) lactic acid (Figure 1). The signs (−) and (+) signify in which direction a chemical induces a rotation of plane-polarized light. In mammalian organisms, only L-LA appears, but if an appropriate strain of bacteria and conditions of process (pH, temperature) are provided, than both isomers or their racemate can be obtained [12,13]. However, the racemic mixture of D-LA and L-LA (i.e., D/L-LA) is not recommended for the pharmaceutical, food, and drink industry by reason of the metabolic problems that D-LA might cause. It is also not suggested in terms of PLA synthesis because the PLA industry usually needs lactic acid with high optical purity (e.g., ~99% L-LA and ~1% D-LA) [13].

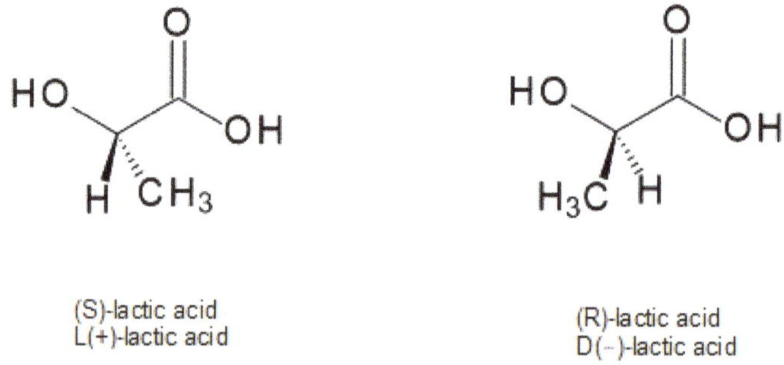

Figure 1. Two lactic acid isomers [13].

Lactic acid can be manufactured on an industrial scale, for example, by the catalytic addition reaction of hydrogen cyanide and acetic aldehyde and then the hydrolysis of the obtained cyanohydrin (Figure 2).

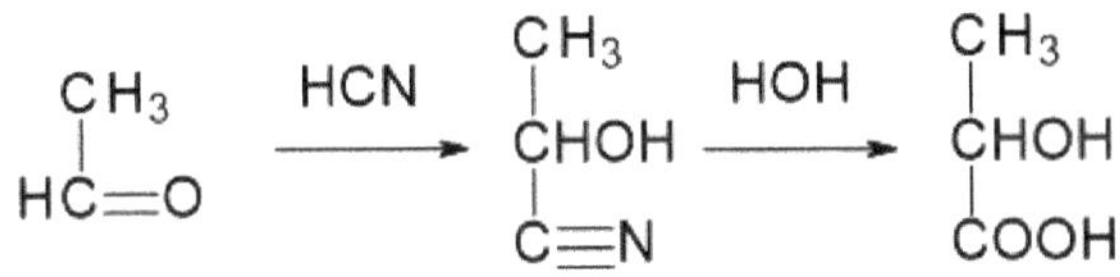

Figure 2. Scheme presenting the method of producing LA from acetic aldehyde using hydrogen cyanide [5].

There is also the possibility of receiving lactic acid as a result of its fermentation from food industry waste products (potatoes, sweetcorn, sugar beet) and proper bacteria (*Streptococcus*, *Pediococcus*, *Lactobacillus*, *Bifidobacterium*) [14–16]. The process, which begins with the hydrolysis of lactose, is presented in Figure 3.

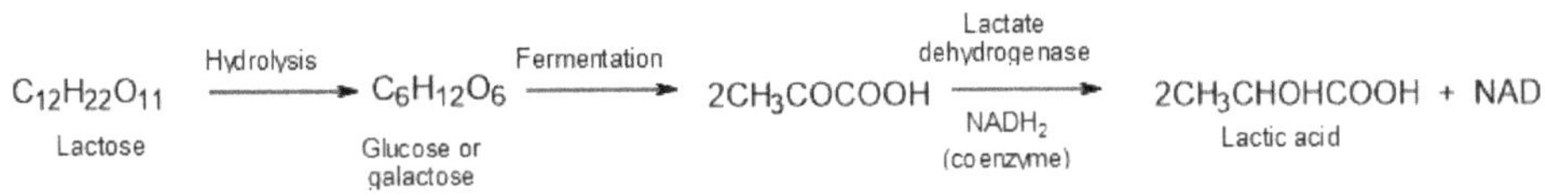

Figure 3. Scheme presenting receiving lactic acid [5].

3. Polylactic Acid (PLA) Chemical Properties

Polylactic acid (PLA) belongs to the family of aliphatic polyesters (Figure 4). It is also known as polylactide [5].

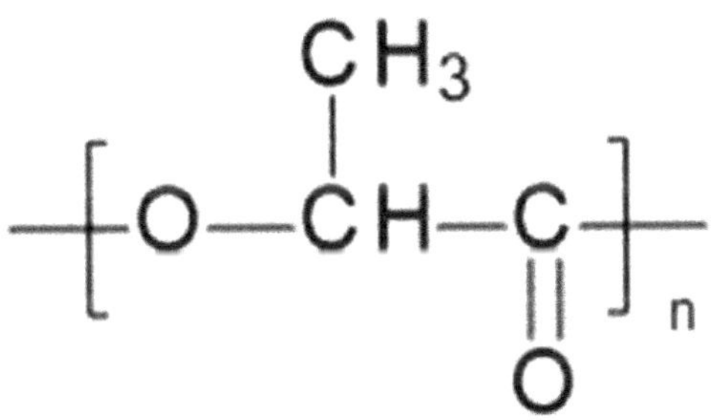

Figure 4. Chemical structure of PLA [5].

Homochiral PLA is isotactic and semi-crystalline (up to 60%). Its glass transition is ca. 55 °C and its melting point is ca. 180 °C [5]. Poly(L lactide) (PLLA) is marked by very good tensile strength (60 MPa), small elongation (3–4%), and high modulus (4.8 GPa) [17]. PLA is modified by adding plasticizers such as polyoxyethylene, polycaprolactone, or citrate esters, and in that way PLA's impact on strength and glass transition is improved [18]. Atactic PLA, made up of heterochiral chains (L,D), is an amorphous polymer [5].

Saeidlou et al. [19] investigated two types of PLA—semi-crystalline PLA (containing 95% L-lactide and 5% D-lactide) and amorphous PLA (containing 82% L-lactide and 18% D-lactide). The results showed that the shear viscosity and the crystallinity increased with increasing L isomer size in the L/D isomer mixture. The L/D isomer ratio in PLA also impacts on glass transition temperatures, as shown in Figure 5. The higher the content of L-stereoisomer is, the higher glass transition temperature occurs.

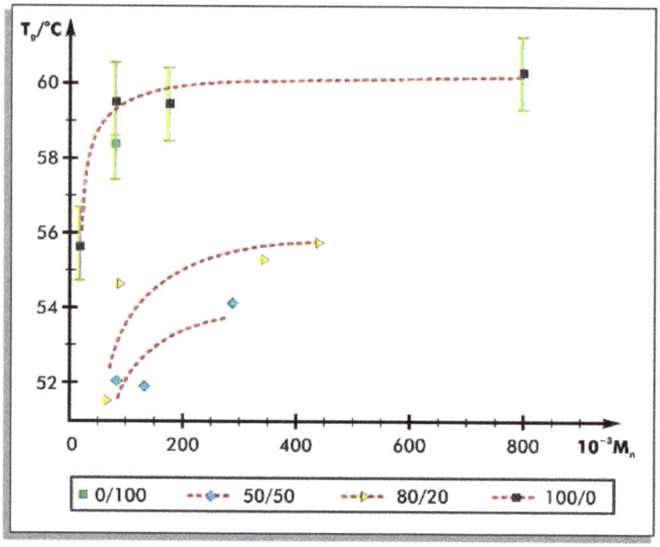

Figure 5. Glass transition temperatures (T_g) of PLA polymers with different L-stereoisomer contents as a function of number-average molecular weights (M_n). Symbols on the legend represent the L/D-isomers ratios [20].

Scientific research has confirmed that PLA is soluble in both polar and nonpolar solvents, e.g., benzene, tetrahydrofuran (THF), dimethyl sulfoxide (DMSO), acetonitrile, and dioxan [21]. PLA is a completely biodegradable material under industrial composting conditions and assimilable by living organisms [5].

Apart from PLA copolymers, which are precisely described in Section 6, various PLA-based composites have been developed in order to improve PLA properties and applications.

Rodenas-Rochina et al. [22] prepared PLA/hydroxyapatite (HA) composites for application in devices created for bone healing. HA micro- or nanoparticles were dispersed in the polymer matrix.

Natural fibers can be used in order to reinforce the PLA matrix. In a recent study, PLA/flax, PLA/jute, and PLA/falx/jute were fabricated. The concentration of natural fibers in individual composites varied (between 0–50%) by weight. PLA/jute and PLA/flax composites with 40% by weight of fibers improved PLA tensile strength the most. The tensile strength of pure PLA (18.77 MPa) increased to 72 MPa and 45 MPa after flax and jute reinforcement in PLA, respectively [23].

PLA/carbon fiber (PLA/CF) composites are believed to have significant applications in biomedical and engineering sectors. The excellent tensile strength and chemical stability of carbon fibers are the main reasons for interest in the production of PLA/CF composites [24].

4. PLA Synthesis—Methods

There are three widely known ways to acquire PLA (Figure 6): lactic acid condensation, lactic acid azeotropic dehydrative condensation, and ring-opening polymerization (ROP) [25]. In this section, every method will be concisely described.

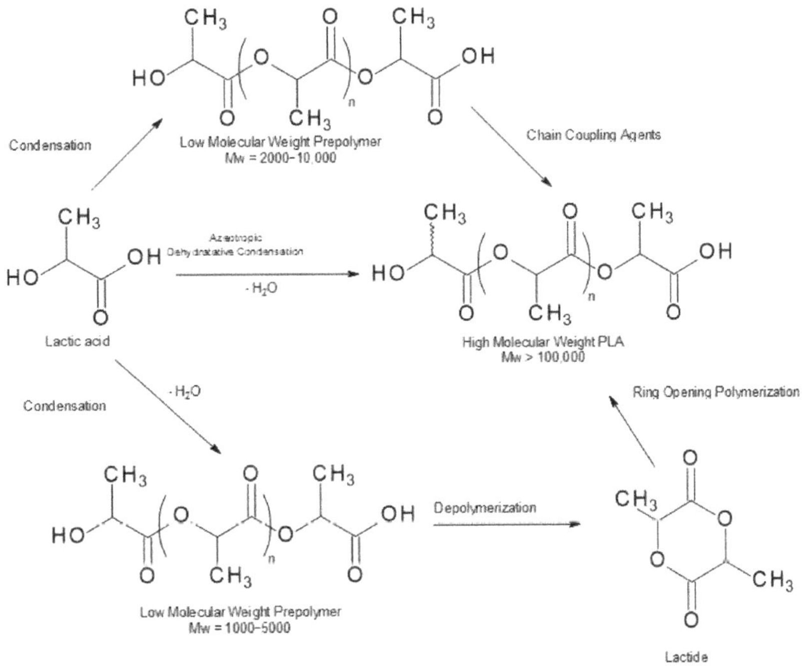

Figure 6. Manufacturing routes to polylactic acid [26].

4.1. Direct Poly-Condensation

The direct poly-condensation process dehydrates lactic acid into oligomers, which are then further polymerized to PLA with concurrent dehydration to avoid the degradation of polymer molecules by moisture. Nonetheless, removal of water produced from the condensation of lactic acid is noticeably demanding during the final stage of polymerization because the diffusion of moisture in the very much viscous polymeric melt is quite slow. The residual water trapped in the PLA melt can limit the achievable molecular weight and the characteristics of PLA. Therefore, the direct poly-condensation process is little used [27].

4.2. Azeotropic Dehydrative Condensation

During the direct LA polycondensation process, there is a difficulty in removing by-produced water. Until 1995, it was believed that a high molecular weight PLA could not be prepared by the polycondensation of LA. Progress was made by Mitsui Chemicals Co. (Tokoy, Japan), because its azeotropic dehydrative polycondensation enabled an increase of the molecular weight of PLA. This method allows the acquisition of PLA with a high molecular weight after a comparatively long reaction time. The dissociated water is removed by means of the so-called azeotropic distillation technique (Figure 7). Solvents with a high boiling point are used for this method [28].

The azeotropic dehydration condensation reaction of LA provides high molecular weight PLA without the use of chain extenders or adjuvants [10]. However, this process has been problematic because of organic solvent usage, which made this method ecologically unattractive [28].

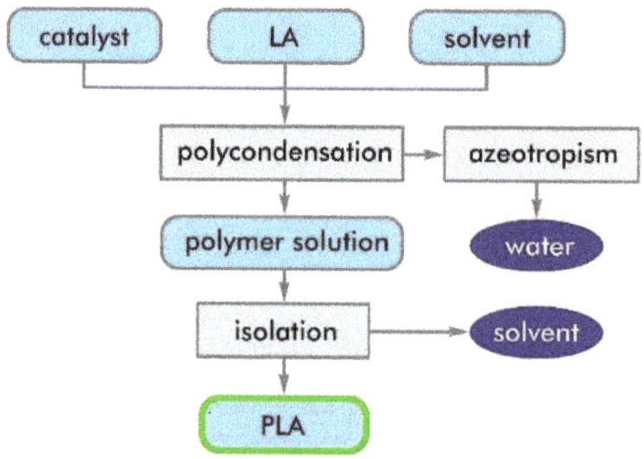

Figure 7. Flow diagram of the azeotropic dehydrative polycondensation of LA [28].

4.3. Ring-Opening Polymerization

Ring-opening polymerization (ROP), catalyzed by organometal catalysts, is a method that consists of converting lactide (the cyclic dimer of lactic) to PLA (Figure 8) [29].

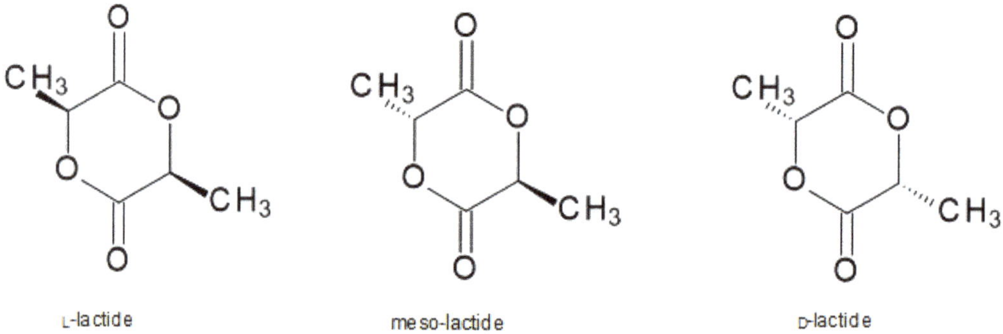

Figure 8. Lactide—cyclic dimers for the ROP process [7].

Initially, lactic acid is dehydrated and poly-condensed into its oligomers at high temperature and under vacuum in order for moisture to be removed. Lactide is then obtained from catalytic depolymerization of these short PLA chains under reduced pressure. The meso-lactide, lactic acid, and residual moisture can be removed from the optically pure D or L form of lactide by diverse means such as crystallization or distillation. Eventually, the purified lactide is polymerized by a ring-opening polymerization reaction into PLA at temperatures above the melting point of lactide and below the degradation temperatures of PLA. The unreacted lactide (around 5%) must be removed from PLA, and the flowing PLA resin is solidified or/and crystallized into pellets. While the ROP of lactide is conducted, there is little/no moisture to be removed from the molten PLA resin [13].

Three different ROP mechanisms for PLA can be specified: cationic, anionic, and coordinative [14,30]. They differ from each other on the grounds that the lactide ring is opened at different positions depending on the polymerization initiator used. A bond between an oxygen and a carbon belonging to an acyl group of atoms (A) or a bond between an oxygen and a carbon of an alkyl group of atoms (B) can be broken. This is presented in Figure 9 [5].

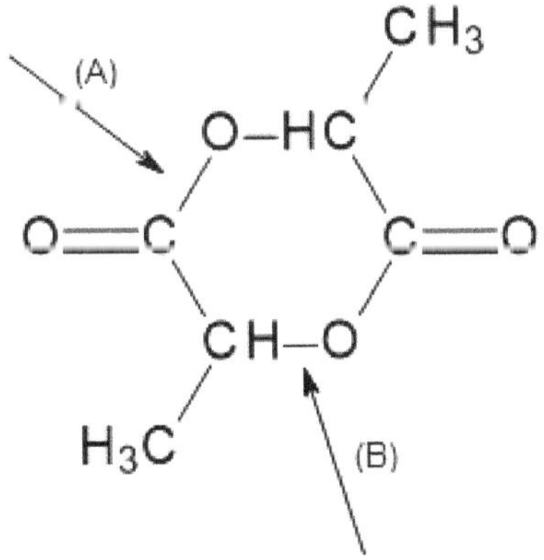

Figure 9. Possible positions of breaking lactide ring [5].

The cationic ROP mechanism for PLA (Figure 10) consists of breaking the alkyl–oxygen bond of the lactide ring. The propagation mechanism starts with the positively charged lactide ring being opened at the alkyl–oxygen bond by an S_N2 attack by the trifluoromethanesulfonate (triflate) anion. The triflate end-group then combines with another molecule of lactide again in an S_N2 manner to obtain a positively charged opened lactide. Polymerization proceeds as the triflate anion again opens the charged lactide [31].

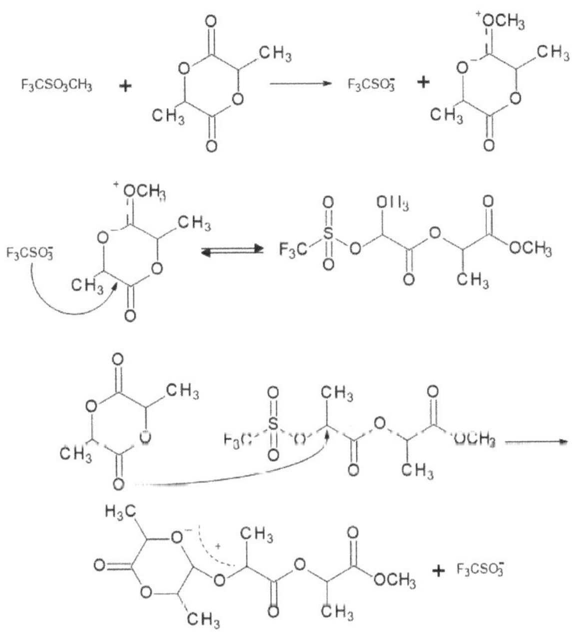

Figure 10. Cationic ring-opening polymerization mechanism for PLA [21].

In this mechanism, the possibility of racemization appears because, during every stage of polymer chain propagation, substitution of a monomer occurs in a chiral center. A low molecular weight polymer is yielded with this method. Hence, it is not used on an industrial scale [5].

In the anionic ROP mechanism for PLA (Figure 11), alkali metals alkoxides (e.g., CH_3OK) act as the initiators of the process. In this process, during the propagation mechanism, the acyl–oxygen bond is cleaved by the initiator's anion attack. In the obtained PLA macromolecules, an oxygen atom of the alkoxide end-group that propagates has a negative charge, so the configuration of these particles cannot be changed. The PLA synthesis with alkali metal alkoxide as an initiator is marked by good reaction rate, few unwanted side reactions, and high efficiency [5]. It has been shown that using primary alkoxides similar to the aforementioned potassium methoxide enables one to obtain well-defined polymers with insignificant racemization, termination, or transesterification. Racemization was lower than 5% when started with 99.9% pure L-lactide [32,33].

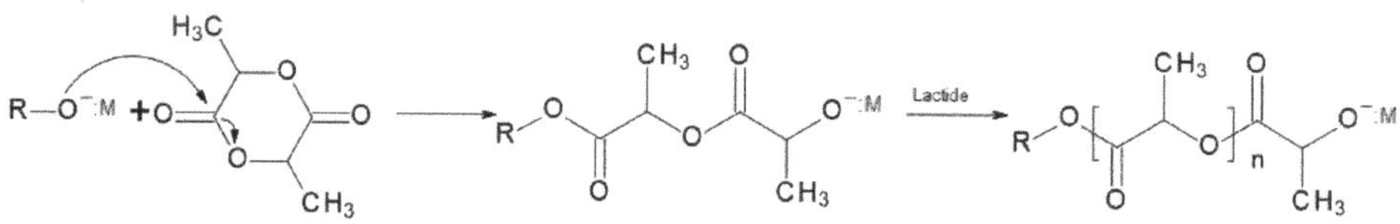

Figure 11. Anionic ring-opening polymerization mechanism for PLA [5].

In coordinative mechanisms (Figure 12) the initiator combines with lactide after previous acyl-oxygen cleavage. In this type of reaction, alkoxides or carboxylates, including covalent bonds between oxygen and copper atoms or unoccupied p or d orbitals, are used as initiators. Zinc, aluminum, titanium, and stannous alkoxides are the most widely known. In industrial conditions in PLA synthesis, tin(II) 2-ethylhexanoate (also known as stannous octoate) is used as a coordinative initiator [5]. This compound will be more precisely described as a catalyst in the next section of this paper.

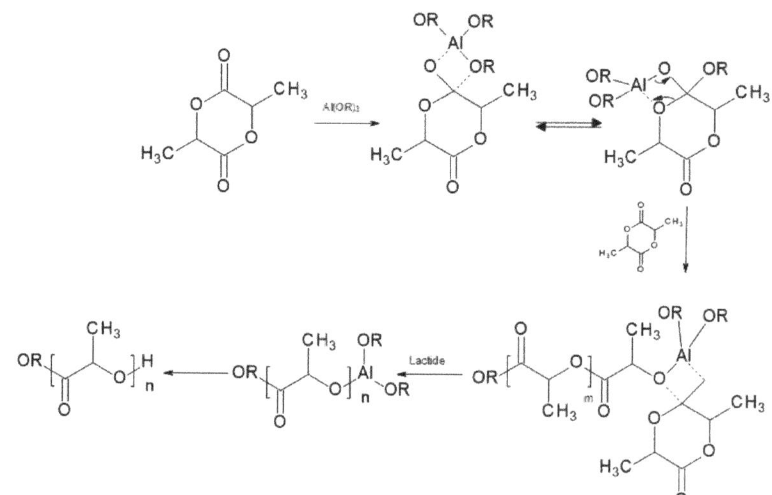

Figure 12. Coordinative ring-opening polymerization mechanism for PLA [5].

5. PLA Synthesis—Catalysts

In this section, the topic of the catalysts used in PLA synthesis will be investigated, and several examples will be described.

When PLA synthesis is conducted, an appropriate catalyst is usually used. During this process, catalysts have an impact on the length of the synthesis, and the choice of catalyst should also depend on the desirable application of the manufactured PLA. PLA polymerization reaction time fluctuates between a few minutes and more than 100 h, depending on the catalyst used, the solvent, and the expected molecular weight [9].

A popular catalyst used in PLA synthesis is stannous octoate—$Sn(Oct)_2$. It is an effective compound that gives high PLA molecular weights. Molecular weight distributions obtained with $Sn(Oct)_2$, compared with other catalysts used in PLA synthesis, are presented in Table 1.

Table 1. Comparison of catalysts used in PLA synthesis.

Polymer	Catalyst	Molecular Weight	Reference
L-PLA	Stannous octoate	M_n < 250,000	[34]
L-PLA	Stannous octoate and compounds of titanium and zirconium	M_n = 40,000–100,000	[9]
L-PLA	Stannous octoate and triphenylamine	M_n = 91,000	[35]
L-PLA	Potassium naphthalenide	M_n < 16,000	[2]
L-PLA	Yttrium tris (2,6-di-tert butyl phenolate)	M_n < 25,000	[36]
D,L-PLA	Lanthanum isopropoxide	M_n = 5300–21,900	[9]

$Sn(Oct)_2$ utility results in PLAs with a low degree of racemization (even at high temperatures). Furthermore, it has low toxicity. $Sn(Oct)_2$ is a catalyst currently used in aliphatic polyester synthesis that produces atactic PLA chains [37]. It should be noted that $Sn(Oct)_2$ is accepted by the FDA (Food and Drug Administration, Silver Spring, MD, USA) for biomedical applications [38]. Figure 13 presents the structure of $Sn(Oct)_2$ and the compounds, which are organometallic catalysts and organocatalysts, used in PLA polymerization [6].

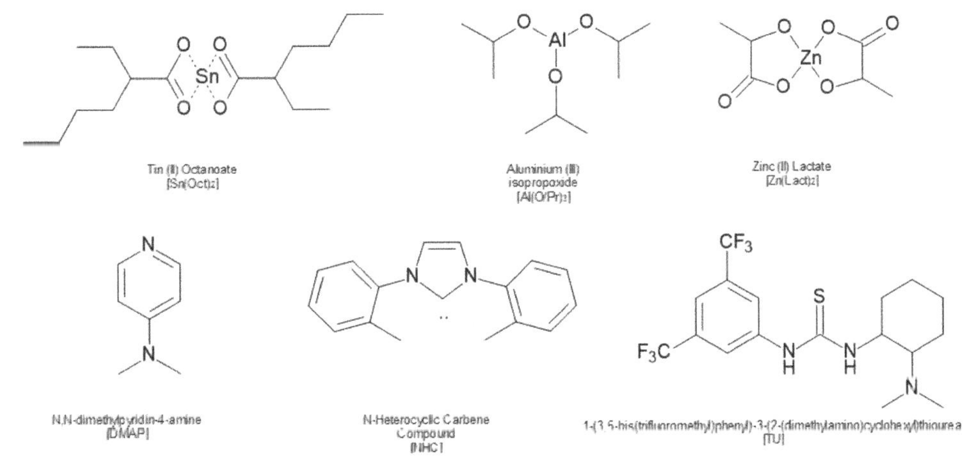

Figure 13. Structure of organometallic catalysts and organocatalysts used in PLA polymerization [6].

5.1. Stannous Octoate

Several studies concerning stannous octoate as a catalyst in PLA synthesis have been made. One of them was carried out on the function of polymerization temperature, time, and concentration of catalyst $Sn(Oct)_2$. During the polymerization of L-lactide, PLA with the highest value of intrinsic viscosity ($M_v = 10^6$) was synthesized when catalyst concentration was low (0.015 wt %) and at the temperature of 100 °C. A nonionic insertion polymerization mechanism was suggested [39].

Hyon et al. [40] used the same catalyst and obtained PLA with the maximum molecular weight at a catalyst concentration of 0.05% at a temperature of 130 °C. Furthermore, at prolonged polymerization, the decrease in M_v and higher polymerization temperature was ascribed to thermal depolymerization of the resultant polylactides.

Other studies have compared the use of tin and zinc compounds. The kinetics and mechanism of L-lactide bulk polymerization using stannous octoate and zinc bis (2,2-dimethyl-3,5-heptanedionate-O,O′) was taken into consideration. Up until 80% conversion, the rate of polymerization using $Sn(Oct)_2$ was higher than that with zinc-containing catalyst. However, at conversions over 80%, the latter catalyst gave the higher rate of polymerization. The accelerating effect on the polymerization was caused by crystallization of the newly formed polymer. It is suggested that the reason why the differences in the rate of polymerizations at high conversion for the two catalysts is observed is that a difference in crystallinity of the newly formed polymer occurs. Additionally, contaminants in the catalyst and the monomer are thought to be the true initiators. Initiation, as well as polymerization, proceeds through a Lewis acid catalyzed transesterification reaction between an activated lactone and a hydroxyl group [41].

The mechanism of the reaction between the lactide and $Sn(Oct)_2$ is recognized as a coordination-insertion mechanism. An alcohol molecule, such as MeOH or the propagating hydrolyzed lactide, which acts as an initiator, exchanges with the octoate ligands. Coordination of lactide to the metal center then occurs. In the next step, the activated nucleophilic alkoxide proceeds by the opening of the lactide. The propagation caused by the generated linear monomer begins, which appears as subsequent lactide coordination and alkoxide insertion until the metal-alkoxide bond is cleaved by termination reactions. Using this method, the PLA obtained includes an ester end group derived from the initiator [42].

5.2. Stannous Octoate and Lewis Base

Stannous octoate can lead to relatively rapid lactide polymerization, but it is also known to have a negative effect on the PLA molecular weight and properties. The back-biting and intermolecular transesterification reactions, which occur not only during the lactide polymerization but also during any further melt processing, are responsible for these complications [43]. A solution for this issue has been proposed. The addition of an equimolar amount of a Lewis base, particularly triphenylphosphine on stannous octoate, considerably improves the lactide polymerization rate in bulk. What is more, the kinetic effect has been accounted for by coordination of the Lewis base onto the metal atom of the initiator, making the insertion of the monomer into the metal alkoxide bond of the initiator easier [9].

5.3. Stannous Octoate and Distannoxane

The copolymerization of lactide (LA) with mevalonolactone (ML) was performed. Copolymerizations with a monomer ratio LA/ML = 10 were prepared in the presence of two different catalysts: stannous octoate or 1-ethoxy-3-chlorotetrabutyldistannoxane (distannoxane). It was found that, in the presence of $Sn(Oct)_2$, branched polymer formation continued by a macromonomer formation step, accompanied by side reactions such as ester exchange and/or alcoholysis. However, copolymerization catalyzed by distannoxane proceeded without side reactions. Due to multiangle laser light scattering and size-exclusion chromatography (MALLS-SEC) analysis and differential scanning calorimetry (DSC), measurement of the formation of branched polymers was also indicated. The following mechanism for the polymerization system using $Sn(Oct)_2$ is proposed. Initially, L-LA reacts with a pendant hydroxy group of ML; L-LA cannot be polymerized using neat stannous octoate, except when in the presence of a protic compound, which is the actual initiator. Side reactions such as the aforementioned ester exchange, alcoholysis, or cyclic oligomer formation can then occur with the extension of the polymerization time. Until L-LA is consumed, polymerization and depolymerization (side reaction) are balanced. This can be an explanation of why the polymerization appears to have stopped in the late phase of this reaction stage. Subsequently, the lactone group of monomeric (unreacted) ML and ML units existing at the initiating terminal of a macromer is ring-opened. Therefore, a considerable increase in the molar portion of the ML units in the copolymer is seen at this point. The hydroxy groups of both macromer ML and monomeric can act as initiating groups [44].

5.4. Bi(III) Acetate and Creatinine

The catalytic activities of four compounds towards the ROP of L,L-lactide were compared: Bi(III) acetate ($Bi(OAc)_3$), creatinine, a $Sn(Oct)_2$-based system, and a system catalyzed by enzymes. In every case, high and moderate molar mass poly(L-lactide)s were obtained. The $Bi(OAc)_3$-based system was akin to $Sn(Oct)_2$ at 140 °C. The reaction mechanism when using bismuth compounds was recognized as a coordination-insertion mechanism. On the other hand, the reactivity of creatinine (following the coordination mechanism) was lower than that of $Bi(OAc)_3$ but highly comparable with that of the enzyme lipase *Pseudomonas fluorescens* [45].

5.5. Aluminum Based Catalysts

During the ring-opening polymerization of D-lactide, catalyst systems based on aluminum alkoxides are reported to give a polylactide marked by controllable molecular weight with narrow dispersion [25]. It was found that $ZnEt_2$ and its complex with aluminum isopropoxide ($Al(OiPr)_3$) gave fast polymerization with low transesterification when D,L-lactide was polymerized in bulk at 150 °C [46]. When $Al(OiPr)_3$ was used in bulk at 100 °C, comparable results were obtained. Moreover, using proton nuclear magnetic resonance (NMR), it was discovered that all chains included isopropoxy ester end-groups and molecular weights correlative to the number of alkoxide groups [47]. This shows that an acyl-oxygen cleavage is a way by which the lactide polymerization is initiated by the aluminum alkoxide. Furthermore, all alkoxide groups of an initiator are active initiating species. It was observed that there was no transesterification at temperatures less than 150 °C, which yields polymers with a narrow molecular weight distribution [47]. The kinetics and the mechanism of the aluminum alkoxide polymerization in solution have been studied [48,49]. One of the findings was that, after an initial induction period, the polymerization is first-order in both monomer and initiator. Despite that, for $Al(OiPr)_3$ in toluene at 70 °C, there are three active sites per aluminum molecule, which is contrary to other lactone polymerizations where the number of active sites is less due to catalyst aggregate formation [48,49]. The mechanism involving the insertion of the lactide into the aluminum-alkoxide bond with lactide acyl-oxygen cleavage is presented in Figure 12 as a ROP coordinative mechanism for PLA.

5.6. Summary

It has been known for many years that high molecular weight PLA can be obtained by the ring-opening polymerization of lactide. An astonishing number of catalysts that have been used in order to initiate this polymerization has been reported. Nevertheless, there is still a need to devise safer, faster, and more stable catalysts. Considering the development of a single-step extrusion process, there are doubts concerning the safety of the catalysts that are left in the final polymer, which is later used in human body applications [9].

6. PLA Copolymers

In order to improve several PLA physical properties, there have been studies concerning PLA modification by creating its copolymers. In this section, such solutions will be described. However, at the outset of this section, nomenclature concerning polymers is presented in Figure 14.

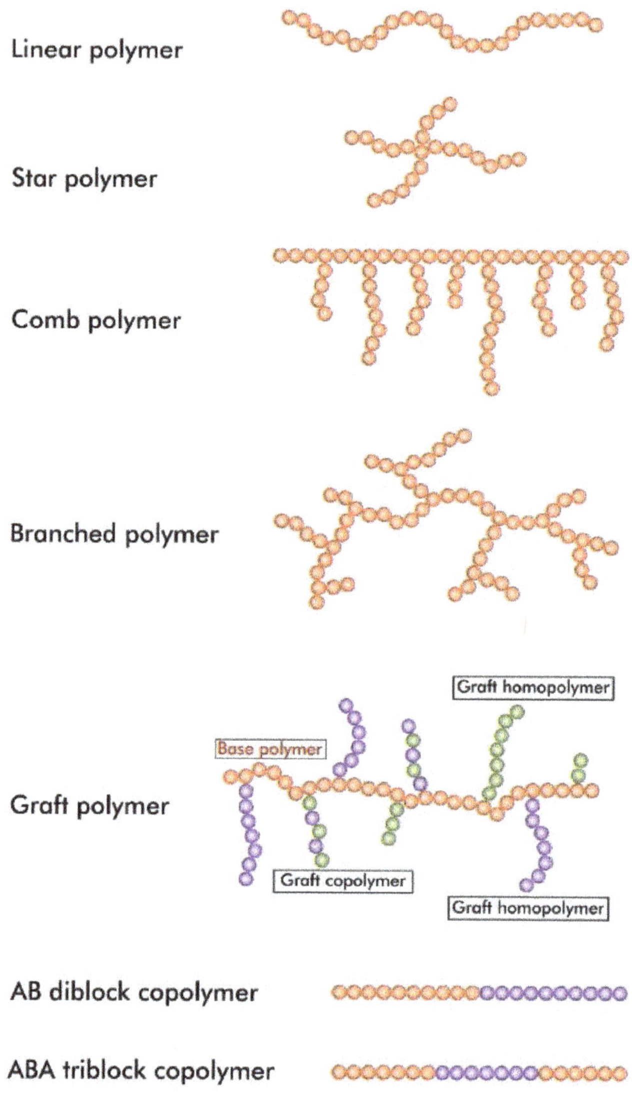

Figure 14. Summary of the selected polymer and copolymer architectures [50–52].

6.1. PLGA Copolymer

Poly(D,L-lactide-co-glycolide) (PLGA) is a linear copolymer of lactic acid (LA) and glycolic acid (GA) [4]. Figure 15 shows its chemical structures. PLGA copolymers prepared at different ratios were investigated, and its degradation time was compared. The results shown in Table 2 indicate that the copolymerization of LA with GA shortens PLA degradation time by as much as three-fold, from 6 to 2 months. Furthermore, the higher the content of polyglycolide (PGA) in the copolymer, the shorter the observed degradation time.

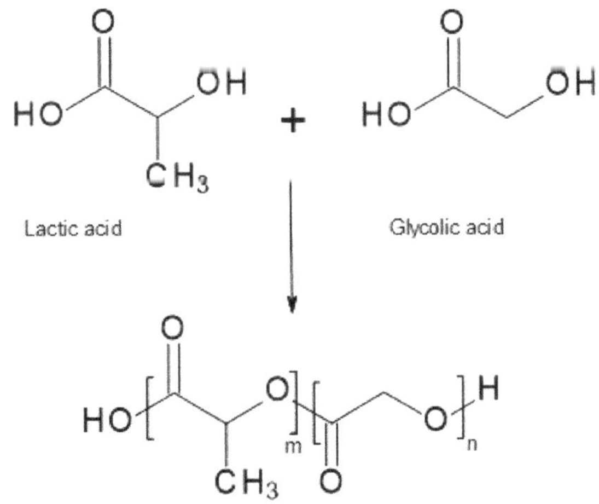

Figure 15. Chemical structure of LA, LG, and PLGA [53].

Table 2. Degradation time of PGA, PLLA, and its copolymers.

Polymer	Degradation Time (Months) [a]	Reference
PGA	1.5–3	[53]
PLLA	6–24	[53,54]
PLGA (LA/GA = 50/50)	1–2	[54]
PLGA (LA/GA = 50/50)	1	[55]
PLGA (LA/GA = 65/35)	1.5	[56]
PLGA (LA/GA = 75/25)	2	[56]
PLGA (LA/GA = 75/25)	4–5	[54]
PLGA (LA/GA = 75/25)	4	[57]
PLGA (LA/GA = 85/15)	5–6	[54]

[a] Time to complete resorption.

There are several ways to obtain PLGA. Solution poly-condensation of LA and GA at a temperature above 120 °C yields low molecular weight PLGA [4]. Using ring-opening polymerization of glycolide and lactide with a metal catalyst (e.g., stannous octoate), high molecular weight PLGA can be prepared. If there is a demand for PLGA with non-possible toxic metallic contaminations, which is favorable for bio-medical applications, enzymatic polymerization should be used [53]. Additionally, it is known that the sequence of PLGA has a significant impact on its degradation rate. Random PLGA degrades quicker than sequenced ones. 4-(dimethylamino) pyridinium p-toluenesulfonate (DPTS) and 1,3-diispropylcarbodiimide (DIC) have been used as catalysts in order to prepare repeating sequence copolymers [58]. Ring-opening polymerization (ROP) and segmer assembly polymerization (SAP) were used. The latter is a method that allows preparation of sequence polymers and gives numerous possibilities for periodic copolymer synthesis. In this approach, sequenced oligomers (segmers) are first prepared and then polymerized. It is a paradigm that shows the convergence of synthetic, organic, and polymer chemistries [59–61]. Figure 16 shows details of the previously described issue.

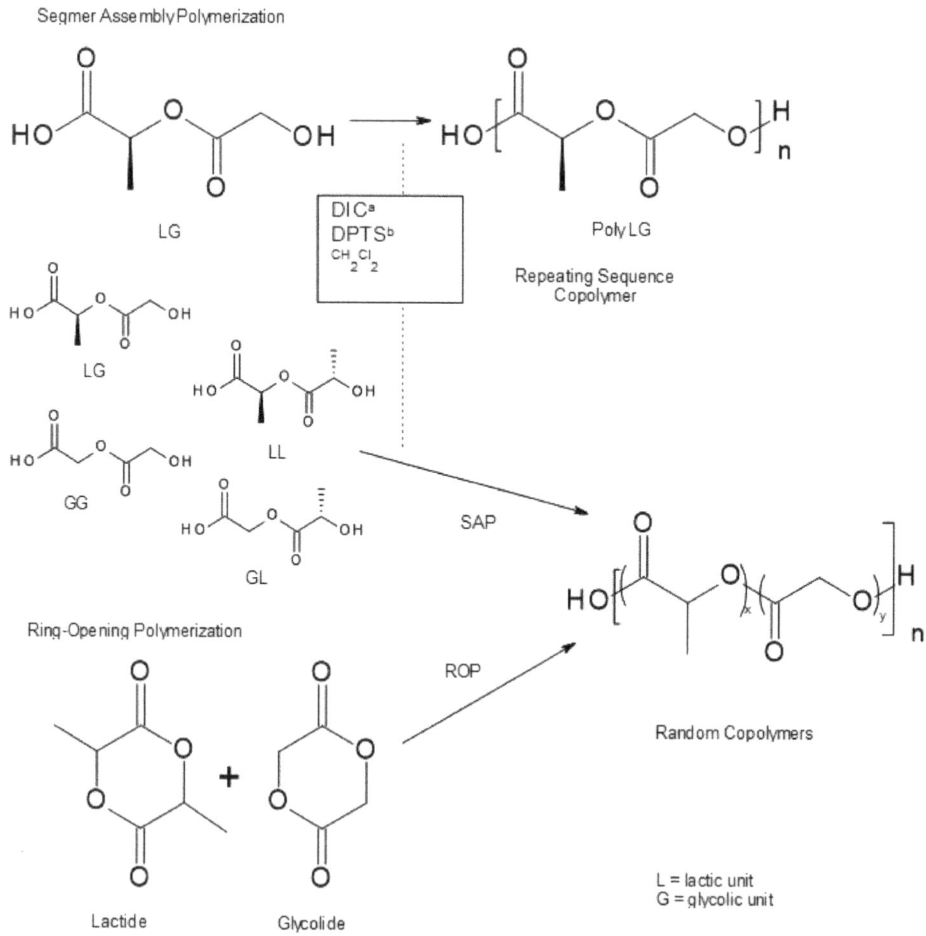

Figure 16. Approaches to sequenced and random PLGAs: (a) DIC = diisopropylcarbodiimide; (b) DPTS = 4-(dimethylamino) pyridinium ptoluenesulfonate. LG, LL, GG and GL are the different dimer units [58].

Because of a suitable time for degradation of PLGA to occur, this biocompatible copolymer can be used, e.g., in drug delivery systems. Depending on the needs, choosing an appropriate LA/GA ratio and the polymer's molecular weight allows the creation of desirable material. Furthermore, the two methods of PLGA synthesis described and the wide variety of available catalysts enable adjustments to suit the needs of the end product.

6.2. Metal-Centered Star-Shaped PLA (Co)Polymers

A star polymer is a polymer composed of branched macromolecules that contain only one common branch unit. Hence, in shape, it looks like a star. If the polymer's arms are chemically different then it is termed a miktoarm star polymer [50]. There are known star-shaped PLA polymers and copolymers with metal ion cores. Fe^{2+}, Fe^{3+}, Eu^{3+}, and Ru^{2+} ions are used as metal centers. Star structures can be produced by the combination of coordination chemistry with controlled or living polymerization. $Sn(Oct)_2$ is commonly used as a catalyst [6]. These materials are designed for a specific role, and because of their stimuli-responsiveness, luminescent materials for drug delivery can be conceived [37]. Luminescent ruthenium tris(bipyridine)-centered star block copolymers, consisting of PLA as the hydrophobic core and poly(acrylic acid) (PAA) as the hydrophilic corona, may provide a multifunctional drug delivery system with the capability of optical imaging. Star copolymers were obtained by the consecutive ROP of D,L-lactide, atom transfer radical polymerization (ATRP) of tert-butyl acrylate, and finally, the tert-butyl end-groups were hydrolyzed [62].

ATRP—atom transfer radical polymerization—is a particular type of a controlled radical polymerization (CRP). ATRP enables entering monomers and cross-linking agents into a polymer chain in a controlled way. This allows the procurement of nearly equally long polymer chains. Their length is determined by the used monomer and initiator ratio. Applying the ATRP method to polymer synthesis allows adjustment of the structure of these polymers to drug delivery requirements. ATRP is based on a redox reaction between dissolved transition metal ions and alkyl halides in order to regulate an equilibrium between their active species (radicals) and dormant species (Figure 17). The initiation stage is faster than the propagation stage in order to provide the concurrent growth of all polymer chains. In the ATRP system, there are four elementary reagents: initiator (P_n-X), activator (transition metal complex; Mt^n/L), deactivator (Mt^{n+1}/L), and macroradicals (P_n*), which grow as a result of adding other monomer particles. Transition metals such as Cu(I), Ni(II), and Fe(II) are the central metals in catalytic complexes. Derivatives of 2,2′-bipyridine, such as N,N,N′,N″,N″-pentamethyldiethylenetriamine (PMDETA), appear as ligands (L). During the reversible redox process, one electron is moved from the transition metal complex (which is in its the lowest oxidation state), and a halogen atom is ripped off a polymer chain (dormant species), which leads to creating a radical and deactivated catalytic complex in a higher oxidation state [63].

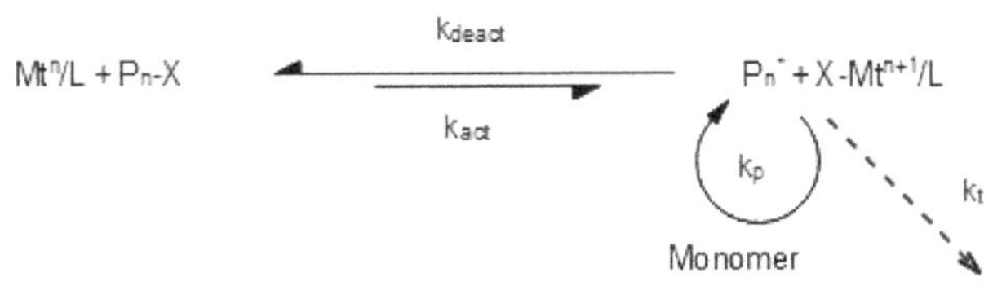

Figure 17. ATRP Equilibrium, k_p—propagation reaction rate constant, k_t—termination reaction rate constant [64].

Implementing metal ions into polymer structures broadens the possibilities of further research. Customizing the synthesis process in terms of the catalyst used, the provided hydrophilicity/hydrophobicity, and the macromolecule structure, gives a potential opportunity to use the obtained copolymers in areas of nanotechnology, biomedicine, and drug delivery systems.

6.3. PEG-PLA Copolymer

PLA and its copolymers can be applied as a base for hydrogel medicinal implants, which can be used as scaffolds for tissue engineering or drug delivery vehicles. For example, the copolymers of lactic acid with poly(ethylene glycol) (PEG) (structure of PEG in Figure 18) form thermo-responsive hydrogels. The physical cross-linking mechanisms of PEG-PLA consist of: ionic or lactic acid segment hydrophobic interactions, chemical bond formation by radical- or photo-cross-linking, and stereocomplexation of D- and L-lactic acid segments [65].

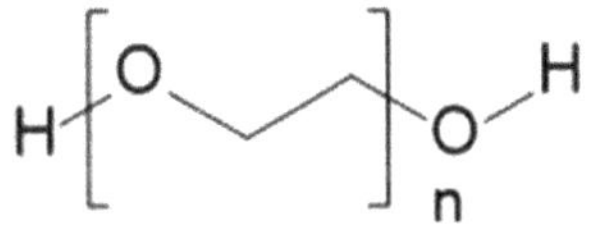

Figure 18. Chemical structure of PEG [66].

Eight-arm star block PEG-b-$(PLLA)_8$ copolymers, functionalized with pyridine, were used in order to form star block PEG-b-PLLA-py metallo-hydrogels in the presence of transition metal ions (Cu^{2+}, Co^{2+}, Mn^{2+}). These PEG-b-PLLA block the copolymer hydrogels, present distinguished biocompatibility and biodegradability, and could be adopted in a broad range of biomedical and industrial applications [67].

With the use of Michael-type addition reaction, chemically cross-linked PEG-b-PLLA hydrogels were prepared. Eight-arm thiol-terminated star PEG (PEG-$(SH)_8$) reacted with acrylated PEG-b-PLLA star block copolymers (PEG-b-$(PLLA_{12})_8$-AC) (Figure 19). Excellent mechanical properties presented by these hydrogels were observed. Furthermore, the degradation time of the formed hydrogels oscillated between a few days to several months and was regulated by the incorporated amount of PEG-$(SH)_8$. Lysozyme was released from the eminently cross-linked PEG-b-$(PLLA_{12})_8$-AC/PEG-$(SH)_8$ hydrogels, predominantly by diffusion [68]. Additionally, miktoarm star polymers, PLLA/PEG, were reported to form thermo-responsive hydrogels in water at high concentrations (22.5 wt %) [69].

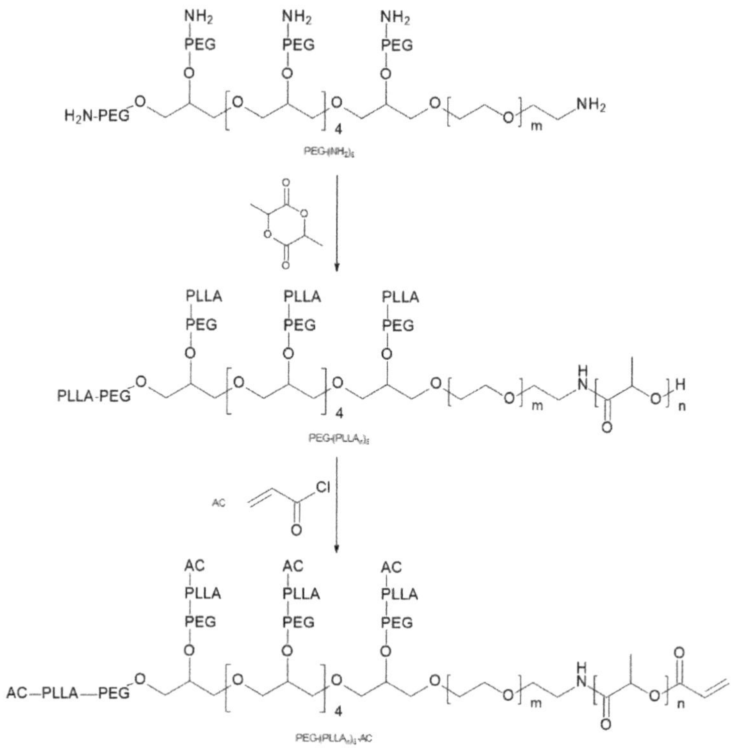

Figure 19. Scheme of PEG-$(PLLA_n)_8$-AC synthesis [68].

In order to introduce stable cross-links and thus improve the mechanical properties of the hydrogels formed by PEG-b-PLA star block copolymers, the stereocomplexation, that is co-crystallization, of PLLA and PDLA blocks can also be exploited. Moreover, enantiomeric PEG-PLA star block copolymers with a central PEG core and outer PLA blocks were shown to gelate faster and form stereocomplexed hydrogels with improved mechanical strength, in comparison with triblock PLA-b-PEG-b-PLA copolymers [6].

Regarding degradation issue, four-arm star PEG-b-PLA polymers are thermally stable at biological conditions. Due to relatively short degradation times, PEG-b-PLA star copolymers might be a superb candidate for drug delivery applications. Because, during hydrolytic degradation, the PLA chain length decreased, lactic acid concentration increased.

The lactic acid concentration of the medium content PLA samples (PEG/PLA = 2.5/0.8) reached a maximum around day 21 (0.256 mg/mL × 10), suggesting total degradation of the PLA chain. The initial concentration was ca. 0.140 mg/mL × 10. The high content PLA sample (PEG/PLA = 2.5/1.6) at day 21 had an acid concentration of 0.449 mg/mL × 10—the initial concentration was ca. 0.200 mg/mL × 10. Using the aforementioned copolymers as a short-term drug release agent was suggested [70].

The solutions presented in this subsection give information concerning other metal ions that can be used in PLA copolymerization processes as they have key importance in the issue of forming a shape of molecule by creating ligands. On the other hand, the presence of star-shaped PEG-$(SH)_8$ is important in terms of degradation and protein release time because of its molecular composition.

6.4. PCL-PLA Copolymer

Polycaprolactone (PCL) is a synthetic thermoplastic polymer. It is a linear semicrystalline polyester that is degraded in a natural environment by bacteria and fungi. In order to enhance its properties, copolymers such as poly(glycolide-co-caprolactone) (PGCL) and poly($_{L}$-lactide-co-ε-caprolactone) (PLCL) have been studied [17].

PCL-PLA copolymers have features of thermoplastic elastomers [71]. A biodegradable PCL-PLA multiblock copolymer was obtained when hexamethylene diisocyanate (HMDI) was added during the chemical reaction. A strong interaction between PCL and PLA was observed. This might indicate that there is a reaction between the PCL hydroxyl, PLA carboxyl, and HMDI isocyanate groups. A copolymer with a mass ratio of PCL:PLA = 80:20 has a Young's modulus of 2.7 ± 0.7 MPa and tensile elongation at break ca. 790%. This composition is desirable in medicine and technology [72].

A PCL-PLA long-chain branched block copolymer was introduced in order to prepare a biodegradable PLA material with enhanced crystallinity, rheological behavior, and mechanical properties. Adding the PCL-PLA copolymer to the neat PLA improved its tensile toughness without ill effect on the above-mentioned properties. Furthermore, PLA/PCL-PLA blend with 15 wt % of the PCL-PLA copolymer had much better elongation at break (210.7%) than neat PLA (7.1%). The studied copolymer was synthesized in the reaction of single hydroxyl-terminated PLA (PLA-OH) with hydroxyl-terminated three-arm star PCL (PCL-3OH) in the presence of HMDI. HMDI was used as the chain-extending agent [73]. The toughening of PLA, while simultaneously preserving its biodegradability and mechanical properties, should be highlighted. Subsequent studies should focus on seeking the most practical mass ratio between PLA and the PCL-PLA copolymer. The bioplastic material described may find numerous technological applications.

Song et al. [74] studied ABA PCL/PLA/PCL block copolymers and AB PCL/PLA (example structure in Figure 20) block copolymers. The physical and mechanical properties of these compounds are intriguing—especially elongation at break. Elongation at break of the PCL/PLA block copolymer with [CL]/[LA] = 72/28 is 380%, while the value for the PCL/PLA/PCL block copolymer with [CL]/[LA] = 75/25 is 600%, without the sacrifice of tensile strength. The suggestion of using ABA block copolymers as a biomedical material with tough membrane forming properties for a sustained release drug delivery system was presented.

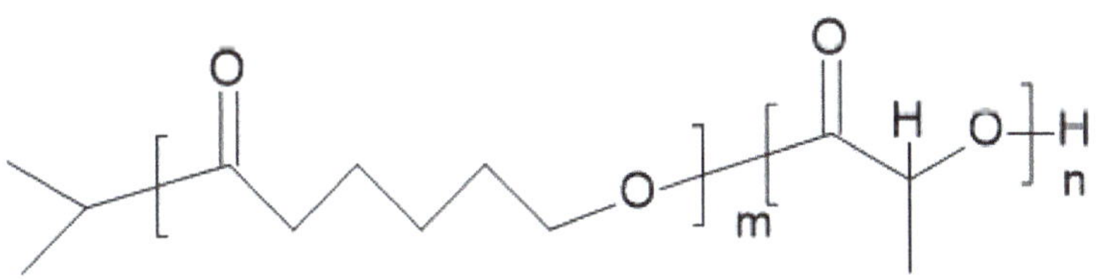

Figure 20. Chemical structure of diPCLPLA block copolymer [75].

Regarding thermal properties, tubular scaffolds made of poly(L-lactide-co-ε-caprolactone) (PLCL) (50:50) random copolymers were synthesized, and DSC analysis showed two potential glass transition temperatures (T_g) of the scaffolds in the 0 to −40 °C region. No crystalline melting peak was observed, which shows that the PLCL random copolymer is amorphous. The DMA profile indicated two T_gs: one at −38 °C and a second at −11 °C. This indicated a phase-separated structure of the PLCL studied. PCL and PLA homopolymers had higher T_g: −60 °C and +50 °C, respectively. Hence, the T_g at −38°C aligns with a phase composed of mainly CL units, and the T_g at −11 °C signifies the other phase containing more LA moieties. Additionally, the scaffolds had 200% higher elongation at break values than their homopolymers, which indicates the advantages of copolymerizing LA and CL to make flexible and soft copolymer scaffolds. FT-IR analysis confirmed the formation of random copolymers [76].

Elongation at break of PCL-PLA copolymers should be noted. The value of this property is much higher for PCL-PLA copolymers than for PCL and PLA homopolymers. This advantage can be used in the packaging and medical industries.

6.5. POSS-PLA Hybrid Copolymer

The physical and mechanical properties of PLA can be greatly improved by developing organic-inorganic hybrid materials. Polyhedral oligomeric silsesquioxane (POSS), consisting of silicon and oxygen atoms arranged in an inner eight-cornered cage with Si atoms positioned at the corners, was synthesized first. The star-shaped POSS-polylactides (POSS-PLAs) with varied PLA arm lengths were then obtained through ring opening polymerization of D,L-lactide. Eventually, the star-shaped POSS-PLA based polyurethanes (POSS-PLAUs) were formed by cross-linking POSS-PLA and polytetramethylene ether (PTMEG) with HMDI. POSS-PLAUs presented superb shape memory properties. POSS-PLAUs with shorter arm length showed faster recovery speed as a result of the higher POSS core content [77].

The ring-opening polymerization of L,L-LA, catalyzed by $Sn(Oct)_2$ was initiated by the functionalized silsesquioxane cages of the regular octahedral structure. As a result, biodegradable hybrid star-shaped POSS-PLA and linear systems with an octasilsesquioxane cage as a core and PLLA arms were given. Biodegradation of the compounds obtained is assumed on the grounds that both lactide blocks and POSS moieties are biodegradable [78]. Biodegradable POSS lactide systems can be applied in biomedical applications. A unique class of inorganic structures presented by POSS can be utilized in the era of hybrid polymer systems with advantageous properties [79].

In order to reduce the brittleness of PDLLA, a highly branched hybrid copolymer based on polyhedral oligomeric silsesquioxane POSS was composed. POSS-OH was used as the core of the toughening material, and the ring-opening polymerization of ε-caprolactone and D,L-lactide was then initiated sequentially to create the highly branched POSS-g-poly(ε-caprolactone)-b-poly(D,L-lactide) (POSS-g-PCL-b-PLA) copolymer with eight PCL-b-PLA arms. Furthermore, POSS-g-PCL-b-PLA/PDLLA nanocomposites were prepared via solution casting. Due to adding the PLA segment, good compatibility and distribution between POSS-g-PCL-b-PLA and the PDLLA matrix were observed. Elongation at break increased, and the yield stress decreased as the POSS-g-PCL-b-PLA content increased. This was due to the core-shell structure of POSS-g-PCL-b-PLA, which considerably improved the toughness of the PDLLA polymer matrix [80].

Synthesis of the star-shaped organic/inorganic hybrid PLLA, based on POSS, was begun from POSS bearing octa(3-hydroxypropyl) moieties [81]. Subsequently, further transformation of POSS-PLA was made. POSS-PLA was changed into the POSS-containing star-shaped organic/inorganic hybrid amphiphilic block copolymers, poly(L-lactide)-block-poly(N-isopropylacrylamide) (POSS(PLLA-b-PNIPAM)), by the reversible addition-fragmentation transfer (RAFT) polymerization of N-isopropylacrylamide (NIPAM) (see Figure 21). Star-shaped POSS-PLLA-b-PNIPAM amphiphilic block copolymers self-assembled into vesicles in an aqueous solution. Hydrophilic PNIPAM blocks and the hydrophobic

POSS core and PLLA created coronas and the vesicular wall, respectively. The temperature dependence of the hydrodynamic radius (R_h) for POSS($PLLA_{12}$–b–$PNIPAM_{119}$)$_8$ block copolymers in aqueous solution was investigated with dynamic light scattering (DLS) measurements. When temperature decreased from 34 °C to 30 °C, the R_h noticeably increased from 53 nm to 93 nm. This shows that the PNIPAM block in the aggregates is temperature responsive. At temperatures below 30 °C, the R_h did not change significantly during the cooling or heating processes, meaning that the phase-transition process of the PNIPAM block is reversible. It can be deduced that, for the cooling process with temperatures below 34 °C, PNIPAM chains began to stretch. The self-assembly morphology of POSS(PLLA–b–PNIPAM) block copolymers was studied by transmission electron microscopy (TEM). Self-assembled vesicular structures of the star-shaped POSS(PLLA–b–PNIPAM) amphiphilic block copolymers in aqueous solution were observed. However, there was a broad disparity in the size of the vesicular aggregates, and the density of the vesicular wall was not uniform. The outer diameter of the vesicles was polydispersed in the range of 20 nm to 35 nm. This size of the vesicles was smaller than the values measured by DLS. This results from the fact that DLS data directly reflects the size of self-assembly aggregates in solution, where the PNIPAM block chains are sufficiently dispersed in water, even although the PNIPAM chains are attached by one end onto the surface of the vesicular wall. The block copolymers described could be exploited in medical and biological fields [81].

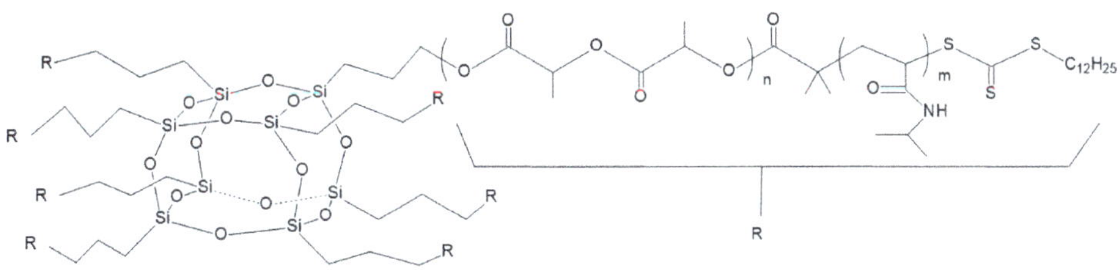

Figure 21. Structure of POSS(PLLA-b-PNIPAM) block copolymer [81].

The compounds presented in this subsection are distinct from the other examples described in this paper on the grounds that they are inorganic–organic hybrid materials. POSS-g-PCL-b-PLA/PDLLA composite desirably decreases the brittleness of linear PDLLA, which is said to be its one of its biggest drawbacks. Moreover, the amphiphilic and self-assembly character of the POSS(PLLA–b–PNIPAM) block copolymer should be noted. The highlighted features of the aforementioned compounds give POSS-PLA hybrid materials considerable prospects for biomedical applications.

6.6. PVA-g-PLA Copolymer

Graft polymers are a subclass of branched polymers. They are composed of blocks connected to the main chain as side-chains, with the chemical constitution of these side-chains differing from those of the main chain [50,51].

The grafting of PLA chains by initiating LA (or LA and GA) polymerization from OH groups of the poly(vinyl alcohol) (PVA) backbone is an example of the "grafting from" method (Figure 22) [82,83]. Such action allows modification of the polymer molecular architecture. This in turn has an impact on the crystallinity and biodegradability of a polymer. Graft polymers can be applied in drug delivery systems.

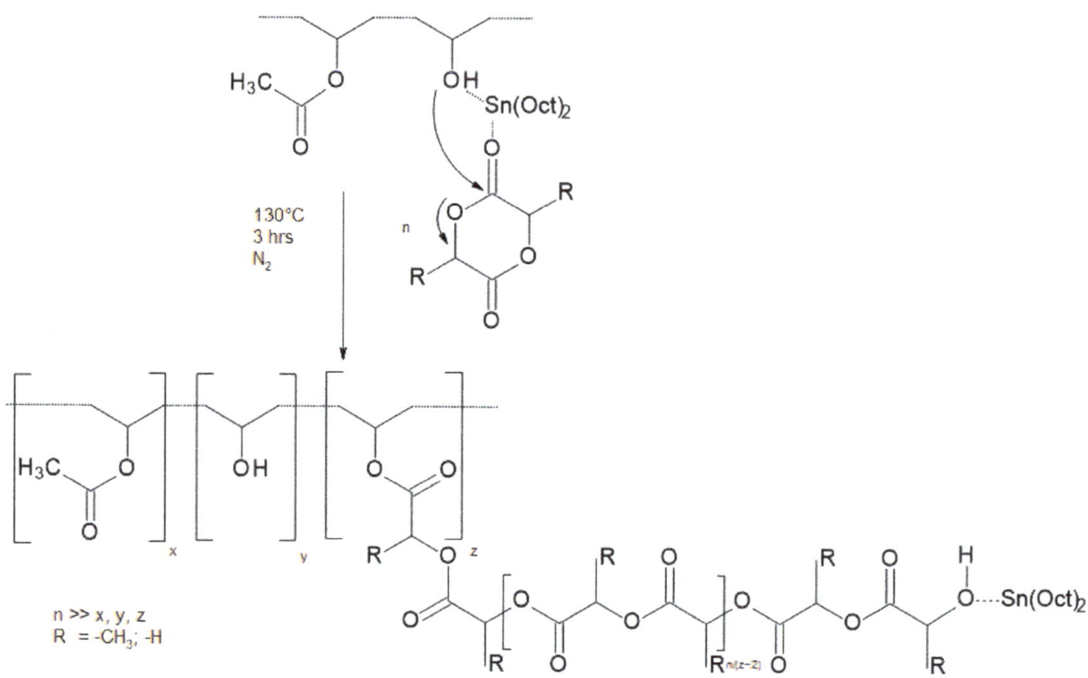

Figure 22. Synthesis of graft polymer PLA (PGA) on PVA by "grafting from" method [82].

Wang et al. [83] performed lactide polymerizations in bulk (130 °C) and used stannous octoate as a catalyst. PVA-g-PLA and PVA-g-P(LA-co-GA) copolymers with M_n in the range of 75,000–275,000 (SEC) were obtained. They can be applied as drug delivery systems with tunable physicochemical properties such as molecular composition, molecular weight, degree of crystallinity, and both melting and glass transition temperatures that could be adapted to the demands of drug delivery. Release of the hydrophilic dextran was investigated by preparing microspheres. The release profiles depended on degradation characteristics and were further modified by introducing sulfonate groups to poly(vinyl alcohol), which resulted in generating negative charges along the PVA backbone.

Sulfonate modified P(VS-VA)-g-PLGA compounds are especially promising because the degree of sulfonate substitution determines degradation time. One of the samples lost 50% of its mass in 8 days. It seems to be very useful in terms of drug delivery systems. Further research concerning degradation and drug release appears to be necessary. Studies concerning the potential applications of P(VS-VA)-g-PLGA in the packaging industry could also be carried out.

6.7. *PLA-Glycidol Copolymer*

It was found that adding glycidol to PLA chains efficiently led to its branching. Branched polymers were formed in a one-pot approach during polymerization performed in bulk at temperatures above the melting point of LA (100–180 °C); 4–70 ratios of lactide/glycidol (LA/GLY) were used with either $BF_3 \cdot Et_2O$ as a Lewis acid or diphenyl phosphate as a protic acid. With the use of diphenyl phosphate as a catalyst, a higher level of lactide conversion was observed, and the products were colorless in contrast to $BF_3 \cdot Et_2O$ usage. The branched structure of the obtained polymers was additionally confirmed with thermal analysis. PLA copolymers were marked by a significant decrease in the melting temperature (T_m) and melting enthalpy (ΔH_m) compared with PLA homopolymer. This resulted in a strongly lowered tendency for crystallization. In the copolymer, where [LA]/[GLY] = 17, T_m was 130 °C and 134 °C and ΔH_m was 4.5 J/kg. By use of the multimodal SEC trace, it is suggested that, at the beginning of the copolymerization, macromolecules of PLA initiated with glycidol molecules were formed, and only later were these macromolecules coupled via terminal ring opening. Initiation with hydroxyl groups occurring simultaneously in macromolecules eventuated in the formation of branches (Figure 23) [84]. It is assumed that a branched structure in the studied system is formed in terms of the activated monomer mechanism of cationic polymerization proceeding in the presence of hydroxyl groups [85].

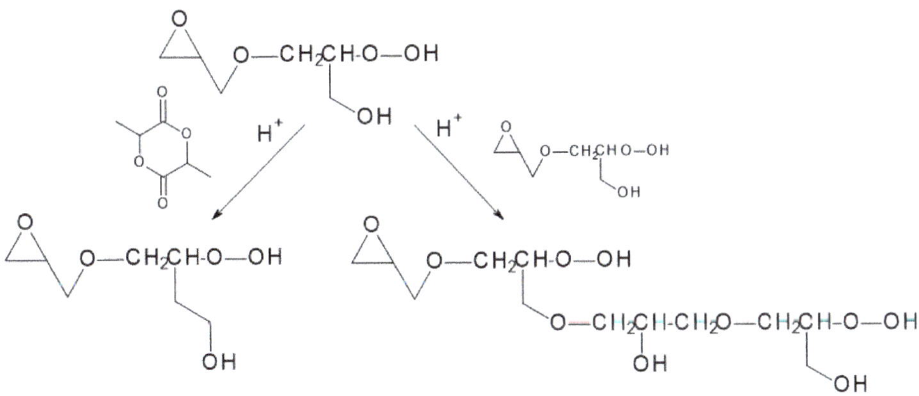

Figure 23. Mechanism of branched polymer formation by ring-opening copolymerization of lactide with glycidol [84].

Besides improving thermal properties, the hydrophilicity of PLA was also markedly enhanced as a result of adding glycidol units, which was confirmed with contact angle measurements. The contact angle of polymer films on a glass plate (with water as a reference liquid) decreased from 95.7° for linear PLA to 62.4° for a copolymer containing 20 mol% glycidol [11]. Better hydrophilicity is important for PLA because degradation increases as material hydrophilicity increases.

Reducing LA-GLY copolymer melting temperature by over 30 °C in comparison with linear PLA might create new possibilities in PLA synthesis, because the temperature of synthesis is one of the key factors. The lower the melting temperature becomes, the weaker are the forces between molecules and, as a result, chemical synthesis is facilitated. Furthermore, carrying out the synthesis at lower temperature is technically more practical. When synthesis is conducted taking melting temperature into consideration, finding the purity of the compound obtained is also simplified. In order to extend the study, other Lewis acids can be tested as catalysts. With regards to the enhanced hydrophilicity of PLA/GLY, copolymer research concerning the degradation properties of these compounds could be desirable in order to analyze the potential biomedical applications of this material.

6.8. PLA-Hydroxyoxetanes Copolymers

The bulk copolymerizations of lactide with 4-membered hydroxyetanes (3-hydroxymethyl-3-methyl-oxetane (HMMOX) and 3-ethyl-3-hydroxymethyl-oxetane (EHMOX) were considered) catalyzed by diphenyl phosphate were completed. As a result, branched polymers with molecular weights in the range of 1800 16,600 (SEC measurement with refractive index (RI) detection and against PSt standards) were obtained. Their molecular weights were depended on the lactide/oxetane ratio in the feed. Due to crude reaction mixture SEC traces, it is revealed that a monomodal distribution was executed, in contradistinction to that of lactide/glycidol copolymers. However, just as in lactide/glycidol copolymers, DSC analyses showed a considerable decrease in crystallinity. Investigating copolymer with [LA]/[OX] = 17, T_m was 134 °C and ΔH_m was 0.5 J/kg. Matrix assisted laser desorption/ionization time of flight (MALDI TOF) analyses carried out for all polymers confirmed the presence of several "branching monomer" units in LA/GLY and LA/OX copolymers [11].

Attempts to increase molecular weights could be made. Additionally, it is worth noting that the authors of studies on LA/OX copolymer reached even lower melting enthalpy than researchers working on the aforementioned LA/GLY compound. Lower melting enthalpy, and hence crystallinity, is essential for PLA compounds because of degradation. Crystalline regions are more resistant to hydrolysis and, as a result, crystalline and semicrystalline polymers are marked by slower degradation rates than amorphous ones [7].

6.9. PLA Copolyesters

The "grafting from" method can be used to prepare graft polyesters, where the polymer backbone is also a polyester, by the application of precisely designed comonomers [11].

Using lactide and a specially synthesized monomer, the cyclodepsipeptide cyclo[Glc-Ser(OBz)] (CGS-OBz) copolyesters were prepared. Polymerization was catalyzed by stannous octoate. When polymerizations were performed with 5–15 mol% cyclo[Glc-Ser(OBz)], copolyesters with molecular weights in the range of 20,000–25,000 were obtained. Afterwards, deprotection of hydroxyl groups was conducted, and polylactide chains were grafted onto this polymer, again using $Sn(Oct)_2$ as a catalyst of polymerization performed in bulk [86]. The applied strategy is presented in the Figure 24.

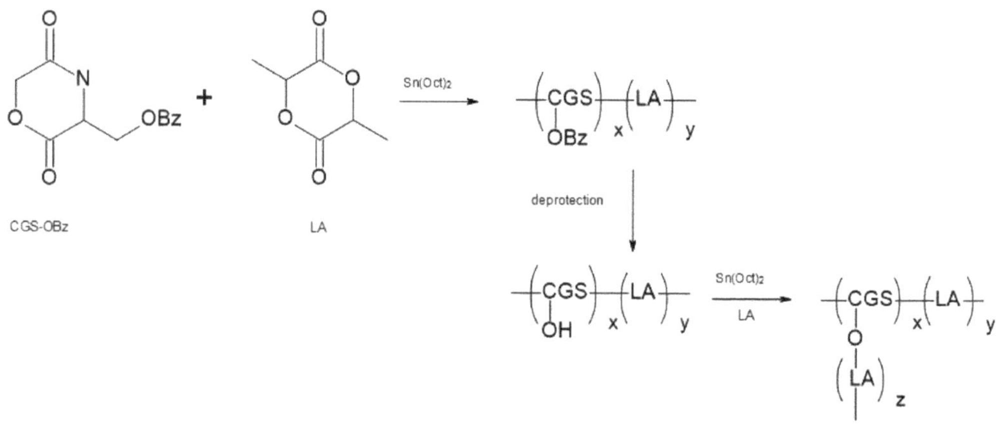

Figure 24. Synthesis of graft polymer PLA onto poly(LA/CGS-OBz) by the generation of initiating groups and the "grafting from" method [86].

In order to scrutinize the crystallinity of PLA-based graft polymers, the DSC analysis method was introduced. Moreover, the copolymers were subjected to hydrolysis with the aim of making a correlation between the polymer architecture and their crystallinity and hydrolytic stability. The result was that the obtained comb-type PLA presented a decrease in crystallinity and an increase in biodegradability compared with linear PLA. Varying the molecular architecture could bring a change in the degradation rate of PLA [86].

Observed correlations between crystallinity, molecular architecture, and the degradability of synthesized copolymers are noticeably detailed, and extending the research on mechanical studies is welcomed. Contact angle measurements would certainly give more data regarding the hydrophilicity of graft comb-type PLA copolymers.

6.10. PLA Copolymers—Summary

In summary, the aim of presenting PLA copolymers in this section was to show that PLA copolymerization is a prospective method for changing several PLA properties. Essentially, copolymerizing PLA with compounds such as PGA, PEG, or glycidol shortens PLA degradation time. In the authors' judgement, this is the main advantage of forming different PLA copolymers. Secondly, creating various PLA copolymers enables the manipulation of PLA hydrophobicity. The ruthenium star block copolymer mentioned has a PLA hydrophobic core and a PAA hydrophilic corona. Additionally, adding glycidol to PLA enhances PLA hydrophilicity, which has an impact on degradation rate and potential biomedical applications (e.g., drug delivery systems). Furthermore, PLA copolymers (e.g., PCL-PLA) are marked by improved toughness and/or elongation at break in comparison with neat PLA. Depending on the needs, diversifying other PLA properties, such as crystallization or contact angle values, makes it possible to compose new PLA copolymers. The fact that both inorganic (e.g., POSS) and organic compounds can be used for PLA copolymerization is worthy of note Many possibilities as regards designing different copolymer architectures also aid this method's advance. As a result, PLA copolymers with broad applications can be obtained—the next section is focused on this issue.

7. PLA Applications

Two main areas of PLA use can be detailed. The first field is represented by products of general use, while the second involves specialistic applications in medicine. The following products made of PLA for general use can be listed: containers, wrapping films, disposable products, and elements of interior furnishings. On the other hand, in medicine, PLA can be applied in the areas of bioresorbent implants, surgical sutures, clamps, clips, surgical masks, or dressings [14,87–91]. PLA copolymers also have special applications.

Studies on the PLA-polyethylene glycol block copolymer (PLA-PEG) and the PLA-p-dioxanone-polyethylene glycol block copolymer (PLA-p-DPEG) have been made. These compounds have been used as carriers for bone morphogenetic proteins (BMPs). BMPs are biologically active molecules able to induce new bone formation, and they are expected to be used clinically in connection with biomaterials, such as bone-graft substitutes to stimulate bone repair. The effect of PLA on the osteoinduction of demineralized bone and the usefulness of PLA as a carrier of BMP were analyzed. It was found that PLA was a satisfactory candidate as a carrier for BMP [92]. Initially, low molecular weight PLA was mixed with BMP in order to form a composite, which was next implanted in the host bone; new bone cells were formed while the degradation of the PLA matrix in the composite was occurring [93]. However, the newly-formed bone was too low in quantity (bone mineral density). For this reason, PLA copolymers with low molecular weight were used to solve these problems [10].

However, in some cases, bone defects occur in positions that require dynamic strength (e.g., the long bone of the leg). In these situations, in order to restore the bone, the BMP/polymer composite has to be combined with a solid biomaterial with good affinity for bone. As an example, a titanium implant featuring a porous surface on which the BMP/polymer composites can be settled is an appropriate material that can be implanted into the bone defect. The new bone formed by the BMP/polymer composites would then firmly surround the titanium implant until the implant and the host bone fused [93]. Thus, even repairing bone defects that require strength would become possible. This process is described in Figure 25 [92]. When the pores of the solid implant are filled with the BMP/polymer composite and implanted, the composite exudes from the pores and forms a layer of bone covering the surface of the biomaterials. This layer of bone may encase the implant and thus improve biological fixation of the biomaterials to the host bone [92].

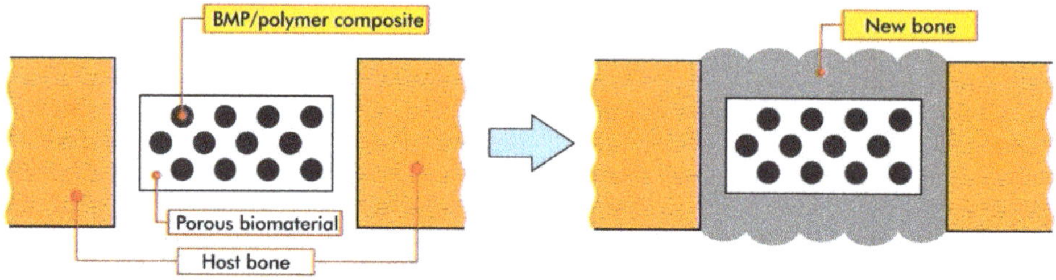

Figure 25. The process of repairing bone defects with the use of BMP/polymer composites [92].

PLA, in view of its biodegradability, is also used in drug delivery systems (DDS) in which the drug can be released continuously for different periods of time up to one year [10]. PLA and their copolymers, in the form of nano-particles, have been used in several applications in the encapsulation process of many drugs, such as dermatotherapy [94], protein (BSA) [95], hormones [96], restenosis [97], antitumor oridonin [98], and psychotic [99]. These nano-particles were received with different methods, such as solvent displacement, solvent evaporation [100], emulsion solvent diffusion [101], and salting out [99].

A lisinopril-conjugated triblock PLA-PEG-PLA copolymer was synthesized by the reaction of PLA-PEG-PLA copolymer with lisinopril (the antihypertensive drug). Subsequently, the lisinopril-conjugated PLA-PEG-PLA was self-assembled into micelles in an aqueous solution. Conjugated micelles were characterized by a better sustained release profile in comparison with the lisinopril-conjugated copolymer and physically loaded micelles [102]. This micellar formulation of drug conjugated amphiphilic copolymers seems to be a significant achievement in controlled drug delivery.

A PLA-PEG-PLA copolymer was used to prepare microspheres containing paclitaxel. The microspheres obtained had a porous structure inside, which advanced drug release. In vitro release was 49% in one month, making drug action time in the body longer. This drug delivery system avoids inconsistent local concentration on account of drug release. Thus, the therapeutic effect is enhanced [90].

Poly(lactic-co-glycolic) acid nanoparticles were used for loading the drug, paclitaxel. The preparation of nanoparticles was conducted by the emulsion solvent evaporation method in the presence of tocopheryl polyethylene glycol succinate as an emulsion agent. The ability of in vitro drug release of the nanoparticles, encapsulation efficiency, and the drug loading efficiency were investigated. The in vitro release mode of drug-loaded nanoparticles appeared to be two-staged—a fast release in the initial stage and a slower release in a second stage. The encapsulation efficiency was 4.84%, and the drug loading efficiency was 67.35% [103]. In order to enhance the abovementioned values, further studies on the porosity of PLGA nanoparticles could be carried out. Changing the LA/GA mass ratio in a copolymer would have a potential impact on its degradation behavior. The two-staged mode of drug release described can be practically applied in particular medical circumstances.

The combination of PLGA with the antibacterial substance totarol was used as a novel coating solution for surgical sutures. Collected data suggests that the biodegradable suture coating obtained has the potential to reduce the risk of surgical site infections (SSIs)—a nosocomial infection that can result in severe complications after surgical intervention. The coating prevents post-operative biofilm formation during the critical phase of wound healing. It has no negative impact on tissue [104].

Regarding tissue engineering, PCL-PLA copolymer is a relevant compound for this type of applications. Recently, a PCL-PLA copolymer nanofiber has been used in the regeneration of damaged tissue [105]. Moreover, an improvement in the mechanical properties of PCL can be achieved by copolymerization with PLA, enabling its use for orthopedic applications, such as the repair of bone defects [106].

Another copolymer that has the potential to be applied in medicine is PVA-g-PLA. Animal experiments featuring thin copolymer films made of this material were conducted. The thickness of the samples was 0.04–0.06 mm. In comparison with the PLA homopolymer, the copolymer obtained had improved hydrophilicity and flexibility. The films showed a satisfying anti-tissue adhesion effect and applicable degradability in the body of the mouse. The film entirely disappeared after 8 weeks of implantation. The films were also biocompatible, as expected, because no inflammation, hematoma, or infection were noticed [107]. These results are very promising regarding future PVA-g-PLA use for preventing post-operative organ tissue adhesion.

A poly(3,4-ethylenedioxythiophene) (PEDOT) polymer and a biocompatible polymer polylactide were synthesized to design graft copolymers. The weight fraction of 3,4-ethylenedioxythiophene (EDOT) was between 5 and 40%. The compound created can be 3D printed using direct melting extrusion methods; 5:95 wt % PEDOT-g-PLA showed the best performance during the printing process as it improved the ability to control its shape. This compound was used as the patterning material for biocompatibility tests on neonatal cardiac cultures. Tissue-like structures made of cardiomyocytes with fibroblast were developed in PEDOT-g-PLA and cardiomyocyte, improving the approved suitable maturation and functionality of these cells. The promising results of the presented study could assist the progress of generating artificial tissue [108].

8. Conclusions

PLA is a biodegradable polymer synthesized from lactic acid. Because PLA is an ecological material, its use could make a positive difference to the worldwide environment.

Regarding PLA synthesis, ring-opening polymerization is a method during which lactide is converted into PLA. The reaction is catalyzed by organometal catalysts. Stannous octoate ($Sn(Oct)_2$) is a compound commonly used as a catalyst. It gives PLA weight distributions (M_n) between 40,000 and 250,000. $Sn(Oct)_2$ efficiency can be improved by adding a Lewis base or distannoxane to the synthesis process. Moreover, stannous octoate can be applied in the biomedical sector as it is accepted by the FDA (Food and Drug Administration, Silver Spring, MD, USA). However, studies concerning new catalysts that can be used in PLA synthesis are necessary because there is a known negative impact of $Sn(Oct)_2$ on PLA properties. In particular, biocompatible catalysts are desirable.

Apart from the many advantages of PLA, such as appropriate biodegradability, durability, and transparency, it also has some drawbacks: brittleness, relatively high melting temperature, and hydrophobicity. Hence, PLA copolymers are composed with the intent of enhancing specific PLA properties. Diverse compounds in terms of composition (e.g., POSS-PLA hybrid copolymers) and polymer architecture (e.g., PVA-g-PLA or PLA copolyesters) have been created. As shown by DSC analysis, graft PLA copolymers are marked by a lower crystallinity and higher degradation rate than linear PLA. The sequence of individual monomers also has an influence; random PLGAs degrade quicker than sequenced ones. A PLA-glycidol copolymer provides significant improvement in terms of thermal properties and better hydrophilicity. Regarding elongation at break and tensile strength, PCL-PLA copolymers should be highlighted. It can be concluded that PLA copolymers enable the creation of ideal materials that can be applied in certain fields, such as medicine, packaging, or technology. Biocompatible PLA copolymers seem to have considerable potential in the field of biomedical applications such as drug delivery systems or tissue engineering. Studies concerning this subject should be continued.

Polylactic acid might be a crucial material for the plastics industry in the near future due to its environmentally friendly character. Several methods of PLA synthesis and the many possibilities for creating new PLA copolymers make conducting new research desirable. Improving PLA properties and searching for new applications for PLA seem to be the two biggest challenges for the development of PLA-based materials.

Author Contributions: Conceptualization, A.M.; methodology, K.S. and A.M.; formal analysis, A.M.; investigation, K.S.; data curation, A.M.; writing—original draft preparation, K.S.; writing—review and editing, A.M.; visualization, K.S. and A.M.; supervision, A.M.; project administration, A.M. Both authors have read and agreed to the published version of the manuscript.

Funding: This research received no external funding.

Institutional Review Board Statement: Not applicable.

Informed Consent Statement: Not applicable.

Data Availability Statement: No new data were created or analyzed in this study. Data sharing is not applicable to this article.

Conflicts of Interest: The authors declare no conflict of interest.

References

1. Shah, A.A.; Hasan, F.; Hameed, A.; Ahmed, S. Biological degradation of plastics: A comprehensive review. *Biotechnol. Adv.* **2008**, *26*, 246–265. [CrossRef]
2. Nampoothiri, K.M.; Nair, N.R.; John, R.P. An overview of the recent developments in polylactide (PLA) research. *Bioresour. Technol.* **2010**, *101*, 8493–8501. [CrossRef]
3. Szaraniec, B.; Stodolak-Zych, E.; Morawska-Chochół, A. Eco-composites based on polylactide and natural fibers. *Przetwórstwo Tworzyw* **2011**, *6*, 534–538.
4. Jem, K.J.; Tan, B. The development and challenges of poly (lactic acid) and poly (glycolic acid). *Adv. Ind. Eng. Polym. Res.* **2020**, *3*, 60–70. [CrossRef]
5. Żenkiewicz, M.; Richert, J. Synthesis, properties and applications of polylactide. *Przetwórstwo Tworzyw* **2009**, *15*, 192–199.
6. Michalski, A.; Brzeziński, M.; Lapienis, G.; Biela, T. Star-shaped and branched polylactides: Synthesis, characterization, and properties. *Prog. Polym. Sci.* **2018**, *89*, 159–212. [CrossRef]
7. Casalini, T.; Rossi, F.; Castrovinci, A.; Perale, G. A Perspective on Polylactic Acid-Based Polymers Use for Nanoparticles Synthesis and Applications. *Front. Bioeng. Biotechnol.* **2019**, *7*, 259. [CrossRef]
8. Fortelny, I.; Ujcic, A.; Fambri, L.; Slouf, M. Phase Structure, Compatibility, and Toughness of PLA/PCL Blends: A Review. *Front. Mater.* **2019**, *6*. [CrossRef]
9. Mehta, R.; Kumar, V.; Bhunia, H.; Upadhyay, S.N. Synthesis of Poly(Lactic Acid): A Review. *J. Macromol. Sci. Part C* **2005**, *45*, 325–349. [CrossRef]
10. Hamad, K.; Kaseem, M.; Yang, H.W.; Deri, F.; Ko, Y.G. Properties and medical applications of polylactic acid: A review. *Express Polym. Lett.* **2015**, *9*, 435–455. [CrossRef]
11. Bednarek, M. Branched aliphatic polyesters by ring-opening (co)polymerization. *Prog. Polym. Sci.* **2016**, *58*, 27–58. [CrossRef]
12. Morrison, R.T.; Boyd, R.M. *Chemia Organiczna*; PWN: Warszawa, Poland, 1997.
13. Jem, K.J.; van der Pol, J.F.; de Vos, S. Microbial Lactic Acid, Its Polymer Poly(lactic acid), and Their Industrial Applications. *Probiotics* **2009**, *14*, 323–346. [CrossRef]
14. Foltynowicz, Z.; Jakubiak, P. Polylactid acid—Biodegradable polymer obtained from vegetable resources. *Polimery* **2002**, *47*, 769–774. [CrossRef]
15. Jankiewicz, M. *Zastosowanie Kwasu Mlekowego i Jego Pochodnych*; Akademia Rolnicza w Poznaniu, PWN: Poznań, Poland, 1995.
16. Schlegel, H.G. *Mikrobiologia Ogólna*; PWN: Warsaw, Poland, 2003.
17. Zaborski, M. Elastomer composites with proecological additives Kompozyty elastomerowe z dodatkami proekologicznymi. *Przem. Chem.* **2017**, *1*, 167–172. [CrossRef]
18. Gołębiewski, J.; Gibas, E.; Malinowski, R. Selected biodegradable polymers—Preparation, properties, applications. *Polimery* **2008**, *53*, 799–807. [CrossRef]
19. Saeidlou, S.; Huneault, M.A.; Li, H.; Park, C.B. Poly(lactic acid) crystallization. *Prog. Polym. Sci.* **2012**, *37*, 1657–1677. [CrossRef]
20. Dorgan, J.R.; Janzen, J.; Clayton, M.P.; Hait, S.B.; Knauss, D.M. Melt rheology of variable L-content poly(lactic acid). *J. Rheol.* **2005**, *49*, 607. [CrossRef]
21. Hartmann, M.H.; Kaplan, D. *Biopolymers from Renewable Resources*; Springer: Berlin/Heidelberg, Germany, 1998; ISBN 978-3-662-03680-8.
22. Rochina, J.R.; Vidaurre, A.; Castilla-Cortázar, I.; Lebourg, M. Effects of hydroxyapatite filler on long-term hydrolytic degradation of PLLA/PCL porous scaffolds. *Polym. Degrad. Stab.* **2015**, *119*, 121–131. [CrossRef]
23. Ejaz, M.; Azad, M.M.; Shah, A.U.R.; Afaq, S.K.; Song, J.-I. Mechanical and Biodegradable Properties of Jute/Flax Reinforced PLA Composites. *Fibers Polym.* **2020**, *21*, 2635–2641. [CrossRef]
24. Murariu, M.; Dubois, P. PLA composites: From production to properties. *Adv. Drug Deliv. Rev.* **2016**, *107*, 17–46. [CrossRef]
25. Garlotta, D. A Literature Review of Poly(Lactic Acid). *J. Polym. Environ.* **2001**, *9*, 63–84. [CrossRef]
26. Lunt, J. Large-scale production, properties and commercial applications of polylactic acid polymers. *Polym. Degrad. Stab.* **1998**, *59*, 145–152. [CrossRef]
27. Ajioka, M.; Enomoto, K.; Suzuki, K.; Yamaguchi, A. Basic Properties of Polylactic Acid Produced by the Direct Condensation Polymerization of Lactic Acid. *Bull. Chem. Soc. Jpn.* **1995**, *68*, 2125–2131. [CrossRef]

28. Ren, J. Synthesis and Manufacture of PLA. In *Biodegradable Poly(Lactic Acid): Synthesis, Modification, Processing and Applications*; Springer: Berlin/Heidelberg, Germany, 2010; pp. 15–37.
29. Henton, D.; Gruber, P.; Lunt, J.; Randall, J. Polylactic Acid Technology. In *Natural Fibers, Biopolymers, and Biocomposites*; CRC Press: Boca Raton, FL, USA, 2005.
30. Gupta, A.; Kumar, V. New emerging trends in synthetic biodegradable polymers—Polylactide: A critique. *Eur. Polym. J.* **2007**, *43*, 4053–4074. [CrossRef]
31. Kricheldorf, H.; Dunsing, R. Polylactones, 8. Mechanism of the cationic polymerization of L,L-dilactide. *Die Makromol. Chem.* **1986**, *187*, 1611–1625. [CrossRef]
32. Jedlinski, Z.; Wałach, W.; Kurcok, P.; Adamus, G. Polymerization of lactones. XII, Polymerization of L-dilactide and L,D-dilactide in the presence of potassium methoxide. *Macromol. Chem. Phys.* **1991**, *192*, 2051–2057. [CrossRef]
33. Kurcok, P.; Matuszowicz, A.; Jedliński, Z.; Kricheldorf, H.R.; Dubois, P.; Jérôme, R. Substituent effect in anionic polymerization of β-lactones initiated by alkali metal alkoxides. *Macromol. Rapid Commun.* **1995**, *16*, 513–519. [CrossRef]
34. Zhang, X.; Macdonald, D.A.; Goosen, M.F.A.; McAuley, K.B. Mechanism of lactide polymerization in the presence of stannous octoate: The effect of hydroxy and carboxylic acid substances. *J. Polym. Sci. Part A Polym. Chem.* **1994**, *32*, 2965–2970. [CrossRef]
35. Jacobsen, S.; Fritz, H.-G.; Degée, P.; Dubois, P.; Jérôme, R. New developments on the ring opening polymerisation of polylactide. *Ind. Crop. Prod.* **2000**, *11*, 265–275. [CrossRef]
36. Stevels, W.M.; Ankoné, M.J.K.; Dijkstra, P.J.; Feijen, J. Well defined block copolymers of ε-caprolactone and L-lactide using Y5(μ-O)(OiPr)13 as an initiator. *Macromol. Chem. Phys.* **1995**, *196*, 1153–1161. [CrossRef]
37. Cameron, D.J.A.; Shaver, M. Aliphatic polyester polymer stars: Synthesis, properties and applications in biomedicine and nanotechnology. *Chem. Soc. Rev.* **2010**, *40*, 1761–1776. [CrossRef] [PubMed]
38. Karidi, K.; Pladis, P.; Kiparissides, C. A Theoretical and Experimental Kinetic Investigation of the ROP of L,L-Lactide in the Presence of Polyalcohols. *Macromol. Symp.* **2013**, *333*, 206–215. [CrossRef]
39. Leenslag, J.; Pennings, A. Synthesis of high-molecular-weight poly(L-lactide) initiated with tin 2-ethylhexanoate. *Die Makromol. Chem.* **1987**, *188*, 1809–1814. [CrossRef]
40. Hyon, S.-H.; Jamshidi, K.; Ikada, Y. Synthesis of polylactides with different molecular weights. *Biomaterials* **1997**, *18*, 1503–1508. [CrossRef]
41. Nijenhuis, A.J.; Grijpma, D.W.; Pennings, A.J. Lewis acid catalyzed polymerization of L-lactide. Kinetics and mechanism of the bulk polymerization. *Macromolecules* **1992**, *25*, 6419–6424. [CrossRef]
42. Montané, X.; Montornes, J.M.; Nogalska, A.; Olkiewicz, M.; Giamberini, M.; Garcia-Valls, R.; Fabregat, M.B.; Jubany, I.; Tylkowski, B. Synthesis and synthetic mechanism of Polylactic acid. In *Chemical Technologies and Processes*; De Gruyter: Berlin, Germany, 2020; Volume 5, pp. 281–296.
43. Gogolewski, S.; Jovanovic, M.; Perren, S.; Dillon, J.; Hughes, M. The effect of melt-processing on the degradation of selected polyhydroxyacids: Polylactides, polyhydroxybutyrate, and polyhydroxybutyrate-co-valerates. *Polym. Degrad. Stab.* **1993**, *40*, 313–322. [CrossRef]
44. Tasaka, F.; Ohya, Y.; Ouchi, T. One-Pot Synthesis of Novel Branched Polylactide Through the Copolymerization of Lactide with Mevalonolactone. *Macromol. Rapid Commun.* **2001**, *22*, 820–824. [CrossRef]
45. Målberg, S.; Basalp, D.; Finne-Wistrand, A.; Albertsson, A.-C. Bio-safe synthesis of linear and branched PLLA. *J. Polym. Sci. Part A Polym. Chem.* **2010**, *48*, 1214–1219. [CrossRef]
46. Bero, M.; Kasperczyk, J.; Jedlinski, Z.J. Coordination polymerization of lactides, 1. Structure determination of obtained polymers. *Die Makromol. Chem.* **1990**, *191*, 2287–2296. [CrossRef]
47. Kricheldorf, H.R.; Berl, M.; Scharnagl, N. Poly(lactones). 9. Polymerization mechanism of metal alkoxide initiated polymerizations of lactide and various lactones. *Macromolecules* **1988**, *21*, 286–293. [CrossRef]
48. Dubois, P.; Jacobs, C.; Jerome, R.; Teyssie, P. Macromolecular engineering of polylactones and polylactides. 4. Mechanism and kinetics of lactide homopolymerization by aluminum isopropoxide. *Macromolecules* **1991**, *24*, 2266–2270. [CrossRef]
49. Jacobs, C.; Dubois, P.; Jerome, R.; Teyssie, P. Macromolecular engineering of polylactones and polylactides. 5. Synthesis and characterization of diblock copolymers based on poly ε-caprolactone and poly(L,L or D,L)lactide by aluminum alkoxides. *Macromolecules* **1991**, *24*, 3027–3034. [CrossRef]
50. Barón, M.; Hellwich, K.-H.; Hess, M.; Horie, K.; Jenkins, A.D.; Jones, R.G.; Kahovec, J.; Kratochvil, P.; Metanomski, W.V.; Mormann, W.; et al. Glossary of class names of polymers based on chemical structure and molecular architecture (IUPAC Recommendations 2009). *Pure Appl. Chem.* **2009**, *81*, 1131–1186. [CrossRef]
51. Jenkins, A.D.; Kratochvil, P.; Stepto, R.F.T.; Suter, U. Glossary of basic terms in polymer science (IUPAC Recommendations 1996). *Pure Appl. Chem.* **1996**, *68*, 2287–2311. [CrossRef]
52. Sherazi, T.A. Graft Polymerization. In *Encyclopedia of Membranes*; Springer: Berlin/Heidelberg, Germany, 2014; pp. 1–2.
53. Gentile, P.; Chiono, V.; Carmagnola, I.; Hatton, P.V. An Overview of Poly(lactic-co-glycolic) Acid (PLGA)-Based Biomaterials for Bone Tissue Engineering. *Int. J. Mol. Sci.* **2014**, *15*, 3640–3659. [CrossRef]
54. Middleton, J.C.; Tipton, A.J. Synthetic biodegradable polymers as orthopedic devices. *Biomaterials* **2000**, *21*, 2335–2346. [CrossRef]
55. Abid, Z.; Mosgaard, M.D.; Manfroni, G.; Petersen, R.S.; Nielsen, L.H.; Müllertz, A.; Boisen, A.; Keller, S.S. Investigation of Mucoadhesion and Degradation of PCL and PLGA Microcontainers for Oral Drug Delivery. *Polymers* **2019**, *11*, 1828. [CrossRef]

56. Vey, E.; Rodger, C.; Booth, J.; Claybourn, M.; Miller, A.F.; Saiani, A. Degradation kinetics of poly(lactic-co-glycolic) acid block copolymer cast films in phosphate buffer solution as revealed by infrared and Raman spectroscopies. *Polym. Degrad. Stab.* **2011**, *96*, 1882–1889. [CrossRef]
57. Mao, H.-Q.; Shipanova-Kadiyala, I.; Zhao, Z.; Dang, W.; Brown, A.; Leong, K.W. Biodegradable poly(terephthalate-co-phosphate)s: Synthesis, characterization and drug-release properties. *J. Biomater. Sci. Polym. Ed.* **2005**, *16*, 135–161. [CrossRef]
58. Li, J.; Stayshich, R.M.; Meyer, T. Exploiting Sequence to Control the Hydrolysis Behavior of Biodegradable PLGA Copolymers. *J. Am. Chem. Soc.* **2011**, *133*, 6910–6913. [CrossRef]
59. Marszałek-Harych, A.; Jędrzkiewicz, D.; Ejfler, J. Bio- and chemocatalysis cascades as a bridge between biology and chemistry for green polymer synthesis. *Cell. Mol. Biol. Lett.* **2017**, *22*, 28. [CrossRef]
60. Zeng, F.-R.; Liang, Y.; Li, Z.-L. Precision Aliphatic Polyesters via Segmer Assembly Polymerization. *Molecules* **2018**, *23*, 452. [CrossRef] [PubMed]
61. Nowalk, J.A.; Fang, C.; Short, A.L.; Weiss, R.M.; Swisher, J.H.; Liu, P.; Meyer, T.Y. Sequence-Controlled Polymers Through Entropy-Driven Ring-Opening Metathesis Polymerization: Theory, Molecular Weight Control, and Monomer Design. *J. Am. Chem. Soc.* **2019**, *141*, 5741–5752. [CrossRef] [PubMed]
62. Johnson, R.M.; Fraser, C.L. Metalloinitiation Routes to Biocompatible Poly(lactic acid) and Poly(acrylic acid) Stars with Luminescent Ruthenium Tris(bipyridine) Cores. *Biomacromolecules* **2004**, *5*, 580–588. [CrossRef]
63. Szcześniak, K.B. *Star Polymers as a Model Delivery System for Drugs and Nucleic Acids: Synthesis, Structure and Dynamics of Cationic Star Polymers Based on Poly (Ethylene Glycol)*; Uniwersytet im. Adama Mickiewicza: Poznan, Poland, 2015.
64. Matyjaszewski, K. Atom Transfer Radical Polymerization (ATRP): Current Status and Future Perspectives. *Macromolecules* **2012**, *45*, 4015–4039. [CrossRef]
65. Basu, A.; Kunduru, K.R.; Doppalapudi, S.; Domb, A.J.; Khan, W. Poly(lactic acid) based hydrogels. *Adv. Drug Deliv. Rev.* **2016**, *107*, 192–205. [CrossRef] [PubMed]
66. Jung, J.; Hu, J.W. Characterization of Polyethylene Oxide and Sodium Alginate for Oil Contaminated-Sand Remediation. *Sustainability* **2017**, *9*, 62. [CrossRef]
67. Buwalda, S.J.; Dijkstra, P.J.; Feijen, J. Poly(ethylene glycol)-poly(L-lactide) star block copolymer hydrogels crosslinked by metal-ligand coordination. *J. Polym. Sci. Part A Polym. Chem.* **2012**, *50*, 1783–1791. [CrossRef]
68. Buwalda, S.J.; Dijkstra, P.J.; Feijen, J. In Situ Forming Poly(ethylene glycol)- Poly(L-lactide) Hydrogels via Michael Addition: Mechanical Properties, Degradation, and Protein Release. *Macromol. Chem. Phys.* **2012**, *213*, 766–775. [CrossRef]
69. Velthoen, I.W.; Tijsma, E.J.; Dijkstra, P.J.; Feijen, J. Thermo-Responsive Hydrogels Based on Branched Poly(L-lactide)-poly(ethylene glycol) Copolymers. *Macromol. Symp.* **2008**, *272*, 13–27. [CrossRef]
70. Salaam, L.E.; Dean, D.; Bray, T.L. In vitro degradation behavior of biodegradable 4-star micelles. *Polymer* **2005**, *47*, 310–318. [CrossRef]
71. Irska, I.; Piesowicz, E.; Rosłaniec, Z. Biodegradable PLA-based thermoplastic elastomers. *Elastomery* **2016**, *20*, 17–25.
72. Zhang, J.; Xu, J.; Wang, H.; Jin, W.; Li, J. Synthesis of multiblock thermoplastic elastomers based on biodegradable poly (lactic acid) and polycaprolactone. *Mater. Sci. Eng. C* **2009**, *29*, 889–893. [CrossRef]
73. Liu, M.-J.; Chen, S.-C.; Yang, K.-K.; Wang, Y.-Z. Biodegradable polylactide based materials with improved crystallinity, mechanical properties and rheological behaviour by introducing a long-chain branched copolymer. *RSC Adv.* **2015**, *5*, 42162–42173. [CrossRef]
74. Song, C.X.; Feng, X.D. Synthesis of ABA triblock copolymers of iε-caprolactone and DL-lactide. *Macromolecules* **1984**, *17*, 2764–2767. [CrossRef]
75. Liu, F.; Zhao, Z.; Yang, J.; Wei, J.; Li, S. Enzyme-catalyzed degradation of poly(l-lactide)/poly(ε-caprolactone) diblock, triblock and four-armed copolymers. *Polym. Degrad. Stab.* **2009**, *94*, 227–233. [CrossRef]
76. Jeong, S.I.; Kim, B.-S.; Lee, Y.M.; Ihn, K.J.; Kim, S.H.; Kim, Y.H. Morphology of Elastic Poly(l-lactide-co-ε-caprolactone) Copolymers and in Vitro and in Vivo Degradation Behavior of Their Scaffolds. *Biomacromolecules* **2004**, *5*, 1303–1309. [CrossRef]
77. Gu, S.Y.; Gao, X.F. Improved shape memory performance of star-shaped POSS-polylactide based polyurethanes (POSS-PLAUs). *RSC Adv.* **2015**, *5*, 90209–90216. [CrossRef]
78. Qiu, Z.; Pan, H. Preparation, crystallization and hydrolytic degradation of biodegradable poly(l-lactide)/polyhedral oligomeric silsesquioxanes nanocomposite. *Compos. Sci. Technol.* **2010**, *70*, 1089–1094. [CrossRef]
79. Rozga-Wijas, K.; Stanczyk, W.A.; Kurjata, J.; Kazmierski, S. Star-Shaped and Linear POSS-Polylactide Hybrid Copolymers. *Materials* **2015**, *8*, 4400–4420. [CrossRef]
80. Wang, T.; Ding, J.; Li, J.; Liu, Y.; Hao, J. Blends of poly(d,l-lactide) with polyhedral oligomeric silsesquioxanes-based biodegradable polyester: Synthesis, morphology, miscibility, and mechanical property. *J. Appl. Polym. Sci.* **2014**, *131*, 9341–9348. [CrossRef]
81. Zhang, W.; Wang, S.; Li, X.; Yuan, J.; Wang, S. Organic/inorganic hybrid star-shaped block copolymers of poly(l-lactide) and poly(N-isopropylacrylamide) with a polyhedral oligomeric silsesquioxane core: Synthesis and self-assembly. *Eur. Polym. J.* **2012**, *48*, 720–729. [CrossRef]
82. Breitenbach, A.; Kissel, T. Biodegradable comb polyesters: Part 1 synthesis, characterization and structural analysis of poly(lactide) and poly(lactide-coglycolide) grafted onto water-soluble poly(vinyl alcohol) as backbone. *Polymer* **1998**, *39*, 3261–3271. [CrossRef]
83. Wang, X.; Xie, X.; Cai, C.; Rytting, E.; Steele, T.; Kissel, T. Biodegradable Branched Polyesters Poly(vinyl sulfonate-covinyl alcohol)-graft Poly(d,l-lactic-coglycolic acid) as a Negatively Charged Polyelectrolyte Platform for Drug Delivery: Synthesis and Characterization. *Macromolecules* **2008**, *41*, 2791–2799. [CrossRef]

84. Basko, M.; Bednarek, M.; Kubisa, P. Cationic copolymerization of L,L-lactide with hydroxyl substituted cyclic ethers. *Polym. Adv. Technol.* **2015**, *26*, 804–813. [CrossRef]
85. Kubisa, P.; Penczek, S. Cationic activated monomer polymerization of heterocyclic monomers. *Prog. Polym. Sci.* **1999**, *24*, 1409–1437. [CrossRef]
86. Tasaka, F.; Ohya, Y.; Ouchi, T. Synthesis of Novel Comb-Type Polylactide and Its Biodegradability. *Macromolecules* **2001**, *34*, 5494–5500. [CrossRef]
87. Duda, A. Poylactide: A polymer of the XXIst century. *Przem. Chem.* **2003**, *82*, 905–907.
88. Bendix, D. Chemical synthesis of polylactide and its copolymers for medical applications. *Polym. Degrad. Stab.* **1998**, *59*, 129–135. [CrossRef]
89. Lin, P.-L.; Fang, H.-W.; Tseng, T.; Lee, W.-H. Effects of hydroxyapatite dosage on mechanical and biological behaviors of polylactic acid composite materials. *Mater. Lett.* **2007**, *61*, 3009–3013. [CrossRef]
90. Li, G.; Zhao, M.; Xu, F.; Yang, B.; Li, X.; Meng, X.; Teng, L.; Sun, F.; Li, Y. Synthesis and Biological Application of Polylactic Acid. *Molecules* **2020**, *25*, 5023. [CrossRef]
91. Giammona, G.; Craparo, E.F. Biomedical Applications of Polylactide (PLA) and Its Copolymers. *Molecules* **2018**, *23*, 980. [CrossRef]
92. Saito, N.; Takaoka, K. New synthetic biodegradable polymers as BMP carriers for bone tissue engineering. *Biomaterials* **2003**, *24*, 2287–2293. [CrossRef]
93. Murakami, N.; Saito, N.; Horiuchi, H.; Okada, T.; Nozaki, K.; Takaoka, K. Repair of segmental defects in rabbit humeri with titanium fiber mesh cylinders containing recombinant human bone morphogenetic protein-2 (rhBMP-2) and a synthetic polymer. *J. Biomed. Mater. Res.* **2002**, *62*, 169–174. [CrossRef]
94. Rancan, F.; Papakostas, D.; Hadam, S.; Hackbarth, S.; Delair, T.; Primard, C.; Verrier, B.; Sterry, W.; Blume-Peytavi, U.; Vogt, A. Investigation of Polylactic Acid (PLA) Nanoparticles as Drug Delivery Systems for Local Dermatotherapy. *Pharm. Res.* **2009**, *26*, 2027–2036. [CrossRef] [PubMed]
95. Cao, H.; Wang, Y.; Fan, Y.; Ma, J. Synthesis of a biodegradable tadpole-shaped polymer via the coupling reaction of polylactide onto mono(6-(2-aminoethyl)amino-6-deoxy)-β-cyclodextrin and its properties as the new carrier of protein delivery system. *J. Control. Release* **2005**, *107*, 158–173. [CrossRef] [PubMed]
96. Matsumoto, J.; Nakada, Y.; Sakurai, K.; Nakamura, T.; Takahashi, Y. Preparation of nanoparticles consisted of poly(l-lactide)–poly(ethylene glycol)–poly(l-lactide) and their evaluation in vitro. *Int. J. Pharm.* **1999**, *185*, 93–101. [CrossRef]
97. Fishbein, I.; Chorny, M.; Rabinovich, L.; Banai, S.; Gati, I.; Golomb, G. Nanoparticulate delivery system of a tyrphostin for the treatment of restenosis. *J. Control. Release* **2000**, *65*, 221–229. [CrossRef]
98. Xing, J.; Zhang, D.; Tan, T. Studies on the oridonin-loaded poly(d,l-lactic acid) nanoparticles in vitro and in vivo. *Int. J. Biol. Macromol.* **2007**, *40*, 153–158. [CrossRef]
99. Leroux, J.-C.; Allémann, E.; De Jaeghere, F.; Doelker, E.; Gurny, R. Biodegradable nanoparticles—From sustained release formulations to improved site specific drug delivery. *J. Control. Release* **1996**, *39*, 339–350. [CrossRef]
100. Fessi, H.; Puisieux, F.; Devissaguet, J.; Ammoury, N.; Benita, S. Nanocapsule formation by interfacial polymer deposition following solvent displacement. *Int. J. Pharm.* **1989**, *55*, R1–R4. [CrossRef]
101. Esmaeili, F.; Ghahremani, M.H.; Ostad, S.N.; Atyabi, F.; Seyedabadi, M.; Malekshahi, M.R.; Amini, M.; Dinarvand, R. Folate-receptor-targeted delivery of docetaxel nanoparticles prepared by PLGA–PEG–folate conjugate. *J. Drug Target.* **2008**, *16*, 415–423. [CrossRef] [PubMed]
102. Danafar, H.; Rostamizadeh, K.; Davaran, S.; Hamidi, M. Drug-conjugated PLA–PEG–PLA copolymers: A novel approach for controlled delivery of hydrophilic drugs by micelle formation. *Pharm. Dev. Technol.* **2016**, *22*, 947–957. [CrossRef]
103. Ling, Y.; Huang, Y. *Preparation and Release Efficiency of Poly (Lactic-Co-Glycolic) Acid Nanoparticles for Drug Loaded Paclitaxel*; Springer: Berlin/Heidelberg, Germany, 2008.
104. Reinbold, J.; Uhde, A.-K.; Müller, I.; Weindl, T.; Geis-Gerstorfer, J.; Schlensak, C.; Wendel, H.-P.; Krajewski, S. Preventing Surgical Site Infections Using a Natural, Biodegradable, Antibacterial Coating on Surgical Sutures. *Molecules* **2017**, *22*, 1570. [CrossRef]
105. Asghari, F.; Samiei, M.; Adibkia, K.; Akbarzadeh, A.; Davaran, S. Biodegradable and biocompatible polymers for tissue engineering application: A review. *Artif. Cells Nanomed. Biotechnol.* **2016**, *45*, 185–192. [CrossRef] [PubMed]
106. Venugopal, J.; Ramakrishna, S. Applications of Polymer Nanofibers in Biomedicine and Biotechnology. *Appl. Biochem. Biotechnol.* **2005**, *125*, 147–158. [CrossRef]
107. Ni, C.; Lu, R.; Tao, L.; Shi, G.; Li, X.; Qin, C. Synthesis of poly(vinyl alcohol-graft-lactic acid) copolymer and its application as medical anti-tissue adhesion thin film. *Polym. Bull.* **2015**, *72*, 1515–1529. [CrossRef]
108. Dominguez-Alfaro, A.; Gabirondo, E.; Alegret, N.; De León-Almazán, C.M.; Hernandez, R.; Vallejo-Illarramendi, A.; Prato, M.; Mecerreyes, D. 3D Printable Conducting and Biocompatible PEDOT- graft -PLA Copolymers by Direct Ink Writing. *Macromol. Rapid Commun.* **2021**, *42*, 2100100. [CrossRef] [PubMed]

materials

MDPI

Article

Investigations on the Influence of Collagen Type on Physicochemical Properties of PVP/PVA Composites Enriched with Hydroxyapatite Developed for Biomedical Applications

Magdalena Głąb [1,*], Anna Drabczyk [1,*], Sonia Kudłacik-Kramarczyk [1,*], Magdalena Kędzierska [2], Agnieszka Tomala [1], Agnieszka Sobczak-Kupiec [1], Dariusz Mierzwiński [1] and Bożena Tyliszczak [1]

[1] Department of Materials Science, Faculty of Materials Engineering and Physics, Cracow University of Technology, 37 Jana Pawła II Av., 31-864 Krakow, Poland; agnieszka.tomala@pk.edu.pl (A.T.); agnieszka.sobczak-kupiec@pk.edu.pl (A.S.-K.); dariusz.mierzwinski@pk.edu.pl (D.M.); bozena.tyliszczak@pk.edu.pl (B.T.)

[2] Department of Chemotherapy, Medical University of Lodz, WWCOiT Copernicus Hospital, 90-001 Lodz, Poland; kameleonmagda6@gmail.com

* Correspondence: magdalena.glab@doktorant.pk.edu.pl (M.G.); anna.drabczyk2@pk.edu.pl (A.D.); sonia.kudlacik-kramarczyk@pk.edu.pl (S.K.-K.)

Citation: Głąb, M.; Drabczyk, A.; Kudłacik-Kramarczyk, S.; Kędzierska, M.; Tomala, A.; Sobczak-Kupiec, A.; Mierzwiński, D.; Tyliszczak, B. Investigations on the Influence of Collagen Type on Physicochemical Properties of PVP/PVA Composites Enriched with Hydroxyapatite Developed for Biomedical Applications. *Materials* **2022**, *15*, 37. https://doi.org/10.3390/ma15010037

Academic Editors: Roser Sabater i Serra and Ángel Serrano-Aroca

Received: 18 October 2021
Accepted: 17 December 2021
Published: 21 December 2021

Abstract: Nowadays, a great attention is directed into development of innovative multifunctional composites which may support bone tissue regeneration. This may be achieved by combining collagen and hydroxyapatite showing bioactivity, osteoconductivity and osteoinductivity with such biocompatible polymers as polyvinylpyrrolidone (PVP) and poly(vinyl alcohol) (PVA). Here PVA/PVP-based composites modified with hydroxyapatite (HAp, 10 wt.%) and collagen (30 wt.%) were obtained via UV radiation while two types of collagen were used (fish and bovine) and crosslinking agents differing in the average molecular weight. Next, their chemical structure was characterized using Fourier transform infrared (FT-IR) spectroscopy, roughness of their surfaces was determined using a stylus contact profilometer while their wettability was evaluated by a sessile drop method followed by the measurements of their surface free energy. Subsequently, swelling properties of composites were verified in simulated physiological liquids as well as the behavior of composites in these liquids by pH measurements. It was proved that collagen-modified composites showed higher swelling ability (even 25% more) compared to unmodified ones, surface roughness, biocompatibility towards simulated physiological liquids and hydrophilicity (contact angles lower than 90°). Considering physicochemical properties of developed materials and a possibility of the preparation of their various shapes and sizes, it may be concluded that developed materials showed great application potential for biomedical use, e.g., as materials filling bone defects supporting their treatments and promoting bone tissue regeneration due to the presence of hydroxyapatite with osteoinductive and osteoconductive properties.

Keywords: polymer-ceramic composites; hydroxyapatite; collagen; swelling capacity; surface roughness; wettability; surface free energy; tissue engineering

1. Introduction

The development of the medical sciences taking shape over the years had a huge impact on the extension of the average human life expectancy. This is undoubtedly a great success but simultaneously a big challenge related to the necessity of the search for effective therapies of treatments of diseases that often accompany the aging of organisms including mainly diseases of the skeletal system [1]. Importantly, a development of the so-called civilization diseases such as e.g., cardiovascular diseases, cancers or osteoporosis is also a problematic issue [2,3]. In the case of osteoporosis, the risk of bone fractures increases significantly as a result of a decrease in their mechanical resistance [4,5]. The patients suffering from this ailment often experience very serious injuries, which in the majority of cases require surgical intervention [6]. Thus, investigations on the development of novel bioactive composite materials that might support the treatment of bone defects and promote faster bone regeneration are very meaningful and constitute one of the biggest challenges of contemporary medicine.

Recently, one of the most popular research topics seems to be studies on the development of composite materials containing a ceramic phase [7–9]. The substance which is the most frequently used for this purpose is hydroxyapatite (HAp) which constitutes a mineral component of bones [10]. Considering its bioactivity [11,12], biocompatibility [13–15], osteoinductivity [16,17] and osteoconductivity [18,19], this substance has a key meaning in preparation of materials designed for bone tissue regeneration. HAp is widely used in bone substitute materials [20,21], in tissue engineering [22,23] and as a drug carrier [24,25].

Composite materials based on the combination of biodegradable synthetic and natural polymers and modified additionally with hydroxyapatite were described by Gordienko et al. The composites containing biologically active substance were analyzed in detail which allowed to conclude e.g., that the developed materials showed no cytotoxicity during in vivo experiments [26]. Next, Shirdar et al. proposed composites based on poly(methyl methacrylate) and modified with hydroxyapatite nanofibers and magnesium phosphate-based two-dimensional nanosheets. The statistical analysis performed indicated the significant impact of both modifiers on the improvement of the compressive strength of developed materials and, importantly, higher viability of cells was reported in the presence of such modified composites compared to unmodified materials [27].

Other interesting studies were described by Liu et al. Here, the main attention was focused on biodegradable composites based on poly(ε-caprolactone) (PCL) and HAp nanoparticles. The materials were prepared via 3D printing. It was observed during in vivo biological analyses that developed scaffolds based on PCL and HAp provided adequate adhesion and penetration of bone cells which, in turn, allowed bone tissue to regenerate [28].

Next, Rodzeń et al. also presented an innovative composite material designed for bone regeneration. As a base material, polyetheretherketone (PEEK) modified with HAp (30 wt.%) was used. In order to obtain a composite, one of the 3D printing methods, i.e., fused filament fabrication, was employed. It was proved that surfaces of developed PEEK/HAp composites supported adhesion and growth of U-2 OS osteoblast-like cells [29]. In other work, composite scaffolds based on HAp and bioglass obtained using a hydrothermal method were characterized. Performed in vitro biological studies confirmed that developed materials showed no cytotoxicity while their biological performance was significantly higher than in the case of unmodified scaffold [30].

Despite numerous studies on the development of innovative composite materials intended for bone tissue regeneration, the methods leading to preparation of such composites still have some limitations or require a lot of work, reagents and are time-consuming. Thus the main purpose of the research was to prepare and characterize composite materials with a wide application potential in regenerative medicine. The potential of developed materials lies in their quick and simple synthesis methodology and an interesting composition in viewpoint of their potential biomedical use which may affect bone regeneration process. Furthermore, the methodology applied gives an opportunity to synthesize composites of various shapes and sizes tailored to a given bone defect.

Furthermore, the presence of collagen of fish or bovine origin increases the innovativeness of the developed composite materials. Collagen is a main component of extracellular matrix which makes it an interesting biopolymer for applications in tissue engineering and regeneration medicine [31,32].

Collagen-based materials show interesting properties due to their good biocompatibility, ability to promote cell proliferation and adhesion, and non-toxicity. Moreover, collagen exhibits hemostatic properties, and provides the reconstructed tissues with an adequate elasticity [33]. In regenerative processes, collagen-based biomatrix imitates scaffolds that fulfill structural and mechanical functions thus favoring the reconstruction of damaged organs and tissues. Importantly, the introduction of collagen into composite biomaterials increases the effectiveness of such materials in regenerative processes, their ability to combine with growth factors, and also enables intracellular transmission and stabilization of cellular components due to which such composite plays a role of a physical structure supporting the regeneration processes [34,35].

The measurement methodology has been chosen so as to verify these composites' usefulness and evaluate their potential for providing adequate conditions for such processes as e.g., cell proliferation. That is why such properties of developed materials were investigated as wettability (it was demonstrated that hydrophilic surfaces favor the bone growth process) or roughness. Materials' roughness affects the interactions between such an implant and the newly formed tissue. This, in turn, has a crucial meaning in viewpoint of potential cell proliferation, and so the healing and regeneration processes.

Thus, this work examined a series of composites based on synthetic polymers such as polyvinylpyrrolidone (PVP) and poly(vinyl alcohol) and, importantly, containing also a ceramic phase (hydroxyapatite) and an additional modifying agent—collagen, while two series were prepared using two type of collagens, i.e., fish and bovine. Developed composites were subsequently analyzed in detail while the main attention was paid to verifying the impact of the type of collagen and the average molecular weight of the crosslinking agent used during the synthesis of composite materials on their physicochemical properties. For this purpose, such analyses as Fourier transform infrared (FT-IR) spectroscopy as well as swelling measurements and incubation in simulated physiological liquids were performed. Moreover, the roughness of prepared materials and the topography of their surfaces were also investigated.

2. Materials and Methods

2.1. Materials

Ammonium phosphate monobasic ($NH_4H_2PO_4$, ACS reagent, ≥98%), calcium nitrate tetrahydrate ($Ca(NO_3)_2{\cdot}4H_2O$, ACS reagent, 99%) and ammonia water (NH_4OH, 25%) were used to obtain hydroxyapatite (HAp). Next, poly(vinyl alcohol) (PVA, crystalline powder, 87–89% hydrolyzed, M_w 13,000–23,000), polyvinylpyrrolidone (PVP, powder, average mol wt. 10,000), diacrylate poly(ethylene glycol) (crosslinking agent, PEGDA, average molecular weight M_n = 700 g/mol and M_n = 575 g/mol) and 2-hydroxy-2-methylpropiophenone (photoinitiator, 97%, d = 1.077 g/mL) were applied during the preparation of composite materials. All mentioned reagents were purchased from Sigma Aldrich (Saint Louis, MO, USA). Fish collagen peptide (molecular weight 2000–5000 Da) and bovine collagen peptide (molecular weight < 3000 Da) were purchased from Xi'an Gawen Biotechnology Co., Ltd. (Xi'an, China).

2.2. Synthesis of Composite Materials

Firstly, HAp powder was prepared via wet precipitation method. Synthesis was performed at room temperature and at constant stirring while ammonium phosphate monobasic (0.36 mol/L) and calcium nitrate tetrahydrate (0.60 mol/L) were used as reagents. The calcium nitrate solution was added dropwise (1 drop/s) into the ammonium phosphate solution while maintaining the alkaline pH of the reaction mixture (pH > 10) by adding ammonia water. pH of the reaction mixture was controlled in a continuous manner via a multifunctional CX-701 pH-meter (Elmetron, Zabrze, Poland). After the reaction, the mixture obtained was allowed to sediment for 24 h. The sediment was washed with distilled water to neutral pH, lyophilized (parameters: T = −50 °C, p = 0.07 mbar, t = 24 h) and used in further works for preparation of polymer-ceramic composites.

PVP/PVA-based composite materials containing HAp were prepared according to procedure described previously in [36]. Briefly, the aqueous solutions of polymers—PVA and PVP—were mixed with a ceramic phase (HAp) and selected collagen, i.e., fish collagen or bovine collagen. Next, an appropriate amount of crosslinking agent PEGDA with an average molecular weight of either 575 g/mol or 700 g/mol and photoinitiator (2-hydroxy-

2-methylpropiophenone) was added. The whole mixtures were placed in vessels of the selected shapes and polymerized under UV lamp (EMITA VP-60, power: 180 W, λ = 320 nm; Famed, Lodz, Poland). The process of photopolymerization was performed for 120 s. Detailed compositions of obtained polymer-ceramic composites are given in Table 1.

Table 1. Compositions of polymer-ceramic composites.

15% PVP, mL	5% PVA, mL	HAp, wt.%	Photoinitiator, mL	Crosslinking Agent, mL	Fish Collagen, wt.%	Bovine Collagen, wt.%	Sample Name
3	7	10	0.05	2.0 (575 g/mol)	-	-	575_HAp
					30	-	575_HAp/Coll-fish
					-	30	575_HAp/Coll-bov
				2.0 (700 g/mol)	-	-	700_HAp
					30	-	700_HAp/Coll-fish
					-	30	700_HAp/Coll-bov

The use of vessels with various shapes allowed to obtain composites of selected shapes and sizes; prepared materials are presented in Figure 1.

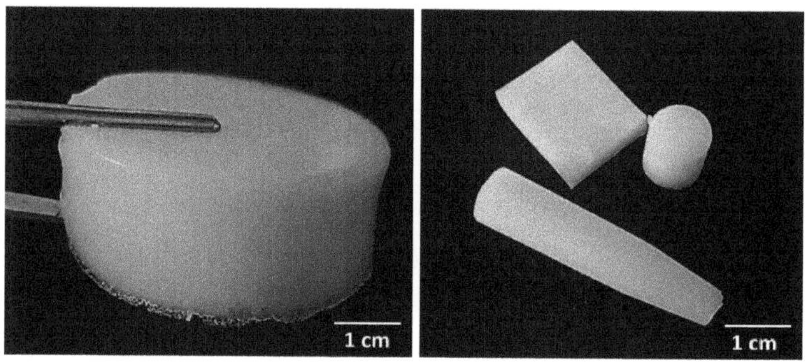

Figure 1. Various shapes of prepared composite materials—sample 575_HAp/Coll-fish (**left**) and samples 700_HAp/Coll-bov (**right**).

Prepared polymer-ceramic composites were subsequently subjected to physicochemical evaluation aimed at determining their properties such as swelling ability, surface hydrophilicity or roughness.

2.3. Characterization of the Chemical Structure of Composite Materials via Fourier Transform Infrared (FT-IR) Spectroscopy

FT-IR spectroscopy was performed to identify the functional groups present in the structures of tested materials while the main attention was paid to group characteristics for collagen. The analysis was conducted for samples without this protein as well as for samples modified with collagen and prepared using crosslinking agents with different average molecular weights. The investigations were carried out at room temperature and using a Thermo Scientific Nicolet iS5 FT-IR spectrophotometer equipped additionally with iD7 ATR (Attenuated Total Reflectance, Loughborough, UK) accessory. FT-IR spectra were recorded within the wavenumber range 4000–500 cm^{-1} (32 scans, resolution 4.0 cm^{-1}).

2.4. Evaluation of the Swelling Properties of Composite Materials

Subsequent study concerned the evaluation of swelling ability of the composites. Considering intended application of developed composites for biomedical purposes, it is significant to determine their behavior in liquids simulating environments occurring in the human body. Thus, the investigations were performed using such liquids as SBF (simulated body fluid), Ringer liquid (isotonic to human blood) and distilled water as a reference liquid. In order to determine this property, prepared materials were dried at 37 °C for 24 h, weighed and placed in 50 mL of the mentioned solutions. After selected time periods—i.e., 1 h, 24 h, 48 h and 72 h—samples were separated from the solutions, an excess of unbound water was removed using a paper towel and samples were weighed again. The swelling ability defined using a swelling ratio α was determined via the following Equation (1):

$$\alpha = \frac{(m - m_0)}{m_0} \quad (1)$$

where:

α—swelling ratio, g/g; m—mass of swollen sample, g/g; and m_0—mass of dry sample (before the study), g.

The study was performed for all prepared samples, i.e., both for unmodified materials and materials containing collagen.

2.5. Incubation Studies

The study was performed to verify the interaction between simulated physiological liquids and prepared composite materials during their long-term immersion in such environments. Samples of the composites (weighing approx. 1.0 g) were placed in 50 mL of selected liquids (the same liquids were used as in the case of swelling studies, i.e., SBF, Ringer liquid and distilled water) whose pH values were verified every two days. The study was conducted for 14 days while samples were incubated at 36.6 °C so at temperature of human body. Both samples with and without collagen were subjected to the research.

2.6. Evaluation of the Wettability of Composites

Wettability of prepared materials was measured by a sessile drop method followed by Drop Shape Analysis system—DSA 10Mk2, Kruss, Germany. Contact angle was determined for polar solvent which was ultra-high quality (UHQ) water (Purelab UHQ, Elga) of resistivity 18 MΩ/cm, and non-polar diodomethane (Sigma Aldrich). Surface free energy was calculated according to the Owens, Wendt, Rabel and Kaelble (OWRK) approach.

The mean contact angle values and deviation were determined from three repetitions measurements at different spots on the sample.

2.7. Assessment of Roughness of Composites

The surface roughness was measured using a stylus contact profilometer TALYSURF6 (Taylor Hobson, San Francisco, CA, USA) Surface roughness was measured according to ISO 4287 [37] with contact stylus acquisition mode in a line distance of 4 mm. The roughness profile including parameters Ra (arithmetic mean of the departures), Rq (kurtosis) and Rsk (skewness) numerically describe the topography of the measured surfaces. The mean value and deviation was determined from at least three repetition measurements at different spots on the sample. The TalyMap Platinum software (Taylor Hobson, San Francisco, CA, USA) was used for 3D topography analysis and export of the surface images.

2.8. Statistical Analysis of the Results of the Investigations

The results of the investigations were subjected to the statistical analysis performed via the two-way analysis of variance (ANOVA) while alpha value 5% was applied. All measurements were carried out three times (n = 3) and are presented as an average value with the standard deviation (SD).

3. Results and Discussion

3.1. Results of FT-IR Spectroscopy

Below in Figure 2 FT-IR, spectra of unmodified and modified composite materials are presented. The study was performed both for samples containing fish collagen (Coll-fish) and bovine collagen (Coll-bov).

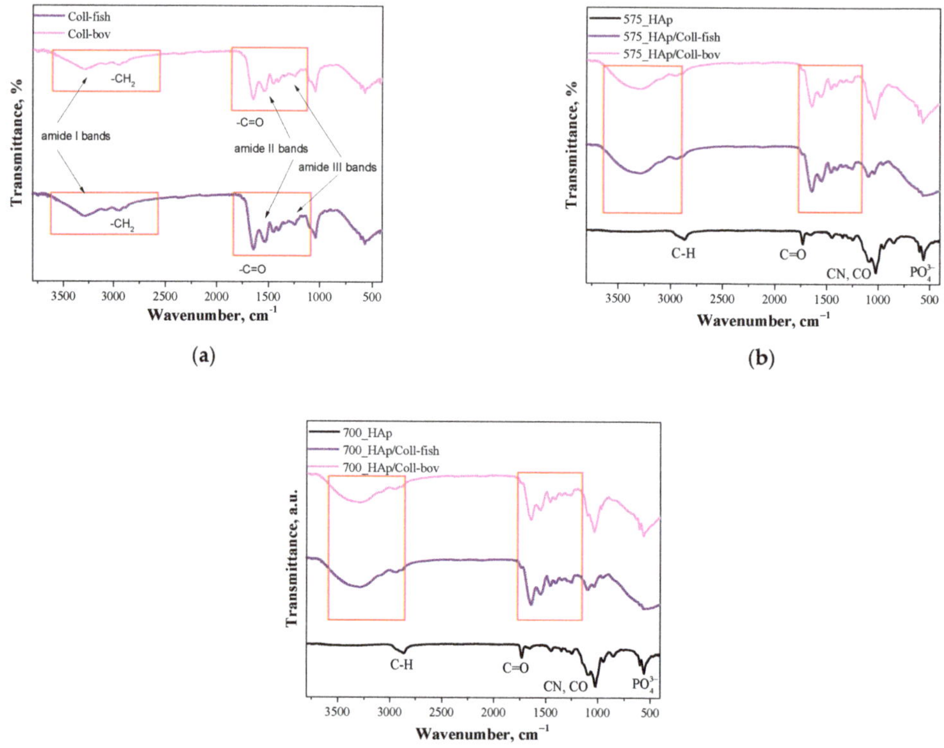

(c)

Figure 2. FT-IR spectra of both used collagens—fish collagen (Coll-fish) and bovine collagen (Coll-bov) (**a**); composites prepared using PEGDA 575 (**b**) and composites prepared using PEGDA 700 (**c**).

In Figure 2a FT-IR, spectra of fish collagen and bovine collagen are presented. In the case of both samples, absorption bands characteristic for functional groups present in the structure of the mentioned proteins were identified. The absorption bands at 3284 cm^{-1} have been assigned to amide I bands corresponding to hydrogen bonds between NH group and carbonyl group of the peptide chain. Next, the absorption bands at approx. 2939 cm^{-1} deriving probably from the asymmetrical stretching vibrations of CH_2 characteristic for amide B bands were identified. In turn, bands within the wavenumber range 1700–1600 cm^{-1} which can be attributed to the stretching vibrations of C=O from peptide bonds (determined also as amide I bands) confirmed the occurrence of secondary structure characteristic for proteins. Moreover, the absorption bands deriving also from amide II and amide II may also be observed on analyzed FT-IR spectra. Next, the absorption bands within the range 1570–1470 cm^{-1} characteristic for amide II which may be assigned for N–H bending vibrations coupled with C–N stretching vibration, were observed; while at 1350–1250 cm^{-1} (amide III), absorption bands characteristic for stretching vibrations of C–H group were identified. Bands marked in Figure 2a were identified analogously as in other works [38–41]. Importantly, the absorption bands most characteristic for collagen identified also on FT-IR spectra of polymer-ceramic composites containing this protein (fish or bovine origin) have been marked with a red frame in Figure 2a–c.

3.2. Studies on Swelling Capacity of Composite Materials

Next, investigations were aimed at determining swelling capacity of composite materials while the main purpose of the study was to verify the potential difference in this property between unmodified materials and materials modified with collagen. Importantly, the impact of the average molecular weight of the crosslinking agent used during the synthesis of composites on their swelling sorption was also discussed. Results of performed studies are presented in Figure 3. In Table 2, the statistical analysis of obtained data is presented.

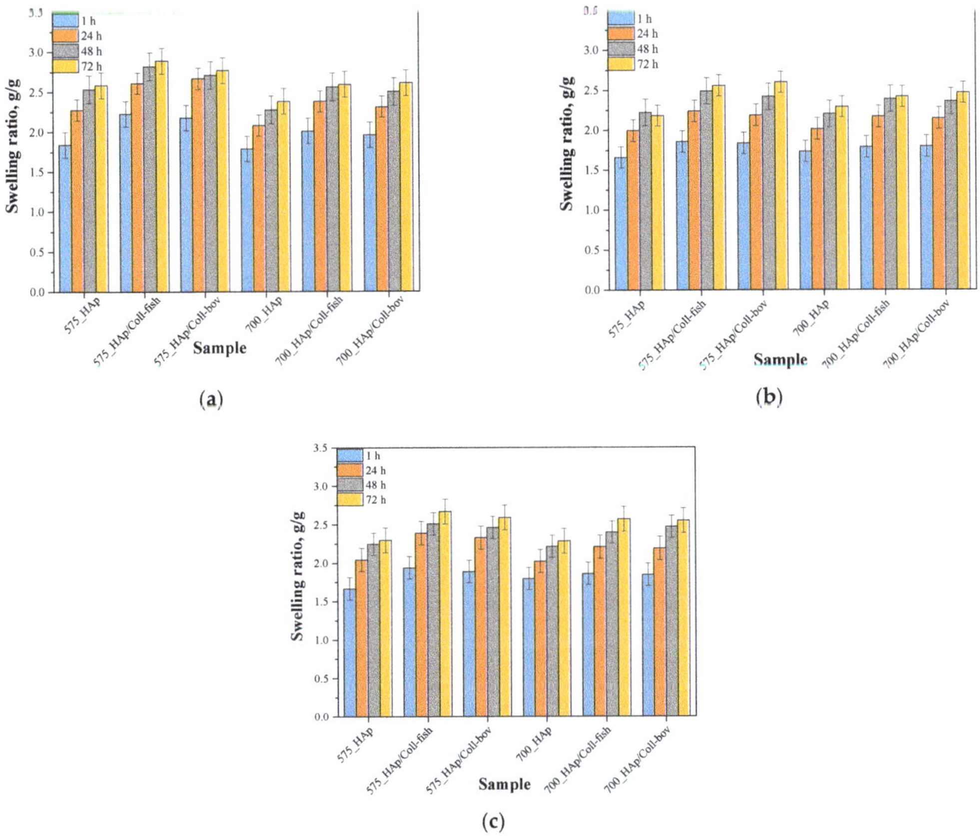

Figure 3. Results of swelling studies of composite materials in distilled water (**a**), SBF (**b**) and Ringer liquid (**c**) (n—number of repetitions, n = 3).

One of the basic properties of hydrogel materials is their ability to absorb liquids without permanent loss of shape and deterioration of mechanical properties. These abilities result mainly from the hydration of such hydrophilic functional groups as –OH, –COOH or –$CONH_2$ occurring in the structure of the polymers and also are due to the presence of capillary forces. Importantly, during such a sorption these materials do not dissolve, which is due to numerous covalent or hydrogen bonds or electrostatic interactions between polymer chains.

Based on the results of performed swelling investigations, it may be concluded that all tested materials show swelling properties. The highest swelling ratios were calculated for samples swelling in distilled water while the sorption in Ringer liquid and in SBF was significantly lower. For example, sample 575_HAp after 1 h swelling showed α = 1.84 g/g in distilled water, α = 1.67 g/g in Ringer liquid and α = 1.66 g/g in SBF. Such results result from the fact that the swelling process depends strictly both on the composition of tested sample and the composition of the liquid in which sorption takes place. Lower values of swelling ratios in SBF and Ringer liquid result probably from the occurrence of divalent ions in these liquids which may, in turn, affect the formation of additional crosslinks between polymer chains thus increasing the crosslinking density of the polymer. This, in turn, leads to a decrease of free spaces between polymer chains available for absorbed liquid and finally to lower swelling properties of samples in such liquids. In the case of distilled water, where any additional ions do not occur, the above-described phenomenon concerning the formation of additional crosslinks does not take place so there is more free space between polymer chains for absorbed liquid.

Table 2. Statistical analysis of obtained data based on the two-way analysis of variance (ANOVA) with repetitions.

Independent Variable	Study	p *
Type of crosslinking agent Type of collagen	1 h swelling in distilled water	1.52923×10^{-15} 2.96059×10^{-14}
Type of crosslinking agent Type of collagen	72 h swelling in distilled water	8.39488×10^{-15} 5.92119×10^{-14}
Type of crosslinking agent Type of collagen	1 h swelling in SBF	0.42331 0.42271
Type of crosslinking agent Type of collagen	72 h swelling in SBF	8.12234×10^{-15} 7.31011×10^{-14}
Type of crosslinking agent Type of collagen	1 h swelling in Ringer liquid	3.97822×10^{-14} 1.54037×10^{-13}
Type of crosslinking agent Type of collagen	72 h swelling in Ringer liquid	6.04203×10^{-14} 1.18424×10^{-13}

* p indicates the statistical significance calculated using the two-way analysis of variance (ANOVA).

Moreover, it was observed that all samples prepared using a crosslinking agent with an average molecular weight 575 g/mol (PEGDA 575) showed higher swelling ability than materials obtained using PEGDA 700. The observed dependence is related to a different structure of crosslinked materials depending on the crosslinking agent used. The highest differences were observed during first 24 h of the swelling. Analogous results were presented in other works [36]. The use of the crosslinking agent PEGDA 575 resulted in the synthesis of polymer matrices consisting of shorter polymer chains compared to the chains occurring in the structure of materials obtained using PEGDA 700. Such a difference affects the swelling properties of tested materials—shorter polymer chains resulted in a formation of more developed and porous structure which may absorb more liquids. Thus, the highest differences in swelling ratios of composites obtained using different crosslinking agents were observed between values calculated after 1 h and after 24 h when the material absorbs liquids mainly on the surface. During the next hours of swelling, a penetration of liquids into the interior of analyzed polymer samples takes place, therefore differences between swelling ratios after 48 h and 72 h compared to α after 24 h were slight. On the other hand, any significant impact of the type of the collagen used on the swelling properties of modified composite materials was not observed. Similar observations concerning the sorption capacity of the materials modified with collagens of various origin were presented by Ghodbane et al. [42]. Nonetheless, the differences between unmodified samples and samples modified with collagen were observed. Both introduction of fish collagen and bovine collagen into the composite materials resulted in an increase in their swelling properties. In general, collagen is responsible for water binding in tissues. Thus, its ability of water sorption translates into the highest swelling ability of composites modified with this protein compared to unmodified materials. These results are consistent with results presented by Bai et al. who stated that as a result of the modification of materials with collagen these materials had a higher amount of hydrophilic functional groups which may interact with water [43]. As a result, materials with collagen show a higher swelling capacity that was also reported in investigations presented here.

3.3. *Results of Incubation of Composite Materials in Selected Simulated Physiological Liquids*

In Tables 3–5, results of 14-day incubation of composite materials are presented. During the study, the change in pH was determined as a function of time while the measurements were also performed for incubation liquids without samples (reference measurements).

Table 3. Results of incubation in distilled water.

Sample	pH Value							
	Day 0	Day 2	Day 4	Day 6	Day 8	Day 10	Day 12	Day 14
Distilled water	7.218 ± 0.216	7.114 ± 0.211	7.015 ± 0.213	7.054 ± 0.209	7.177 ± 0.214	6.912 ± 0.211	7.036 ± 0.216	6.991 ± 0.208
575_HAp	7.132 ± 0.215	7.141 ± 0.212	7.053 ± 0.209	6.990 ± 0.210	7.012 ± 0.215	6.984 ± 0.208	7.013 ± 0.213	7.181 ± 0.214
700_HAp	7.041 ± 0.211	7.028 ± 0.208	6.986 ± 0.210	6.944 ± 0.208	6.972 ± 0.211	6.954 ± 0.209	7.021 ± 0.208	7.047 ± 0.211
575_HAp/Col-fish	7.183 ± 0.215	6.604 ± 0.198	6.694 ± 0.201	6.756 ± 0.205	6.833 ± 0.202	6.757 ± 0.199	6.655 ± 0.203	6.842 ± 0.201
700_HAp/Col-fish	7.144 ± 0.214	6.557 ± 0.194	6.645 ± 0.201	6.723 ± 0.192	6.782 ± 0.203	6.710 ± 0.198	6.681 ± 0.203	6.796 ± 0.196
575_HAp/Col-bov	7.164 ± 0.215	6.564 ± 0.198	6.656 ± 0.197	6.715 ± 0.202	6.793 ± 0.204	6.511 ± 0.195	6.612 ± 0.197	6.809 ± 0.203
700_HAp/Col-bov	7.148 ± 0.214	6.534 ± 0.202	6.625 ± 0.198	6.687 ± 0.203	6.764 ± 0.195	6.486 ± 0.197	6.582 ± 0.194	6.773 ± 0.200

Table 4. Results of incubation in SBF.

Sample	pH Value							
	Day 0	Day 2	Day 4	Day 6	Day 8	Day 10	Day 12	Day 14
SBF	7.485 ± 0.222	7.491 ± 0.226	7.542 ± 0.225	7.493 ± 0.224	7.438 ± 0.224	7.501 ± 0.227	7.493 ± 0.222	7.570 ± 0.225
575_HAp	7.526 ± 0.226	7.466 ± 0.221	7.546 ± 0.223	7.496 ± 0.226	7.406 ± 0.227	7.566 ± 0.225	7.516 ± 0.223	7.636 ± 0.224
700_HAp	7.584 ± 0.227	7.484 ± 0.223	7.434 ± 0.222	7.444 ± 0.222	7.451 ± 0.223	7.474 ± 0.225	7.474 ± 0.224	7.444 ± 0.223
575_HAp/Col-fish	7.576 ± 0.228	7.455 ± 0.226	7.425 ± 0.227	7.435 ± 0.224	7.505 ± 0.229	7.465 ± 0.227	7.465 ± 0.226	7.435 ± 0.226
700_HAp/Col-fish	7.585 ± 0.224	7.484 ± 0.227	7.633 ± 0.223	7.646 ± 0.224	7.511 ± 0.226	7.572 ± 0.227	7.572 ± 0.226	7.546 ± 0.225
575_HAp/Col-bov	7.557 ± 0.226	7.413 ± 0.223	7.443 ± 0.223	7.453 ± 0.224	7.423 ± 0.227	7.483 ± 0.226	7.483 ± 0.227	7.453 ± 0.225
700_HAp/Col-bov	7.694 ± 0.231	7.496 ± 0.229	7.423 ± 0.227	7.431 ± 0.226	7.405 ± 0.224	7.461 ± 0.225	7.463 ± 0.222	7.434 ± 0.225

Based on the above presented results, it may be concluded that in the case of all tested samples, pH values measured in the course of their incubation varied, but only slightly. Rapid jumps in pH values could indicate the degradation of polymer matrices or the release of potential unreacted reagents such as crosslinking agent or photoinitiator from their interior. In the case of tested composite materials, such changes were not observed, which, in turn, may be evidence of the preparation of properly-crosslinked materials which show biocompatibility with tested environments. The only big difference (compared to the other results) was observed in the case of samples modified with collagen. This was probably related to the method of its preparation. The collagen hydrolysates introduced into the polymer matrices were obtained as a result of an enzymatic hydrolysis performed in an acidic environment which, in turn, may translate into the mentioned pH decrease during the incubation of such modified materials. However, these changes were slight and, importantly, the least noticeable for samples immersed in SBF, which may probably be evidence of well buffering properties of prepared composites in this liquid.

Table 5. Results of incubation in Ringer solution.

Sample	pH Value							
	Day 0	**Day 2**	**Day 4**	**Day 6**	**Day 8**	**Day 10**	**Day 12**	**Day 14**
Ringer solution	5.749 ± 0.172	5.881 ± 0.176	5.941 ± 0.174	5.820 ± 0.175	5.842 ± 0.173	5.841 ± 0.172	5.850 ± 0.176	6.051 ± 0.181
575_HAp	5.654 ± 0.169	5.794 ± 0.173	5.854 ± 0.175	5.834 ± 0.175	5.854 ± 0.181	5.754 ± 0.176	5.864 ± 0.174	6.064 ± 0.176
700_HAp	5.751 ± 0.169	5.891 ± 0.164	5.851 ± 0.166	5.831 ± 0.167	5.851 ± 0.164	5.751 ± 0.166	5.861 ± 0.169	6.061 ± 0.170
575_HAp/Col-fish	5.634 ± 0.171	5.474 ± 0.175	5.534 ± 0.172	5.474 ± 0.181	5.434 ± 0.174	5.534 ± 0.175	5.544 ± 0.174	5.584 ± 0.176
700_HAp/Col-fish	5.641 ± 0.169	5.481 ± 0.164	5.541 ± 0.166	5.421 ± 0.167	5.441 ± 0.164	5.541 ± 0.169	5.551 ± 0.167	5.451 ± 0.168
575_HAp/Col-bov	5.625 ± 0.167	5.465 ± 0.168	5.425 ± 0.165	5.405 ± 0.162	5.425 ± 0.166	5.525 ± 0.167	5.475 ± 0.164	5.435 ± 0.169
700_HAp/Col-bov	5.621 ± 0.168	5.561 ± 0.165	5.621 ± 0.168	5.501 ± 0.166	5.421 ± 0.168	5.521 ± 0.169	5.531 ± 0.171	5.541 ± 0.167

3.4. Results of Investigations on Wettability of Composites

In order to verify the hydrophilic/hydrophobic nature of the surface of prepared composite materials, the values of their contact angles were determined. The study was conducted using two measuring liquids, i.e., distilled water and diiodomethane. Results of the research are shown in Figure 4 while in Table 6 the statistical analysis of obtained data is presented.

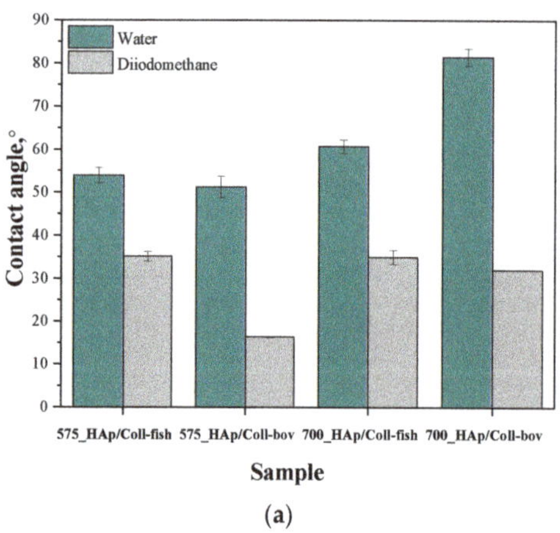

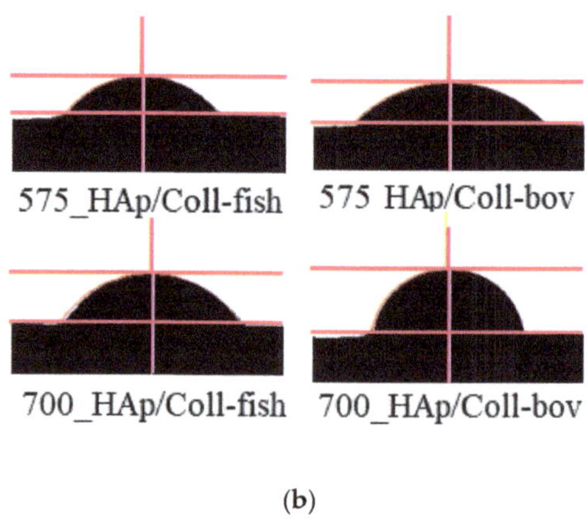

Figure 4. Contact angles of composite materials modified with collagen (fish or bovine one) determined both for distilled water and diiodomethane (**a**) and example images of distilled water-wetted composites (**b**) (n—number of repetitions, n = 3).

Table 6. Statistical analysis of obtained data based on the two-way analysis of variance (ANOVA) with repetitions.

Independent Variable	Measuring Liquid	p *
Type of crosslinking agent Type of collagen	Distilled water	0.00157 0.01485
Type of crosslinking agent Type of collagen	Diiodomethane	1.85299×10^{-4} 0.00591

* p indicates the statistical significance calculated using the two-way analysis of variance (ANOVA).

Furthermore, the analysis performed using two mentioned measuring liquids, i.e., both polar and non-polar one, allowed to determine the surface free energies of analyzed composite materials, whose values are presented in Table 7.

Table 7. Surface free energies determined for tested composites.

Sample *	Surface Free Energy		
	Polar, mJ/m^2	Dispersive, mJ/m^2	Total Free Energy, mJ/m^2
575_HAp/Coll-fish	21.28	26.59	47.87
575_HAp/Coll-bov	18.71	33.74	52.45
700_HAp/Coll-fish	14.90	29.74	44.64
700_HAp/Coll-bov	1.40	43.66	45.06

* for samples HAp_ 575 and HAp_ 700 values of contact angles could not be determined.

Proper dental orthopedic implantation depends strictly on the osteointegration degree, i.e., on the possibility of formation of stable connection between the implant and the tissue. This, in turn, depends on many factors including e.g., the physicochemistry of the implant surface. One of the most important parameters of such a surface is its hydrophilicity. Thus, it was important to perform investigations aimed at determining the contact angles of obtained composite materials which, in turn, allowed to define the hydrophilic or hydrophobic nature of their surfaces. For example, it was demonstrated that hydrophilic surfaces affect the formation of the environment conducive to bone growth [44]. It is assumed that the mentioned dependency is related to the rapid spread of serum on the hydrophilic surface of the implant which, in turn, provides a good substrate for bioactive substances [45]. These substances may significantly affect the early adhesion of cells, their proliferation and differentiation [46,47]. The materials whose contact angle with polar liquid is less than 90° are defined as hydrophilic. Thus, it may be concluded that all tested composites are hydrophilic in nature. This is a significant advantage of developed materials when considering their potential use for biomedical use including tissue engineering.

In the case of unmodified samples, i.e., 575_HAp and 700_HAP, determining the contact angle was not possible. The drops of the measuring liquid spread over the surface of these material at the very first moment of the measurement. Thus, the images showing the first contact of a drop with unmodified materials were not recorded which, in turn, made it impossible to determine the contact angles of these samples.

In the case of the rest of tested samples, it may be concluded that composites prepared using PEGDA 575 showed lower contact angles than samples obtained using PEGDA 700, indicating better surface wettability. Values of contact angles of samples synthesized using PEGDA 700 were significantly higher, which indicated a less hydrophilic nature of their surfaces. Here, due to the use of the crosslinking agent with a higher average molecular weight during the synthesis, prepared materials consisted of long polymer chains. As a result, functional groups deriving from collagen as well as functional groups from the structure of PVP and PVA responsible for interactions with polar measuring liquid (distilled water) may be "embedded" between these long chains ("trapped" inside the polymer matrix). On the other hand, the polymer matrix obtained using PEGDA 575 consisted of short polymer chains, thereby the mentioned hydrophilic functional groups may be more exposed and thus more available for interactions with measuring liquid via e.g., hydrogen bonds.

3.5. Results of the Evaluation of the Surface Roughness of Composite Materials

The roughness of tested materials is another important parameter affecting the adhesion and growth of cells, therefore the next investigations concerned the determination of parameters describing the surface roughness. The study was performed via the profile method according to the ISO 4287 standard [37]. Values characterizing the roughness of tested materials are presented in Table 8, and statistical analysis of obtained data is presented in Table 9.

Table 8. Roughness parameters according to ISO 4287 standard [37].

Sample Name	R_a, µm	R_q, µm	R_{sk}, µm
575_HAp	8.18 ± 1.36	9.87 ± 1.79	−0.09 ± 0.17
575_HAp/Coll-fish	9.63 ± 0.06	11.71 ± 0.06	−0.07 ± 0.03
575_HAp/Coll-bov	9.02 ± 0.53	11.15 ± 0.38	−0.02 ± 0.32
700_HAp	8.42 ± 1.49	10.15 ± 1.56	−0.11 ± 0.31
700_HAp/Coll-fish	9.17 ± 0.96	11.23 ± 1.18	−0.11 ± 0.23
700_HAp/Coll-bov	7.98 ± 0.68	10.45 ± 1.57	−0.12 ± 0.47

Table 9. Statistical analysis via the two-way analysis of variance (ANOVA) with repetitions.

Independent Variable	p *		
	R_a	R_q	R_{sk}
Type of crosslinking agent	0.08072	0.19422	0.89284
Type of collagen	0.21614	0.48395	0.94069
Addition of Coll-fish (compared to unmodified material)	0.25956	0.26533	0.95593
Addition of Coll-bov (compared to unmodified material)	0.43402	0.04523	0.86937

* p indicates the statistical significance calculated using the two-way analysis of variance (ANOVA).

In Figures 5 and 6, roughness profiles and surface topography of analyzed materials are shown.

The roughness and the surface topography of composite materials designed for application in implantology and tissue regeneration are one of the most important factors which may promote the osteointegration. The roughness of the surface had a great impact on the potential interactions between newly formed bone tissue and the implant. The proliferation of cell as well as the formation of extracellular matrix at the interface between the implant and the healthy tissue is a key aspect in the bone reconstruction process. For example, in previous studies it was proved that rough and bioactive surface of the implant may provide an improved adhesion and proliferation of human osteoblasts [48]. In the case of all tested materials, roughness parameter R_a was within the range 7.98–9.63 µm, which indicated a relatively high roughness of analyzed composites. This, in combination with the fact that analyzed materials were modified with a ceramic phase, i.e., hydroxyapatite, which provides osteoinductive and osteoconductive properties, allowed to conclude that developed materials showed properties conducive to bone tissue regeneration due to the roughness of their surface and simultaneous bioactivity provided via a presence of a ceramic phase.

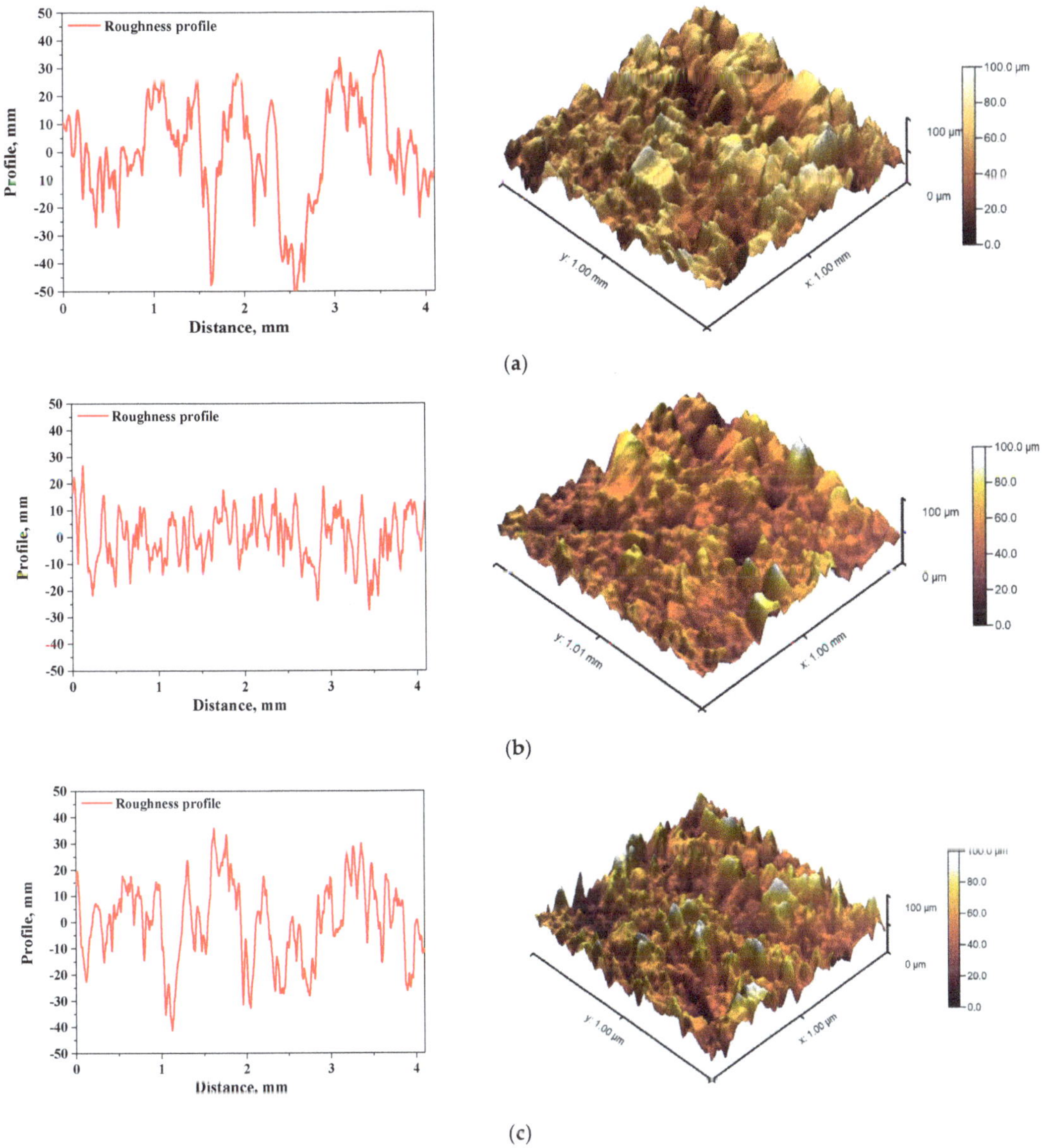

Figure 5. The roughness profile (left) and 3D image of sample 575_HAp (**a**), 575_HAp/Coll-fish (**b**) and 575_HAp/Coll-bov (**c**).

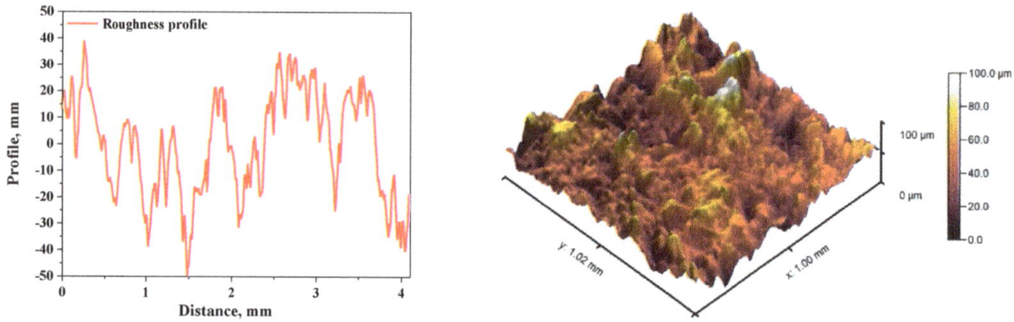

(**a**)

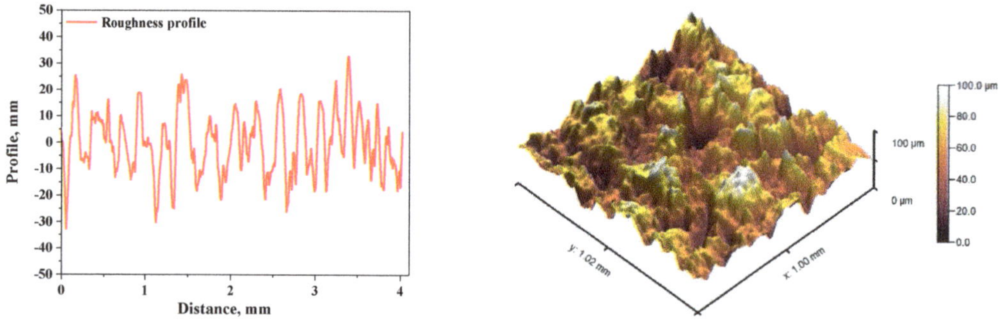

(**b**)

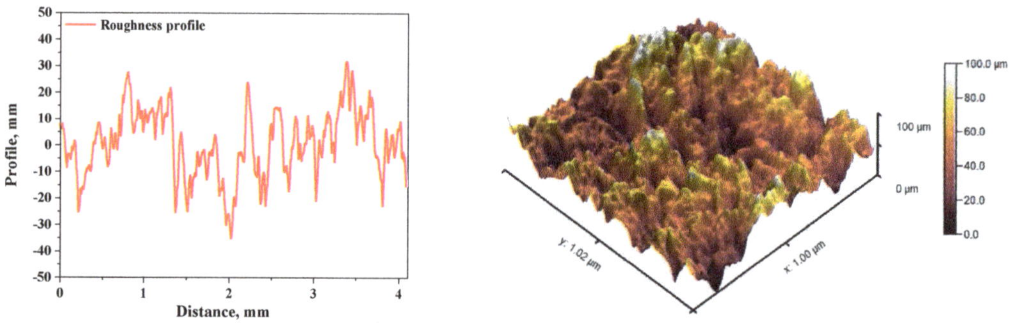

(**c**)

Figure 6. The roughness profile (left) and 3D image of sample 700_HAp (**a**), 700_HAp/Coll-fish (**b**) and 700_HAp/Coll-bov (**c**).

4. Conclusions

Spectroscopic analysis confirmed the occurrence of absorption bands characteristic for the modifier used, i.e., collagen. This, in turn, confirmed the effectiveness of the synthesis methodology applied which led to obtaining modified composites.

Next, it was proved that all tested composite materials showed swelling properties in simulated physiological liquids while the highest sorption was observed in distilled water (swelling ratio $\alpha = 2.0$ g/g after 1 h). Swelling in SBF and Ringer liquid was adequately

11.0% and 8.5% lower, which was a result of the presence of numerous ions in these solutions which may increase the crosslinking density of such materials.

Impact of the modification of composites on their physicochemical properties was demonstrated. It was proved that modification of polymer–ceramic composites with collagen resulted in an increase of swelling ratios by 0.5–1.0 g/g compared to unmodified composites. Next, considering the results of wettability investigations, it was demonstrated that values of contact angles determined for polymer–ceramic composites modified with collagen were lower than 90°, which indicated the hydrophilic nature of the surfaces of developed materials. Furthermore, performed statistical analysis proved that the both the type of the crosslinking agent used and the type of collagen had a statistically significant impact on such properties as wettability and swelling ability in selected conditions.

During 14-day incubation of composites in simulated physiological liquids, any rapid pH changes were not observed. Such changes might indicate the potential degradation of immersed liquids.

In the case of studies on roughness of composite materials, any statistically significant impact of both the average molecular weight of the crosslinking agent used and the type of collagen introduced into the polymer matrix, was not demonstrated. All composites showed similar surface roughness, which was due to a ceramic phase (hydroxyapatite) present in all tested materials.

Considering such advantages of developed polymer–ceramic composites modified with collagen as their quick and simple synthesis, a possibility of the preparation of their various shapes and sizes depending on the vessel in which the polymerization is performed as well as their properties desirable in viewpoint of bone tissue regeneration, it may be concluded that developed materials showed great application potential and should be subjected to more advanced experiments.

Author Contributions: Conceptualization, M.G., A.D., S.K.-K. and B.T.; methodology, M.G., A.D., S.K.-K. and B.T.; software, M.G., A.D., S.K.-K. and B.T.; validation, M.G., A.D. and S.K.-K.; formal analysis, M.G., A.D. and S.K.-K.; investigation, M.G., A.D., S.K.-K., M.K., A.T. and B.T.; resources, M.G., A.D., S.K.-K., M.K. and B.T.; data curation, M.G., A.D. and S.K.-K.; writing—original draft preparation, M.G., A.D. and S.K.-K.; writing—review and editing, M.G., A.D. and S.K.-K.; visualization, B.T.; supervision, D.M. and B.T.; project administration, A.S.-K. and B.T.; funding acquisition, A.S.-K. and B.T. All authors have read and agreed to the published version of the manuscript.

Funding: This research was funded by the Foundation for Polish Science, grant number POIR.04.04.00–00–16D7/18 ("Multifunctional biologically active composites for applications in bone regenerative medicine"). The authors gratefully acknowledge the financial support of the TEAM-NET program of the Foundation for Polish Science which is financed by the European Union under the European Regional Development Fund.

Institutional Review Board Statement: Not applicable.

Informed Consent Statement: Not applicable.

Data Availability Statement: The data presented in this study are available on request from the corresponding authors.

Acknowledgments: The research work was carried out within the SMART-MAT Functional Materials Science Club at the Faculty of Materials Engineering and Physics of the Cracow University of Technology.

Conflicts of Interest: The authors declare no conflict of interest.

References

1. Iorio, S.; Badino, P.; Gorini, I.; Aliverti, M. Osteoarchaeology and the History of Medicine in our experience. *Acta Biomed.* **2019**, *90*, 353–364.
2. Isbit, J. Preventing diseases of civilization. *J. Pediatr. Surg.* **2018**, *53*, 1261. [CrossRef]
3. Ensrud, K.; Crandall, C.J. Osteoporosis. *Ann. Intern. Med.* **2017**, *167*, ITC17–ITC32. [CrossRef]
4. Aspray, T.J.; Hill, T.R. Osteoporosis and the Ageing Skeleton. *Subcell. Biochem.* **2019**, *91*, 453–476.

5. Kendler, D.L.; Body, J.J.; Brandi, M.L.; Cannata-Andia, J.; Cannata-Ortiz, M.J.; Maghraouri, A.E.; Guglielmi, G.; Hadji, P.; Pierroz, D.D.; de Villiers, T.J.; et al. Osteoporosis management in hematologic stemcelltransplant recipients: Executive summary. *J. Bone Oncol.* **2021**, *28*, 100361. [CrossRef]
6. Coughlan, T.; Dockety, F. Osteoporosis and fracture risk in older people. *Clin. Med.* **2014**, *14*, 187–191. [CrossRef]
7. Kumar, R.; Singh, R.; Hashmi, M.S.J. Polymer- Ceramic composites: A state of art review and future applications. *Adv. Mater. Process. Technol.* **2020**, 1–14. [CrossRef]
8. Victor, S.R.; Muthu, J. Polymer Ceramic Composite Materials for Orthopedic Applications—Relevance and Need for Mechanical Match and Bone Regeneration. *J. Mech.* **2014**, *2*, 1–10.
9. Misra, S.K.; Boccaccini, A.R. Biodegradable and bioactive polymer/ceramic composite scaffolds. In *Tissue Engineering Using Ceramics and Polymers, Woodhead Publishing Series in Biomaterials*; Boccaccini, A.R., Gough, J.E., Eds.; Woodhead Publishing: Sawston, UK, 2007; pp. 72–92.
10. Shaban, N.Z.; Kenawy, M.Y.; Taha, N.A.; Abd El-Latif, M.M.; Ghareeb, D.A. Synthesized Nanorods Hydroxyapatite by Microwave Assisted Technology for In Vitro Osteoporotic Bone Regeneration through Wnt/β-Catenin Pathway. *Materials* **2021**, *14*, 5823. [CrossRef]
11. Fraile-Martínez, O.; García-Montero, C.; Coca, A.; Álvarez-Mon, M.A.; Monserrat, J.; Gómez-Lahoz, A.M.; Coca, S.; Álvarez-Mon, M.; Acero, J.; Bujan, J.; et al. Applications of Polymeric Composites in Bone Tissue Engineering and Jawbone Regeneration. *Polymers* **2021**, *13*, 3429. [CrossRef]
12. Faksawat, K.; Limsuwan, P.; Naemchanthara, K. 3D printing technique of specific bone shape based on raw clay using hydroxyapatite as an additive material. *Appl. Clay Sci.* **2021**, *214*, 106269. [CrossRef]
13. Salimi, M.N.; Anuar, A. Characterizations of Biocompatible and Bioactive Hydroxyapatite Particles. *Proc. Eng.* **2013**, *53*, 192–196. [CrossRef]
14. Fang, J.; Li, X.; Fang, L.; Lu, X.; Ren, F. A strong, tough, and osteoconductive hydroxyapatite mineralized polyacrylamide/dextran hydrogel for bone tissue regeneration. *Acta Biomater.* **2019**, *88*, 503–513. [CrossRef]
15. Li, C.; Qin, W.; Lakshmanan, S.; Ma, X.; Sun, X.; Xu, B. Hydroxyapatite based biocomposite scaffold: A highly biocompatible material for bone regeneration. *Saudi J. Biol. Sci.* **2020**, *27*, 2143–2148. [CrossRef]
16. Cheng, L.; Ye, F.; Yang, R.; Lu, X.; Shi, Y.; Li, L.; Fan, H.; Bu, H. Osteoinduction of hydroxyapatite/β-tricalcium phosphate bioceramics in mice with a fractured fibula. *Acta Biomater.* **2010**, *6*, 1569–1574. [CrossRef]
17. Wei, X.; Zhang, X.; Yang, Z.; Li, L.; Sui, H. Osteoinductive potential and antibacterial characteristics of collagen coated iron oxide nanospheres containing strontium and hydroxyapatite in long term bone fractures. *Arab. J. Chem.* **2021**, *14*, 102984. [CrossRef]
18. Chang, B.S.; Lee, C.K.; Hong, K.S.; Youn, H.J.; Ryu, H.S.; Chung, S.S.; Park, K.W. Osteoconduction at porous hydroxyapatite with various pore configurations. *Biomaterials* **2000**, *21*, 1291–1298. [CrossRef]
19. Wang, Q.; Tang, P.; Ge, X.; Li, P.; Lv, C.; Wang, M.; Wang, K.; Fang, L.; Lu, X. Experimental and simulation studies of strontium/zinc-codoped hydroxyapatite porous scaffolds with excellent osteoinductivity and antibacterial activity. *Appl. Surf. Sci.* **2018**, *462*, 118–126. [CrossRef]
20. Kien, P.T.; Quan, T.N.; Tuyet Anh, L.H. Coating Characteristic of Hydroxyapatite on Titanium Substrates via Hydrothermal Treatment. *Coatings* **2021**, *11*, 1226. [CrossRef]
21. Qayoom, I.; Teotia, A.K.; Meena, M.; Singh, P.; Mishra, A.; Singh, S.; Kumar, A. Enhanced bone mineralization using hydroxyapatite-based ceramic bone substitute incorporating *Withania somnifera* extracts. *Biomed. Mater.* **2020**, *15*, 055015. [CrossRef]
22. Rahman, S.U. Hydroxyapatite and tissue engineering. In *Handbook of Ionic Substituted Hydroxyapatites*; Woodhead Publishing: Sawston, UK, 2019; pp. 383–400.
23. Panda, S.; Biswas, C.K.; Paul, S. A comprehensive review on the preparation and application of calcium hydroxyapatite: A special focus on atomic doping methods for bone tissue engineering. *Ceram. Int.* **2021**, *47*, 28122–28144. [CrossRef]
24. Lara-Ochoa, S.; Ortega-Lara, W.; Guerrero Beltrán, C.E. Hydroxyapatite Nanoparticles in Drug Delivery: Phys. Applications. *Pharmaceutics* **2021**, *13*, 1642. [CrossRef]
25. Ribeiro, T.P.; Monteiro, F.J.; Laranjeira, M.S. PEGylation of iron doped hydroxyapatite nanoparticles for increased applicability as MRI contrast agents and as drug vehicles: A study on thrombogenicity, cytocompatibility and drug loading. *Eur. Polym. J.* **2020**, *137*, 109934. [CrossRef]
26. Gordienko, M.; Karakatenko, E.; Menshutina, N.; Koroleva, M.; Gilmutdinova, I.; Eremin, P. Composites Composed of Hydrophilic and Hydrophobic Polymers, and Hydroxyapatite Nanoparticles: Synthesis, Characterization, and Study of Their Biocompatible Properties. *J. Funct. Biomater.* **2021**, *12*, 55. [CrossRef]
27. Shirdar, M.R.; Taheri, M.M.; Qi, M.-L.; Gohari, S.; Farajpour, N.; Narayanan, S.; Foroozan, T.; Sharifi-Asl, S.; Shahbazian-Yassar, R.; Shokuhfar, T. Optimization of the Mechanical Properties and the Cytocompatibility for the PMMA Nanocomposites Reinforced with the Hydroxyapatite Nanofibers and the Magnesium Phosphate Nanosheets. *Materials* **2021**, *14*, 5893. [CrossRef] [PubMed]
28. Liu, F.; Kang, H.; Liu, Z.; Jin, S.; Yan, G.; Sun, Y.; Li, F.; Zhan, H.; Gu, Y. 3D Printed Multi-Functional Scaffolds Based on Poly(ε-Caprolactone) and Hydroxyapatite Composites. *Nanomaterials* **2021**, *11*, 2456. [CrossRef] [PubMed]
29. Rodzeń, K.; McIvor, M.J.; Sharma, P.K.; Acheson, J.G.; McIlhagger, A.; Mokhtari, M.; McFerran, A.; Ward, J.; Meenan, B.J.; Boyd, A.R. The Surface Characterization of Fused Filament Fabricated (FFF) 3D Printed PEEK/Hydroxyapatite Composites. *Polymers* **2021**, *13*, 3117. [CrossRef]

30. Ebrahimi, S.; Hanim, Y.U.; Sipaut, C.S.; Jan, N.B.A.; Arshad, S.E.; How, S.E. Fabrication of Hydroxyapatite with Bioglass Nanocomposite for Human Wharton's-Jelly-Derived Mesenchymal Stem Cell Growing Substrate. *Int. J. Mol. Sci.* **2021**, *22*, 9637. [CrossRef]
31. Mbese, Z.; Alven, S.; Aderibigbe, B.A. Collagen-Based Nanofibers for Skin Regeneration and Wound Dressing Applications. *Polymers* **2021**, *13*, 4368. [CrossRef]
32. Gallo, N.; Natali, M.L.; Curci, C.; Picerno, A.; Gallone, A.; Vulpi, M.; Vitarelli, A.; Ditonno, P.; Cascione, M.; Sallustio, F.; et al. Analysis of the Physico-Chemical, Mechanical and Biological Properties of Crosslinked Type-I Collagen from Horse Tendon: Towards the Development of Ideal Scaffolding Material for Urethral Regeneration. *Materials* **2021**, *14*, 7648. [CrossRef]
33. Mobaraki, M.; Bizari, D.; Soltani, M.; Khshmohabat, H.; Raahemifar, K.; Akbarzade Amirdehi, M. The Effects of Curcumin Nanoparticles Incorporated into Collagen-Alginate Scaffold on Wound Healing of Skin Tissue in Trauma Patients. *Polymers* **2021**, *13*, 4291. [CrossRef]
34. Arun, A.; Malrautu, P.; Laha, A.; Luo, H.; Ramakrishna, S. Collagen Nanoparticles in Drug Delivery Systems and Tissue Engineering. *Appl. Sci.* **2021**, *11*, 11369. [CrossRef]
35. Frolova, A.; Aksenova, N.; Novikov, I.; Maslakova, A.; Gafarova, E.; Efremov, Y.; Bikmulina, P.; Elagin, V.; Istranova, E.; Kurkov, A.; et al. A Collagen Basketweave from the Giant Squid Mantle as a Robust Scaffold for Tissue Engineering. *Mar. Drugs* **2021**, *19*, 679. [CrossRef]
36. Głąb, M.; Kudłacik-Kramarczyk, S.; Drabczyk, A.; Walter, J.; Kordyka, A.; Godzierz, M.; Bogucki, R.; Tyliszczak, B.; Sobczak-Kupiec, A. Hydroxyapatite Obtained via the Wet Precipitation Method and PVP/PVA Matrix as Components of Polymer-Ceramic Composites for Biomedical Applications. *Molecules* **2021**, *26*, 4268. [CrossRef]
37. ISO 4287:1997. *Geometrical Product Specifications (GPS)—Surface Texture: Profile Method—Terms, Definitions and Surface Texture Parameters*; International Organization for Standarization: Geneva, Switzerland, 1997.
38. Riaz, T.; Zeeshan, R.; Zarif, F.; Ilyas, K.; Muhammad, N.; Safi, S.Z.; Rahim, A.; Rizvi, S.A.; Rehman, I.U. FTIR analysis of natural and synthetic collagen. *App. Spec. Rev.* **2018**, *53*, 703–746. [CrossRef]
39. Sripriya, R.; Kumar, R. A Novel Enzymatic Method for Preparation and Characterization of Collagen Film from Swim Bladder of Fish Rohu (*Labeo rohita*). *Food Nutr. Sci.* **2015**, *6*, 1468–1478.
40. Chinh, N.T.; Manh, V.Q.; Trung, V.Q.; Lam, T.D.; Huynh, M.D.; Tung, N.Q.; Trinh, N.D.; Hoang, T. Characterization of Collagen Derived From Tropical Freshwater Carp Fish Scale Wastes and Its Amino Acid Sequence. *Nat. Prod. Comm.* **2019**, *14*, 1–12. [CrossRef]
41. Belbachir, K.; Noreen, R.; Gouspillou, G.; Petbois, C. Collagen types analysis and differentiation by FTIR spectroscopy. *Analytical Bio. Chem.* **2009**, *395*, 829–837. [CrossRef] [PubMed]
42. Ghodbane, S.A.; Dunn, M.G. Physical and mechanical properties of cross-linked type I collagen scaffolds derived from bovine, porcine, and ovine tendons. *J. Biomed. Mat. Res. Part A* **2016**, *104*, 2685–2692. [CrossRef] [PubMed]
43. Bai, Z.; Dan, W.; Yu, G.; Wang, Y.; Chen, Y.; Huang, Y.; Yang, C.; Dan, N. Tough and tissue-adhesive polyacrylamide/collagen hydrogel with dopamine-grafted oxidized sodium alginate as crosslinker for cutaneous wound healing. *RSC Adv.* **2018**, *8*, 42123–42132. [CrossRef]
44. Eriksson, C.; Nygren, H.; Ohlson, K. Implantation of hydrophilic and hydrophobic titanium discs in rat tibia: Cellular reactions on the surfaces during the first 3 weeks in bone. *Biomaterials* **2004**, *25*, 4759. [CrossRef]
45. Huang, Q.; Lin, L.; Yang, Y.; Hu, R.; Vogler, E.A.; Lin, C. Role of trapped air in the formation of cell-and-protein micropatterns on superhydrophobic/superhydrophilic microtemplated surfaces. *Biomaterials* **2012**, *33*, 8213. [CrossRef] [PubMed]
46. Olivares-Navarrete, R.; Hyzy, S.L.; Hutton, D.L.; Erdman, C.P.; Wieland, M.; Boyan, B.D.; Schwartz, Z. Direct and indirect effects of microstructured titanium substrates on the induction of mesenchymal stem cell differentiation towards the osteoblast lineage. *Biomaterials* **2010**, *31*, 2728–2735. [CrossRef] [PubMed]
47. Wennerberg, A.; Galli, S.; Albrektsson, T. Current knowledge about the hydrophilic and nanostructured SLActive surface. *Clin. Cosmet. Investig. Dent.* **2011**, *3*, 59–67. [CrossRef] [PubMed]
48. Aparicio, C.; Gil, F.J.; Planell, J.A. Human-osteoblast proliferation and differentiation on grit-blasted and bioactive titanium for dental applications. *J. Mater. Sci. Mat. Med.* **2002**, *13*, 1105–1111. [CrossRef]

materials MDPI

Article

Engineered Highly Porous Polyvinyl Alcohol Hydrogels with Poly(3-hydroxybutyrate-co-3-hydroxyvalerate) and Graphene Nanosheets for Musculoskeletal Tissue Engineering: Morphology, Water Sorption, Thermal, Mechanical, Electrical Properties, and Biocompatibility

José Luis Aparicio-Collado [1], Qiqi Zheng [1], José Molina-Mateo [1], Constantino Torregrosa Cabanilles [1], Ana Vidaurre [1,2], Ángel Serrano-Aroca [3,*] and Roser Sabater i Serra [1,2,4,*]

[1] Centre for Biomaterials and Tissue Engineering, Universitat Politècnica de València, 46022 València, Spain; joapcol@upvnet.upv.es (J.L.A.-C.); jmmateo@fis.upv.es (J.M.-M.); ctorregr@fis.upv.es (C.T.C.); vidaurre@fis.upv.es (A.V.)
[2] Biomedical Research Networking Centre in Bioengineering, Biomaterials and Nanomedicine (CIBER-BBN), 46022 València, Spain
[3] Biomaterials and Bioengineering Lab, Centro de Investigación Traslacional San Alberto Magno, Universidad Católica de Valencia San Vicente Mártir, 46001 València, Spain
[4] Department of Electrical Engineering, Universitat Politècnica de València, 46022 València, Spain
* Correspondence: angel.serrano@ucv.es (A.S.-A.); rsabater@die.upv.es (R.S.i.S.)

Citation: Aparicio-Collado, J.L.; Zheng, Q.; Molina-Mateo, J.; Torregrosa Cabanilles, C.; Vidaurre, A.; Serrano-Aroca, Á.; Sabater i Serra, R. Engineered Highly Porous Polyvinyl Alcohol Hydrogels with Poly(3-hydroxybutyrate-co-3-hydroxyvalerate) and Graphene Nanosheets for Musculoskeletal Tissue Engineering: Morphology, Water Sorption, Thermal, Mechanical, Electrical Properties, and Biocompatibility. *Materials* **2023**, *16*, 3114. https://doi.org/10.3390/ma16083114

Academic Editor: Leif Kari

Received: 14 March 2023
Revised: 12 April 2023
Accepted: 13 April 2023
Published: 15 April 2023

Abstract: Electroactive composite materials are very promising for musculoskeletal tissue engineering because they can be applied in combination with electrostimulation. In this context, novel graphene-based poly(3-hydroxybutyrate-co-3-hydroxyvalerate)/polyvinyl alcohol (PHBV/PVA) semi-interpenetrated networks (semi-IPN) hydrogels were engineered with low amounts of graphene (G) nanosheets dispersed within the polymer matrix to endow them with electroactive properties. The nanohybrid hydrogels, obtained by applying a hybrid solvent casting–freeze-drying method, show an interconnected porous structure and a high water-absorption capacity (swelling degree > 1200%). The thermal characterization indicates that the structure presents microphase separation, with PHBV microdomains located between the PVA network. The PHBV chains located in the microdomains are able to crystallize; even more after the addition of G nanosheets, which act as a nucleating agent. Thermogravimetric analysis indicates that the degradation profile of the semi-IPN is located between those of the neat components, with an improved thermal stability at high temperatures (>450 °C) after the addition of G nanosheets. The mechanical (complex modulus) and electrical properties (surface conductivity) significantly increase in the nanohybrid hydrogels with 0.2% of G nanosheets. Nevertheless, when the amount of G nanoparticles increases fourfold (0.8%), the mechanical properties diminish and the electrical conductivity does not increase proportionally, suggesting the presence of G aggregates. The biological assessment (C2C12 murine myoblasts) indicates a good biocompatibility and proliferative behavior. These results reveal a new conductive and biocompatible semi-IPN with remarkable values of electrical conductivity and ability to induce myoblast proliferation, indicating its great potential for musculoskeletal tissue engineering.

Keywords: carbon-based nanocomposite; conductive cell substrate; semi-IPN hydrogel; graphene nanosheets; poly(3-hydroxybutyrate-co-3-hydroxyvalerate); polyvinyl alcohol

1. Introduction

Tissue engineering (TE) aims to engineer artificial biocompatible structures which mimic the in vivo environment of a specific tissue, allowing its regeneration and healing. This strategy combines engineered biomaterials, specific cell populations, and bioactive molecules to induce differentiation in different human tissues [1–4]. A wide variety of polymeric biomaterials can be used to generate artificial scaffolds for TE, both of natural and synthetic origin. Natural polymers are those derived from living organisms, which present excellent biocompatibility, biodegradability, and biological performance [5,6]. Nevertheless, their lack of strong mechanical properties and other tissue microstructural features might limit their applications [7]. On the other hand, synthetic polymers are artificially produced, and their properties, such as mechanical strength, chemical stability, or microstructural topography, can be modeled to match specific requirements for different tissues to enhance cell adhesion, proliferation, and differentiation [8].

Hydrogels are of special interest in TE, since they are hydrophilic crosslinked structures able to retain large amounts of water/fluids without being dissolved [9]. Several polymers have been employed for hydrogel preparation, with poly (vinyl alcohol) (PVA) being one of the most interesting. PVA is a synthetic polymer with good biocompatibility and hydrophilic behavior produced from vinyl acetate and approved for biomedical applications by the American Food and Drug Administration (FDA). Its molecular structure can be modified by crosslinking its -OH groups with different methodologies (e.g., chemical crosslinking with glutaraldehyde, physical crosslinking by freeze–thaw, etc.) to form hydrogels with excellent water-sorption capacities, widely used in different applications such as drug delivery, wound dressing, and tissue engineering [10–13]. However, its low mechanical properties and lack of cell adhesive motifs require surface modification or combination with other polymers to produce novel biomaterials with enhanced properties and broader applications [14,15]. Semi-interpenetrated polymer networks (semi-IPNs) represent an interesting approach to combine different natural and/or synthetic polymers, forming polymeric composites in which only one component is crosslinked, and the other/s remains entangled into the crosslinked matrix in a linear or branched conformation [10,16]. With a semi-IPN structure based on crosslinked PVA and a second polymer that will improve its drawbacks, the properties of the hydrogel can be maintained while enhancing mechanical properties and bioactivity (in terms of cell adhesion).

Poly(3-hydroxybutyrate-co-3-hydroxyvalerate) (PHBV) is a natural aliphatic biopolyester found in different bacteria and archaea as an internal carbon source and energy storage [17]. It is also approved by the FDA, and its excellent mechanical properties, tunable degradation, and cell adhesion capacity make PHBV one of the most studied polyhydroxyalkanoates (PHAs) in the field of TE [18–21]. Its brittleness, null antimicrobial activity, and no water sorption due to its hydrophobic nature evidence that a suitable application for PHBV would be its use in combination with hydrophilic and more flexible polymers [22].

In addition, conductive biomaterials have been used in TE to generate electroactive substrates able to stimulate regeneration in electrically active tissues (bone, nerve, heart, muscle) even without external electrical stimulation [23–26]. Conductive polymers, such as polyaniline (PANI), polypyrrole (PPy), poly(3,4-ethylenedioxythiophene (PEDOT), etc., and carbon nanomaterials are typically used in combination with natural or synthetic biomaterials [27,28]. Graphene (G), a characteristic 2D carbon nanomaterial, is of particular interest in the biomedical field since it presents remarkable conductivity, excellent mechanical properties, good thermal stability, and extended surface area [29,30]. Graphene itself and some of its derivates such as graphene oxide (GO) and reduced graphene oxide (rGO) have been proposed as a new approach to enhance regeneration of different electrosensitive tissues [30–33]. Hurtado et al. recently reported that the incorporation of graphene nanoplatelets into semi-IPNs of calcium alginate and PHBV significantly increased its antiviral activity against a surrogate of SARS-CoV-2, while showing a good compatibility with human keratinocyte HaCaT cells [22]. Several approaches have developed conductive PVA hydrogels for tissue engineering applications in combination with different conductive nanoparticles. For example, Wang et al. [34] described how the combination of PVA with GO and rGO increased cell attachment in comparison with pure PVA. Moreover, the mechanical and electrical properties of PVA were significantly enhanced after the addition of few-layer graphene [35]. In addition, the combination of hydrophobic polymers such as

PHBV or PCL with graphene nanosheets have resulted in different nanocomposites with enhanced conductivity, mechanical properties, and good cell adhesion [36,37]. Nevertheless, the lack of hydrophilicity limits their application in tissue engineering. To our knowledge, there is no evidence in the literature about the combination of PVA/PHBV materials with G nanosheets to obtain novel nanohybrid hydrogels with electroactive properties for muscle tissue engineering.

In previous work from our research group, we developed and patented a novel semi-IPN based on PHBV/PVA with incorporated PPy nanoparticles to enhance the network's conductivity [10,38]. The obtained nanocomposite presented a homogeneous compact structure with enhanced thermal and conductive properties, but it lacked a 3D porous structure, and the conductivity values were quite low, even with high nanoparticles content (up to 15% wt/wt). Therefore, the door was left open to further study different approaches to modify this novel hydrogel in terms of morphology, swollen properties, and electrical conductivity for its use as electroactive artificial extracellular matrix for TE.

In the present work, we developed a new method to generate PHBV/PVA semi-IPNs with a porous 3D structure, incorporating G nanosheets to generate new 3D polymeric networks (PHBV/PVA/G) with enhanced conductivity despite using ultra-low graphene concentrations (0.2% and 0.8% wt/wt) to mimic the physiological electrical properties of musculoskeletal tissue and avoid potential cytotoxic effects. The morphological and physicochemical characterization was performed by electronic microscopy, Fourier transform infrared spectroscopy, swelling assay, thermal and mechanical analysis, and surface electrical conductivity. Finally, its biocompatibility was assessed with murine myoblasts (C2C12 cell line).

2. Materials and Methods

2.1. Materials and Reagents

PVA (Mw 13,000–23,000 g/mol, 87–89% hydrolyzed), 1-methyl-2-pyrrolidinone (NMP) (Mw 99.13 g/mol), and G nanosheets were supplied from Sigma Aldrich-Merck, St. Louis, MO, USA. PHBV, with 2% wt of 3-hydroxyvalerate (Mw 410,000 g/mol) was purchased from Goodfellow (Huntingdon, UK). Chloroform (Mw 119.4 g/mol, 99.9% pure), glutaraldehyde (GA) (25 % wt/wt solution), methanol (Mw 32 g/mol, 99.8% pure), and sulfuric acid (Mw 98 g/mol, 95–98% pure) were supplied by Scharlab, Sentmenat, Spain. All reagents were used as received.

2.2. Preparation of Semi-IPN PHBV/PVA/G Hydrogels

The materials' synthesis was adapted from a previous work by the same research group in which a similar goal was shared: development of semi-IPN hydrogels based on PHBV/PVA, ratio 30/70 wt/wt, with conductive properties provided by filler nanoparticles. Nanoparticles of the conductive polymer PPy were added as conductive filler [10]. In this work, the aim is to obtain highly porous 3D structures with electroactive properties by using very low amounts of G nanosheets. A new method to prepare the materials was developed, based on a hybrid solvent casting–freeze-drying process.

PVA was dissolved in NMP (5% wt/wt) for 2 h at 150 °C, while PHBV was dissolved in chloroform (3% wt/wt) for 2 h at 50 °C (both with continuous stirring of 500 rpm for PHBV and 300 rpm for PVA). Then, PHBV and PVA solutions were mixed in a 30/70 PHBV/PVA ratio and the crosslinking solution (4% wt/wt GA with respect to the total PVA content) was added and stirred for 30 min at 50 °C prior to solvent casting, in order to crosslink the PVA chains, forming a hydrogel where PHBV remains entangled within the PVA matrix. This solution was composed of GA (crosslinker, 25% GA solution), methanol (quencher, diluted 50% in MilliQ water), acetic acid (pH controller, diluted 10% in MilliQ water), and sulfuric acid (catalyst, diluted 10% in MilliQ water) mixed in a 2:2:3:1 volumetric ratio [10]. Then, the solution was transferred to Petri dishes for solvent casting (24 h at 60 °C), followed by three consecutive washings with MilliQ water (three times) and three immersions in MilliQ water after 1 h, 3 h, and 24 h to remove any traces of solvent and keep the swollen state of the samples. Then, the samples were frozen at −80 °C for 24 h and subsequently lyophilized for 72 h at −80 °C under vacuum (0.1 mbar) to preserve the porous structure.

Crosslinked PVA and pristine PHBV were used as reference. Crosslinked PVA was obtained following the same procedure but skipping the mixing step with PHBV. PHBV samples were obtained by solvent casting; the PHBV–chloroform solution was poured into a Petri dish, allowing solvent evaporation for 24 h at room temperature, followed by drying at 60 °C under vacuum to constant weight to completely remove all traces of solvent.

A previous step was carried out to prepare the semi-IPNs of PHBV/PVA containing G nanoparticles (in the form of nanosheets). The nanoparticles (0.2% and 0.8% wt/wt relative to the total polymeric mass) were dispersed in the specific amount of NMP needed to dissolve PVA in an ultrasonic bath for 6 h, after which PVA was dissolved in the NMP–G suspension. Then, the semi-IPN PHBV/PVA/G substrates were prepared following the same protocol described for the semi-IPN without G nanosheets.

Table 1 provides detailed information about the sample compositions of the study.

Table 1. Notation of the samples.

Identification	Description
PHBV	PHBV 3% wt/wt film (solvent casting)
c-PVA	PVA 5% wt/wt film crosslinked with 4% wt/wt GA (solvent casting–freeze-drying)
c-PHBV/PVA	30% PHBV/70% PVA semi-IPN crosslinked with 4% wt/wt GA (solvent casting–freeze-drying)
c-PHBV/PVA 0.2% G	30% PHBV/70% PVA semi-IPN crosslinked with 4% wt/wt GA + 0.2% G nanosheets (solvent casting–freeze-drying)
c-PHBV/PVA 0.8% G	30% PHBV/70% PVA semi-IPN crosslinked with 4% wt/wt GA + 0.8% G nanosheets (solvent casting–freeze-drying)

Figure 1 shows the preparation process of the hydrogel scaffolds and the schematic diagram of the proposed PVA–PHBV–G system.

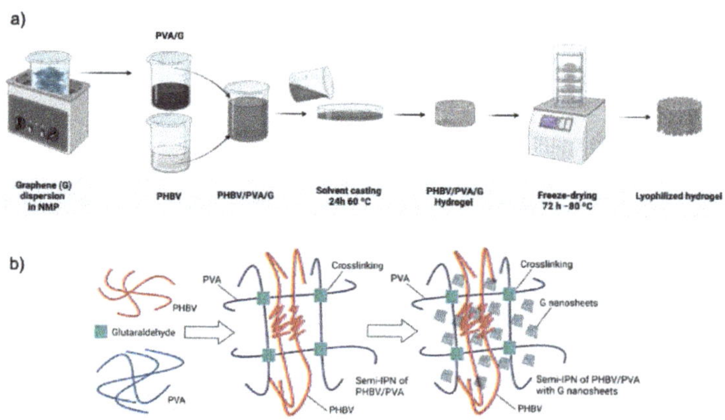

Figure 1. (**a**) Preparation process of the hydrogel scaffolds and (**b**) schematic diagram of the proposed PVA–PHBV–graphene hydrogel system. Created with Biorender.com.

2.3. *Morphological and Physicochemical Characterization*

2.3.1. Electron Microscopy

The surface and cross-section morphology of the crosslinked samples were analyzed by SEM (Zeiss ULTRA 55 Field Emission Scanning Electron Microscope (FESEM) (Carl Zeiss Microscopy, Jena, Germany)) with an accelerating voltage of 1.5–3.0 kV. The samples were coated with a platinum layer using a sputter coating (EM MED020, Leica, Wetzlar,

Germany). The cross-section was observed in samples previously immersed in liquid nitrogen and cryofractured.

2.3.2. Fourier Transform Infrared Spectroscopy (FTIR)

Fourier transform infrared (FTIR, Bruker ALPHA II Compact FT-IR Spectrometer, MA, USA) was used to study the surface functional groups (transmittance mode). The experiments were carried out at room temperature using 32 scans over a range of 4000–400 cm^{-1} at a resolution of 2 cm^{-1}.

2.3.3. Swelling Assay

Swelling experiments were performed gravimetrically in crosslinked samples. Circular freeze-dried samples (11 mm diameter) were immersed in MilliQ water at 37 °C until equilibrium after 24 h. Redundant surface water was removed using filter paper. The experiments were performed in triplicate to ensure reproducibility.

Samples were weighted before (W_0) and after (W_1) swelling, and the swelling degree (W_{eq}) was calculated as follows:

$$W_{eq}(\%) = \frac{W_1 - W_0}{W_0} \cdot 100 \quad (1)$$

2.3.4. Differential Scanning Calorimetry (DSC)

DSC analysis was carried out in a PerkinElmer DSC 8000 (Pekin Elmer, Waltham, MA, USA) under a flowing nitrogen atmosphere (20 mL/min). After erasing the effects of any previous thermal history by heating at 220 °C for 5 min, the samples were subjected to a cooling scan down to −20 °C, followed by a heating scan from that temperature up to 220 °C, both at 20 °C/min. The glass transition temperature, T_g, was calculated from the heating scan as the inflexion point of the specific heat capacity, C_p, vs. temperature, which coincides with a maximum in the temperature derivative (dc_p/dT).

The degree of crystallinity, X_c(%), was calculated according to

$$X_c(\%) = \frac{\Delta H_m/\omega}{\Delta H^0_{comp}} \cdot 100 \quad (2)$$

where ΔH_m is the melting enthalpy of the sample, ΔH^0 is the melting enthalpy for the 100% crystalline component of the semi-IPN, and ω is the component weight fraction.

2.3.5. Thermogravimetric Analysis (TGA)

TGA was used to study the thermal decomposition kinetics. Vacuum-dried samples (5–10 mg weight) were heated from 30 to 600 °C at a rate of 10 °C/min using a Mettler Toledo TGA 2 (SF) system (Mettler Toledo, Columbus, OH, USA). The mass of the samples was constantly measured as a function of temperature.

2.3.6. Mechanical Properties

Dynamic mechanical analysis was performed on a DMA 8000 (PerkinElmer, Waltham, MA, USA) at a frequency of 1 Hz on circular samples (11 mm diameter) in an immersion bath (MilliQ water) in compression mode. The storage modulus (E') and loss modulus (E'') were measured in the temperature range 36 to 38 °C at a heating rate of 1 °C/min.

2.3.7. Electrical Conductivity

To study the conductivity of the developed hydrogels, a flat structure was needed to ensure good contact with the probes. Thus, the hydrogels were first hydrated, subsequently placed between glass covers, and dried in a vacuum until constant weight to remove any water residue.

The electrical surface resistance (R_S) of flat circular samples (10 mm diameter) was measured on a T2001A3-EU four-point probe system-EU plug (Ossila Limited, Sheffield, UK). The electrical conductivity (σ) was calculated according to the following expression:

$$\frac{1}{R_S \times l} \quad (3)$$

where l is the film thickness, measured with a digital caliper (Acha, Spain). The measurements were performed in triplicate to ensure reproducibility.

2.4. Cytotoxicity Assessment

The cytotoxicity of the hydrogels was tested by performing a direct MTS assay. This colorimetric test is based on the reduction of the tetrazolium salt (3-[4,5, dimethylthiazol-2-yl]-5-[3-carboxymethoxy-phenyl]-2-[4-sulfophenyl]-2H-tetrazolium) (MTS) by the mitochondrion NADPH-dependent dehydrogenases, only active in viable cells. This metabolic process results in the generation of a soluble formazan dye whose absorbance can be quantified, providing information on the viability of cells cultured on the material.

Murine myoblasts (C2C12 cell line) were cultured on plasma-treated 24-well culture plates at 20.000 cells/cm^2 with growth media (DMEM high glucose, 10% fetal bovine serum (FBS) and 1% penicillin/streptomycin antibiotic mix (P/S)) for 48 h at 37 °C and 5% CO_2 to allow cell confluence. Cell substrates (c-PHBV/PVA, c-PHBV/PVA 0.2% G, and c-PHBV/PVA 0.8% G) were first sterilized with 3 consecutive washings (5 min) with 70% ethanol and one last washing (10 min) with MilliQ water and then placed above cells in direct contact (3 biological replicates with 3 technical replicates for each biological replicate, material, and time). These conditions were kept for 24, 48, and 72 h. At every time point, the materials were carefully removed, and the culture medium was replaced by DMEM without phenol red with an MTS 1:10 dilution and left incubating for 2 h to allow MTS metabolization by cultured cells. After that, the resulting formazan-dyed media absorbance was measured in a microplate reader (Victor Multilabel Plate Reader, Perkin Elmer, MA, USA) at 490 cm^{-1}. Cell viability was calculated as follows:

$$\text{Cellviability}(\%) = \frac{\text{ODtest}}{\text{ODcontrol}} * 100 \quad (4)$$

where OD corresponds to the optical density of the tested material and OD control is the optical density of the negative control. The negative control (cell life) consisted of C2C12 seeded in growth media without any material. C2C12 cells seeded in growth media, with 2% triton X-100 solution inoculation 1 h before MTS inoculation (at every time point), were used as positive control (cell death).

2.5. Statistical Analysis

Statistical analysis was carried out by one-way ANOVA tests on all samples using GraphPad Prism 8.0 software, with three replicates per condition unless noted. Data are presented as mean ± standard deviation. If significant differences were noted between samples, Tukey tests were used to perform pairwise comparations with a 95% confidence level ($p < 0.05$).

3. Results and Discussion

3.1. Microstructure and FTIR Analysis of the Nanohybrid Hydrogels

The morphology of the samples is depicted in Figure 2 (surface) and Figure 3 (cross-section). The combined technique (solvent casting–freeze-drying) produced a highly interconnected porous structure with pores between 0.5 and 10 μm (crosslinked PVA (Figures 2a and 3a) is included as reference). After the addition of 30% of the hydrophobic polymer PHBV, the porous structure of the swollen samples is maintained, although a decrease in pore size and the presence of small threads (more noticeable in the surface images can be observed (Figures 2b and 3b). The reduction in the pore size indicates a decrease in the swollen capacity, which is analyzed below. However, the semi-IPN shows a homogeneous structure, and the small threads can be produced by traces of PHBV or PVA not integrated into the structure of the semi-IPN during the solvent evaporation.

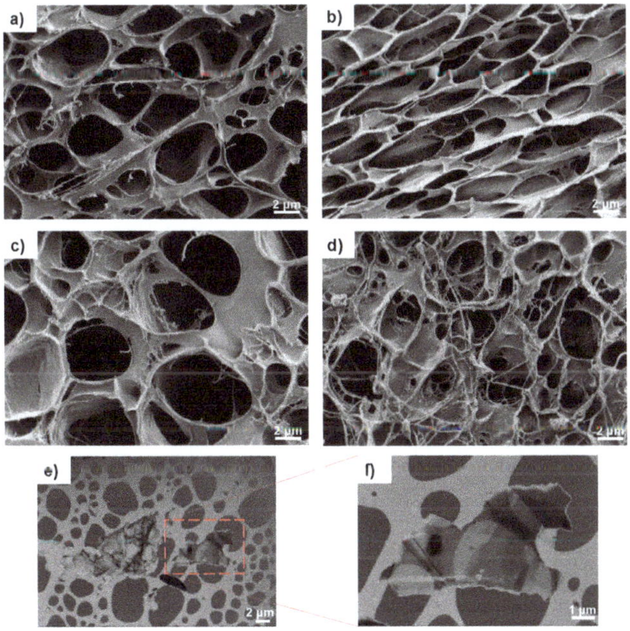

Figure 2. (**a**) High-resolution FESEM surface images of the hydrogels: (**a**) c-PVA, (**b**) c-PHBV/PVA, (**c**) c-PHBV/PVA 0.2% G, and (**d**) c-PHBV/PVA 0.8% G. (**e**,**f**) Aggregated and detailed single-form HRFESEM pictures of graphene nanosheets on a TEM grid previously dispersed in THF. To preserve the porous structure, the samples were immersed in MilliQ water for 24 h, frozen in liquid nitrogen, and stored at −80 °C until freeze-drying for 24 h. Magnification (**a**–**d**): 5 kX, (**e**): 3.5 kX, (**f**): 10.8 kX.

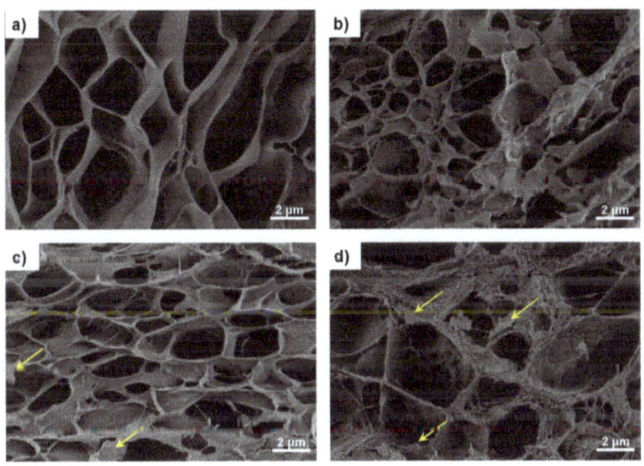

Figure 3. High-resolution FESEM cross-section pictures of the hydrogels: (**a**) c-PVA, (**b**) c-PHBV/PVA, (**c**) c-PHBV/PVA 0.2% G, and (**d**) c-PHBV/PVA 0.8% G. Cross-sections were obtained after cryogenic cut with liquid nitrogen. Magnification: 7 kX.

The addition of G nanosheets does not alter the porous structure, and the mixture of the solvents water and chloroform, which are miscible, favors the dispersion of G nanosheets, as observed in Figure 3c,d (see the yellow arrows in the cross-section images). The methodology used to prepare the samples results in highly porous structures, both in the surface (Figure 2c,d) and the cross-section (Figure 3c,d). It has been reported that nano/micropores promote cell adhesion by increasing the surface area and facilitating nutrient diffusion. In addition, microporous structures (a few microns) play an important role at the cellular level, improving the biological performance of artificial cell substrates [39].

FTIR spectra of the semi-IPN PHBV/PVA hydrogel with and without G nanosheets are shown in Figure 4. Pristine PVA, crosslinked PVA (c-PVA), and PHBV are included as reference. The spectra of neat PVA and c-PVA show the major vibrational peaks: the stretching of OH groups (band between 3600 and 3200 cm^{-1}) related to the intermolecular and intramolecular hydrogel bonds, the vibrational band related to the stretching of the C–H bond in the alkyl group (2840–3000 cm^{-1}), and the stretching of the C=O and C–O bonds (1750–1735 cm^{-1}). PVA crosslinking with bifunctional crosslinker GA produces the formation of acetal bridges between the hydroxyl groups in PVA and the aldehyde groups of GA [40,41]. The reduction in the intensity of the OH band (3600–3200 cm^{-1}) in the c-PVA sample evidences the crosslinking by the formation of acetal. The PHBV spectrum shows the characteristic -C–O–C- stretching vibration (800 and 1050 cm^{-1}) and a band related to the C=O stretching at 1719 cm^{-1} [42]. The spectrum of the semi-IPN PHBV/PVA reveals the peaks related to both PVA and PHBV. The reduction of the intensity of the OH band (3600–3200 cm^{-1}) is indicative of the reduced number of the hydroxyl groups after the addition of the hydrophobic PHBV [10]. As graphene does not have functional groups [43], the FTIR spectra of the hydrogels with G nanosheets do not show differences to that of the semi-IPN without nanoparticles, confirming that there is no interaction between the graphene nanosheets and the polymer matrix.

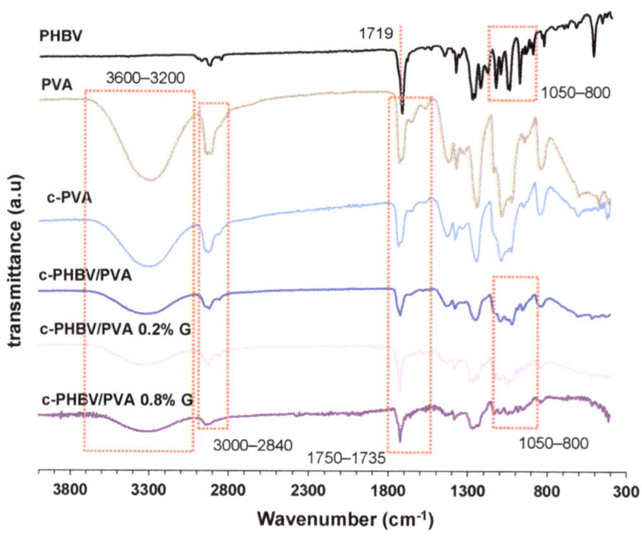

Figure 4. Fourier transform infrared spectroscopy (FTIR) spectra in the region 4000–300 cm^{-1} of the semi-IPN with and without G nanoparticles. Neat PVA, crosslinked PVA (c-PVA), and neat PHBV are also included as reference samples.

3.2. *Physical Characterization*

3.2.1. Swelling Properties

The equilibrium swelling degree of the semi-IPN PHBV/PVA with and without G nanosheets is included in Figure 5. As expected, the swelling capacity of the semi-

IPN c-PHBV/PVA decreased significantly ($p < 0.05$) due to the presence of 30% of the hydrophobic PHBV compared to crosslinked PVA (reference), although due to its high porous structure, the swelling degree is still higher than 1200%. The addition of 0.2% and 0.8% of G nanosheets does not have a significant impact on the swelling capacity compared with the semi-IPN without G nanoparticles, although statistically significant differences remain with respect to crosslinked PVA ($p < 0.05$ and $p < 0.01$ for the nanocomposites with 0.2% and 0.8% of G nanosheets, respectively). This behavior indicates that low percentages of G nanosheets (<1%) distributed within the polymer matrix, despite being highly hydrophobic, are not enough to affect the swelling capacity of the nanocomposites. The highly porous structure is maintained after the addition of G nanosheets, allowing swelling degrees higher than 1100%.

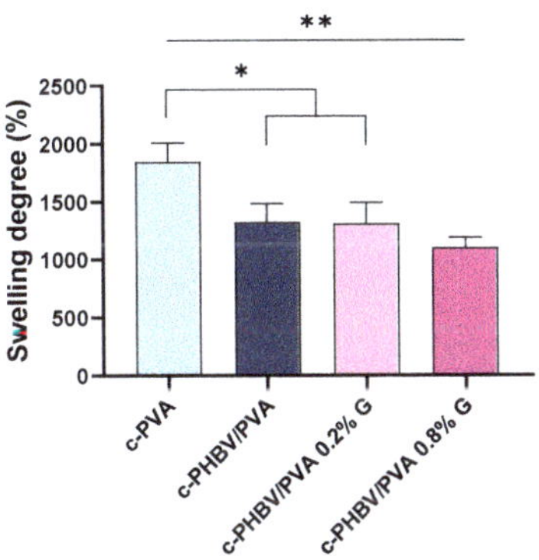

Figure 5. Equilibrium swelling degree of the engineered hydrogels after 24 h immersed in MilliQ water. Graph shows mean ± standard deviation. (*) and (**) indicate significant differences ($p < 0.05$ and $p < 0.01$, respectively). Crosslinked PVA (c-PVA) is included as reference.

3.2.2. Thermal Behavior and Thermal Degradation

Figure 6 shows the DSC scan (on heating) of the material system, and the characteristic thermal parameters obtained from the thermogram (glass transition temperature (T_g), cold crystallization temperature ($T_{c(cold)}$), melting temperature (T_m), melting enthalpy (ΔH_m), degree of crystallinity (X_c(%)) are collected in Table 2. Crosslinked PVA and neat PHBV (flat sample) are included as references. Neat PHBV shows the glass transition process around 0 °C (T_g), an exotherm peak related to cold crystallization ($T_{c(cold)} \approx 45$ °C) followed by a multiple peak associated with the melting process ($T_m \approx 155$ °C). As expected, and consistent with previous results, no melting process was observed in crosslinked PVA, pointing out that crystallization is prevented by the crosslinking to which the sample was subjected. The glass transition process of crosslinked PVA can be observed in the interval between 45 and 65 °C [10]. The semi-IPNs with and without G nanosheets show different processes identified in both pristine PHBV and crosslinked PVA. The glass transition of the PHBV chains can be observed in the composites (see the arrows in Figure 4). In addition, the thermogram shows a cold crystallization process in the same interval as the crosslinked PVA glass transition, followed by a small melting process. The cold crystallization, between 30 and 50 °C, and the melting, from 130 to 170 °C, can be observed in the same intervals as the PHBV cold crystallization and melting, respectively, suggesting that PHBV chains are involved in these processes. These results indicate that the combined methodology used to prepare the hybrid hydrogels induces phase separation between the hydrophilic PVA and the hydrophobic PHBV.

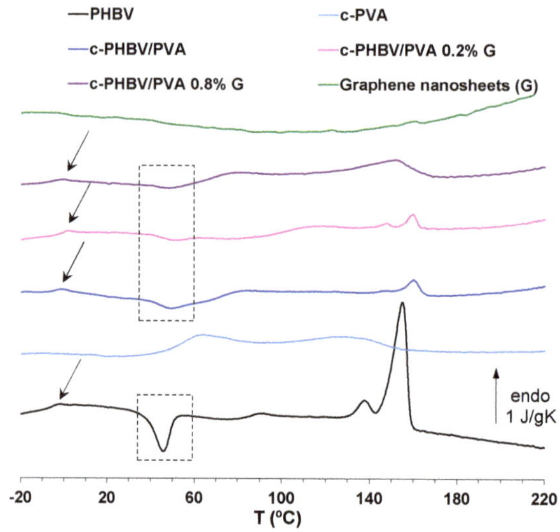

Figure 6. DSC thermograms at a rate of 20 °C/min of c-PHBV/PVA and c-PHBV/PVA with graphene nanosheets (0.2% and 0.8%). Normalized heat flow on heating. Neat PHBV, crosslinked PVA (c-PVA), and G nanosheets are included as reference.

Table 2. Glass transition temperature (T_g), cold crystallization temperature ($T_{c(cold)}$), melting temperature (T_m), melting enthalpy (ΔH_m), degree of crystallinity (X_c(%)), and decomposition temperature at which weight loss was 50% ($T_{d-50\%}$).

Sample	T_g (°C)	$T_{c(cold)}$ (°C)	T_m (°C)	ΔH_m (J/g)	X_c (%)	$T_{d-50\%}$ (°C)
PHBV	−3.6	46	156	21.2	16.1	276.3
c-PVA	57	-		-	-	352.8
c-PHBV/PVA	−3.5	50	162	1.0	2.6	334.5
c-PHBV/PVA 0.2% G	−2	50,5	160	1.8	4.7	331.1
c-PHBV/PVA 0.8% G	−3.1	48	156	4.8	12.3	339.2

However, the semi-IPN morphology (at micrometric scale) is homogeneous, as shown in the microscopy images (Figures 1 and 2), which indicate that the semi-IPN presents microphase separation, with microdomains of PHBV highly intertwined with the PVA network. As mentioned above, in a previous study, where PHBV/PVA semi-IPN with PPy nanoparticles were prepared using a different methodology (solvent casting in two phases at different temperatures), the thermal analysis (DSC results) showed a single glass transition process located between those of the neat components, indicating a good miscibility of the blend, with no phase separation [10]. Using this new method to obtain hydrogels with higher porosity, the microphase separation might have occurred during the washing and swelling in water applied between the solvent evaporation and the freeze-drying, causing PHBV chains (of hydrophobic character) to clump together in the presence of water. The PHBV chains inside the microdomains, despite the restriction imposed by the PVA network, can reorganize and crystallize, albeit to a lesser extent than pristine PHBV, as shown in Figure 6. The degree of crystallinity of the PHBV, X_c(%), was calculated according to equation 2. The enthalpy ΔH_m was obtained from the area of the melting peak, taking a baseline from 125 to 170 °C (Figure 4). Considering the low percentage of hydroxyvalerate in the copolymer PHBV (2%), it can be assumed that the crystallization process is related to poly-hydroxybutyrate (PHB); therefore, $\Delta H^0_{comp} = \Delta H^0_{PHBV} \approx \Delta H^0_{PHB} = 132\ J/g$.

The degree of crystallinity of PHBV microdomains in the semi-IPN decreases significantly compared to pristine PHBV (from 16.1 to 2.6%). The procedure used in this study, despite the microphase separation, produces highly porous hydrogels with a good stability in aqueous environments. The thermal behavior of the semi-IPN after the addition of G nanosheets is not significantly changed. The cold crystallization process and melting of the PHBV microdomains can be observed, although the crystallinity of these microdomains increased from 2.6% for the semi-IPN PHBV/PVA without G nanosheets to 4.7% and 12.3% after the addition of 0.2% and 0.8% of G nanoparticles, respectively. These results indicate that the G nanosheets, distributed within the semi-IPN, act as nucleating agent for PHBV. The increase of G nanosheets from 0.2 to 0.8% (four times higher), raises the crystallinity of the PHBV microdomains, although to a lesser extent (2.6-fold), suggesting the presence of aggregates when the quantity of nanoparticles increases. The structure of the PVA–PHBV–G hydrogel system is shown in Figure 1b.

The thermal degradation, analyzed by thermogravimetry for temperatures up to 600 °C, is shown in Figure 7. PHBV presents a one-stage degradation at temperatures above 250 °C, related to the hydrolysis and chain scission that produce crotonic acid [44]. Crosslinked PVA shows a three-stage profile, related to moisture vaporization (50–130 °C), followed by two processes at higher temperatures related to the dehydration of hydroxyl groups and hydrocarbon products degradation (above 250 °C) and, finally, the breakage of the main chain (above 400 °C) [45,46]. As expected, the degradation profile of the semi-IPN PHBV/PVA is located between those of the components, closer to crosslinked PVA (the major component of the semi-IPN) and with a similar behavior. The decomposition temperature at which weight loss was 50% ($T_{d-50\%}$), included in Table 2, increases from 276.3 °C for pristine PHBV to 334.5 °C for the semi-IPN, close to 352.8 °C obtained for the crosslinked PVA sample. The addition of G nanosheets does not significantly change the degradation profile, which is similar to that of the semi-IPN without G nanosheets, with $T_{d-50\%}$ in the same temperature range (331.1–339 °C). However, at temperatures above 450 °C, a residual weight that increases with G nanosheets content can be observed, which suggests that G nanosheets contribute to improve thermal stability at high temperature, as previously reported [47].

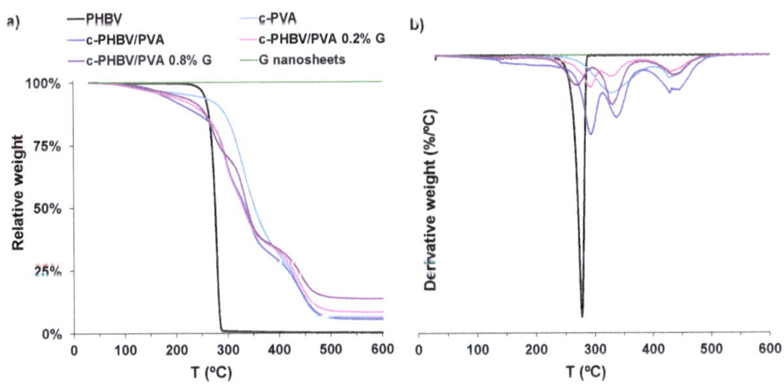

Figure 7. (**a**) Thermogravimetric analysis of the semi-IPN with and without G nanosheets. (**b**) Derivative of the weight loss. Neat PHBV and crosslinked PVA (c-PVA) are included as reference.

3.2.3. Mechanical and Electrical Properties

The measurement of mechanical properties by dynamical mechanical analysis is a first approach to evaluate the applications of engineered scaffolds [48–50]. The complex modulus, storage modulus (E'), and loss modulus (E''), obtained in a wet environment (immersion bath) at 37 °C to simulate physiological conditions, are shown in Figure 8. Statistically significant differences were found between all the samples ($p < 0.001$). As expected, both E' and E'' increased from 7.52×10^4 Pa and 7.26×10^3 Pa, respectively, for crosslinked PVA (considered as reference) to 1.02×10^5 Pa and 2.08×10^4 Pa for the semi-IPN, due to the presence of 30% of PHBV, that possesses higher mechanical strength. This enhancement indicates that the PHBV chains, located between the PVA network, reinforce the structure of the composite, despite the microphase separation between both components (as stated by the thermal characterization). The addition of small percentages of G nanosheets significantly increases the mechanical properties, as reported elsewhere [51,52]. In this study, the increase of the storage modulus after the addition of only 0.2% of G nanosheets increased almost fourfold (from 1.02×10^5 Pa for the semi-IPN to 3.87×10^5 Pa for the nanocomposite), due to the interfacial interactions between the nanoparticles and the polymer matrix, which also suggest a good dispersion within the matrix. However, after incorporating 0.8% of G nanosheets, the enhancement in the storage module decreases until 2.02×10^5, although it is still above of that of the semi-IPN without nanofiller. The loss modulus of the c-PHBV/PVA 0.8 G also decreases, with a value slightly lower than that of the semi-IPN without nanoparticles. These results suggest that, at this concentration, G nanosheets, due to strong van der Waals forces, tend to agglomerate, which results in a weaker interaction with the matrix and a less effective stress transfer [51,53]. Yang et al. prepared hybrid membranes based on polydimethyl siloxane–G nanosheets and reported that when the nanofiller content reached 0.5% wt, the agglomeration of the nanoparticles reduced the interfacial area, hindering the stress transfer across the G nanosheets–polymer matrix interface [54], in good agreement with the obtained results.

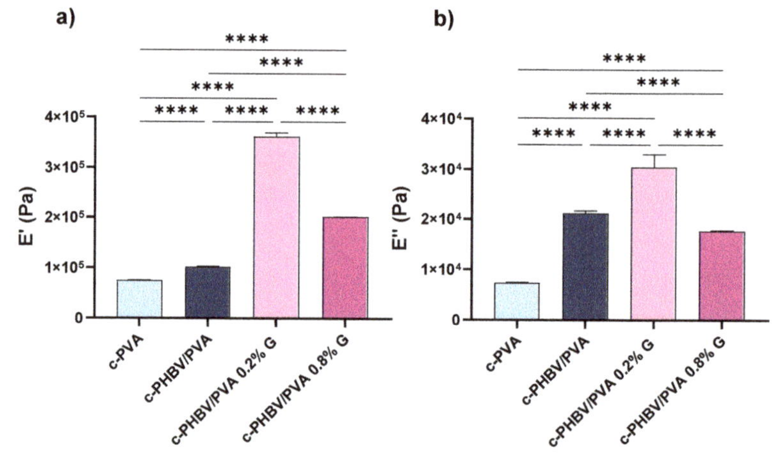

Figure 8. Dynamical mechanical analysis (compression mode). (**a**) Storage modulus (E') and (**b**) loss modulus (E'') at 37 °C for c-PHBV/PVA and nanohybrid hydrogels with 0.2% and 0.8% of G nanosheets. Crosslinked PVA (c-PVA) is included as reference. Graph shows mean ± standard deviation; (****) indicates significant differences ($p < 0.001$).

The electrical surface conductivity, obtained by equation (3) and depicted in Figure 9, shows a statistically significant increase ($p < 0.001$) after the addition of G nanosheets compared to the semi-IPN without nanoparticles as expected. Significant differences were also found between nanocomposites with 0.2% and 0.8% of G nanosheets ($p < 0.001$). The conductivity increases more than 600% after the addition of 0.2% of nanofiller (from 5.0 to 30.7 mS/m). When 0.8% of G nanosheets are added, the surface conductivity increases to 53.2 mS/m; however, increasing the nanofiller content four times results in a 1.7-fold enhancement of surface conductivity (compared to 0.2% G nanosheets). Thus, the noncorrelation between the nanofiller content and conductivity may be related to the presence of nanoparticles aggregates, which is consistent with the previously analyzed thermal behavior and mechanical performance.

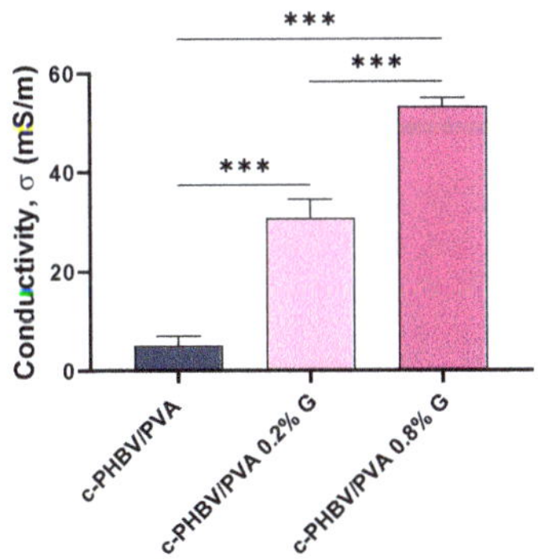

Figure 9. Electrical surface conductivity of the semi-IPNs with and without graphene nanosheets. Graph shows mean ± standard deviation; (***) indicates significant differences ($p < 0.001$).

A particular challenge in musculoskeletal tissue engineering is to engineer constructs that mimic the properties of the native tissue. Mechanical and electrical signaling cues (skeletal muscle is considered an electrosensitive tissue) are considered as important factors for successful muscle regeneration, together with topography and porosity [55,56]. The mechanical properties of the semi-IPN with and without G nanosheets are in the range of those reported for the elastic modulus of skeletal muscle tissue (30–8000 kPa) [57]. Regarding electrical properties, the values of surface conductivity (between 5.03 to 53.25 mS/m), which may be effective in the transference of electrical cues, are also in the same range as the skeletal muscle tissue ($\sigma \approx 10^{-3}$ S/m) [58] and other engineered conductive cells substrates prepared for musculoskeletal regeneration that have shown an improved cell response [59,60].

3.3. Biocompatibility

The biocompatibility of the biomaterial was assessed for C2C12 murine myoblast exposed to direct contact with the materials (semi-IPN PHBV/PVA without G nanosheets and with 0.2% and 0.8% of nanoparticles). Both pristine components, PHBV and PVA, are biocompatible polymers, widely used for biomedical applications [14,61–63]. On the other hand, it has been reported that carbon nanomaterials may induce cytotoxic effects in a dose-dependent manner [64], although the shape (layer or nanotube, for example) and the processing techniques to prepare the biomaterials play a decisive role [65,66]. The cell viability after 24, 48, and 72 h are shown in Figure 10.

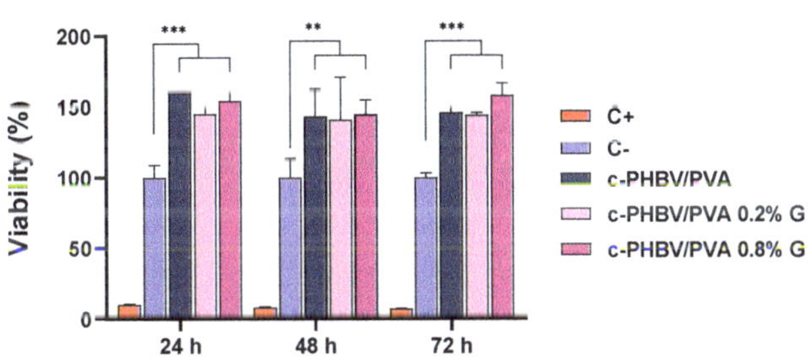

Figure 10. Cell viability after 24, 48, and 72 days from MTS assay. Semi-IPN c-PHBV/PVA and nanohydrogels with 0.2% and 0.8% of graphene nanosheets. Positive control: growth medium with latex extract as cytotoxic agent. Negative control: growth medium. Graph shows mean ± standard deviation; (**) and (***) indicate significant differences ($p < 0.01$ and $p < 0.001$, respectively).

All the materials show viability values over the negative control (cells seeded with growth medium and without any materials, considered as 100% viability) in the evaluated times. Statistically significant differences were found for all the evaluated times ($p < 0.001$ for 24 and 72 h, and $p < 0.01$ for 48 h) between the cell substrates (with and without nanoparticles). The semi-IPN shows values higher than 140% for the three times assessed, with significant differences compared to the negative control, indicating cell proliferative activity, even after only 24 h of culture. The addition of 0.2% and 0.8% of G nanosheets maintains high values of cell viability, in agreement with previous results, where small amounts of G nanosheets entangled in polymeric matrices were not cytotoxic [37,67] and promoted proliferation [68]. The high values of viability in the evaluated times imply that the nanohybrid hydrogels do not lose the proliferative capacity.

4. Conclusions

The engineered conductive hydrogels developed in this study are safe in terms of cytotoxicity and possess high bioaffinity with C2C12 cells, thus promoting cell proliferation. These porous hydrogels, developed by a hybrid solvent casting–freeze-drying method, and 0.2% of G nanosheets possess interconnected porous structure, high water-absorption capacity, thermal stability, and suitable mechanical and conductive properties (in the range of skeletal muscle tissue). Although further physicochemical and biological studies are needed to explore the applications of these novel electroactive semi-IPN PHBV/PVA/G hydrogels, the results of this study postulate them as promising candidates for electroactive tissue regeneration applications, particularly muscle tissue engineering.

Author Contributions: Conceptualization, R.S.i.S. and Á.S.-A.; methodology, R.S.i.S., Á.S.-A. and J.L.A.-C.; validation, R.S.i.S. and J.L.A.-C.; formal analysis, J.L.A.-C. and Q.Z.; investigation, J.L.A.-C., Q.Z., J.M.-M., C.T.C. and A.V.; writing—original draft preparation, J.L.A.-C. and R.S.i.S.; writing—review and editing, J.L.A.-C., R.S.i.S., Á.S.-A., J.M.-M., C.T.C. and A.V.; supervision, R.S.i.S. and Á.S.-A.; funding acquisition, R.S.i.S., J.M.-M. and Á.S.-A. All authors have read and agreed to the published version of the manuscript.

Funding: This research was funded by Spanish Ministry of Science and Innovation (MCINN, Agencia Estatal de Investigación/FEDER funds), grants RTI2018-097862-B-C21 (awarded to R.S.i.S. and J.M.-M.) and PID2020-119333RB-I00/AEI/10.13039/501100011033 (awarded to Á.S.-A.). Á.S.-A. also acknowledges the Fundación Universidad Católica de Valencia San Vicente Mártir through the Grant 2020-231-006UCV. The CIBER-BBN initiative is funded by the VI National R&D&I Plan 2008-2011, Iniciativa Ingenio 2010, Consolider Program. CIBER actions are financed by the Instituto de Salud Carlos III with assistance from the European Regional Development.

Institutional Review Board Statement: Not applicable.

Informed Consent Statement: Not applicable.

Data Availability Statement: Data will be made available on request.

Acknowledgments: The authors give thanks for technical and human support provided by the Microscopy Services at Universitat Politècnica de València.

Conflicts of Interest: The authors declare no conflict of interest.

References

1. Pina, S.; Ribeiro, V.P.; Marques, C.F.; Maia, F.R.; Silva, T.H.; Reis, R.L.; Oliveira, J.M. Scaffolding Strategies for Tissue Engineering and Regenerative Medicine Applications. *Materials* **2019**, *12*, 1824. [CrossRef] [PubMed]
2. Boni, R.; Ali, A.; Shavandi, A.; Clarkson, A.N. Current and Novel Polymeric Biomaterials for Neural Tissue Engineering. *J. Biomed. Sci.* **2018**, *25*, 90. [CrossRef] [PubMed]
3. Di Marzio, N.; Eglin, D.; Serra, T.; Moroni, L. Bio-Fabrication: Convergence of 3D Bioprinting and Nano-Biomaterials in Tissue Engineering and Regenerative Medicine. *Front. Bioeng. Biotechnol.* **2020**, *8*, 326. [CrossRef] [PubMed]
4. Sharma, K.; Mujawar, M.A.; Kaushik, A. State-of-Art Functional Biomaterials for Tissue Engineering. *Front. Mater.* **2019**, *6*. [CrossRef]
5. Choi, G.; Cha, H.J. Recent Advances in the Development of Nature-Derived Photocrosslinkable Biomaterials for 3D Printing in Tissue Engineering. *Biomater. Res.* **2019**, *23*, 18. [CrossRef] [PubMed]

6. Mohammadinejad, R.; Kumar, A.; Ranjbar-Mohammadi, M.; Ashrafizadeh, M.; Han, S.S.; Khang, G.; Roveimiab, Z. Recent Advances in Natural Gum-Based Biomaterials for Tissue Engineering and Regenerative Medicine: A Review. *Polymers* **2020**, *12*, 176. [CrossRef]
7. Ullah, S.; Chen, X. Fabrication, Applications and Challenges of Natural Biomaterials in Tissue Engineering. *Appl. Mater. Today* **2020**, *20*, 100656. [CrossRef]
8. Place, E.S.; George, J.H.; Williams, C.K.; Stevens, M.M. Synthetic Polymer Scaffolds for Tissue Engineering. *Chem. Soc. Rev.* **2009**, *38*, 1139–1151. [CrossRef]
9. Yang, D. Recent Advances in Hydrogels. *Chem. Mater.* **2022**, *34*, 1987–1989. [CrossRef]
10. Luis Aparicio-Collado, J.; Novoa, J.J.; Molina-Mateo, J.; Torregrosa-Cabanilles, C.; Serrano-Aroca, Á.; Sabater i Serra, R. Novel Semi-Interpenetrated Polymer Networks of Poly(3-Hydroxybutyrate-Co-3-Hydroxyvalerate)/Poly (Vinyl Alcohol) with Incorporated Conductive Polypyrrole Nanoparticles. *Polymers* **2021**, *13*, 57. [CrossRef]
11. Gajra, B.; Pandya, S.S.; Vidyasagar, G.; Rabari, H.; Dedania, R.R.; Rao, S. Poly Vinyl Alcohol Hydrogel and Its Pharmaceutical and Biomedical Applications: A Review. *Int. J. Pharm. Res.* **2012**, *4*, 20–26.
12. Boso, D.; Maghin, E.; Carraro, E.; Giagante, M.; Pavan, P.; Piccoli, M. Extracellular Matrix-Derived Hydrogels as Biomaterial for Different Skeletal Muscle Tissue Replacements. *Materials* **2020**, *13*, 2483. [CrossRef] [PubMed]
13. Gull, N.; Khan, S.M.; Butt, O.M.; Islam, A.; Shah, A.; Jabeen, S.; Khan, S.U.; Khan, A.; Khan, R.U.; Butt, M.T.Z. Inflammation Targeted Chitosan-Based Hydrogel for Controlled Release of Diclofenac Sodium. *Int. J. Biol. Macromol.* **2020**, *162*, 175–187. [CrossRef] [PubMed]
14. Kumar, A.; Han, S.S. PVA-Based Hydrogels for Tissue Engineering: A Review. *Int. J. Polym. Mater. Polym. Biomater.* **2017**, *66*, 159–182. [CrossRef]
15. Sionkowska, A. Current Research on the Blends of Natural and Synthetic Polymers as New Biomaterials: Review. *Prog. Polym. Sci.* **2011**, *36*, 1254–1276. [CrossRef]
16. Crosby, C.O.; Stern, B.; Kalkunte, N.; Pedahzur, S.; Ramesh, S.; Zoldan, J. Interpenetrating Polymer Network Hydrogels as Bioactive Scaffolds for Tissue Engineering. *Rev. Chem. Eng.* **2022**, *38*, 347–361. [CrossRef]
17. Rivera-Briso, A.L.; Serrano-Aroca, Á. Poly(3-Hydroxybutyrate-Co-3-Hydroxyvalerate): Enhancement Strategies for Advanced Applications. *Polymers* **2018**, *10*, 732. [CrossRef]
18. Pryadko, A.; Surmeneva, M.A.; Surmenev, R.A. Review of Hybrid Materials Based on Polyhydroxyalkanoates for Tissue Engineering Applications. *Polymers* **2021**, *13*, 1738. [CrossRef]
19. Gheibi, A.; Khoshnevisan, K.; Ketabchi, N.; Derakhshan, M.A.; Babadi, A.A. Application of Electrospun Nanofibrous PHBV Scaffold in Neural Graft and Regeneration: A Mini-Review. *Nanomed. Res. J.* **2016**, *1*, 107–111. [CrossRef]
20. Rivera-Briso, A.L.; Aparicio-Collado, J.L.; Serra, R.S.I.; Serrano-Aroca, Á. Graphene Oxide versus Carbon Nanofibers in Poly(3-Hydroxybutyrate-Co-3-Hydroxyvalerate) Films: Degradation in Simulated Intestinal Environments. *Polymers* **2022**, *14*, 348. [CrossRef]
21. Janowski, G.; Frącz, W.; Bąk, Ł. The Mechanical Properties Prediction of Poly [(3-Hydroxybutyrate)-Co-(3-Hydroxyvalerate)] (PHBV) Biocomposites on a Chosen Example. *Materials* **2022**, *15*, 7531. [CrossRef]
22. Hurtado, A.; Cano-Vicent, A.; Tuñón-Molina, A.; Aparicio-Collado, J.L.; Salesa, B.; i Serra, R.S.; Serrano-Aroca, Á. Engineering Alginate Hydrogel Films with Poly(3-Hydroxybutyrate-Co-3-Valerate) and Graphene Nanoplatelets: Enhancement of Antiviral Activity, Cell Adhesion and Electroactive Properties. *Int. J. Biol. Macromol.* **2022**, *219*, 694–708. [CrossRef] [PubMed]
23. Li, Y.; Wei, L.; Lan, L.; Gao, Y.; Zhang, Q.; Dawit, H.; Mao, J.; Guo, L.; Shen, L.; Wang, L. Conductive Biomaterials for Cardiac Repair: A Review. *Acta Biomater* **2022**, *139*, 157–178. [CrossRef] [PubMed]
24. Dong, R.; Ma, P.X.; Guo, B. Conductive Biomaterials for Muscle Tissue Engineering. *Biomaterials* **2020**, *229*, 119584. [CrossRef]
25. Zarrintaj, P.; Zangene, E.; Manouchehri, S.; Amirabad, L.M.; Baheiraei, N.; Hadjighasem, M.R.; Farokhi, M.; Ganjali, M.R.; Walker, B.W.; Saeb, M.R.; et al. Conductive Biomaterials as Nerve Conduits: Recent Advances and Future Challenges. *Appl. Mater. Today* **2020**, *20*, 100784. [CrossRef]
26. Chen, C.; Xi, Y.; Weng, Y. Progress in the Development of Graphene-Based Biomaterials for Tissue Engineering and Regeneration. *Materials* **2022**, *15*, 2164. [CrossRef]
27. Gajendiran, M.; Choi, J.; Kim, S.J.; Kim, K.; Shin, H.; Koo, H.J.; Kim, K. Conductive Biomaterials for Tissue Engineering Applications. *J. Ind. Eng. Chem.* **2017**, *51*, 12–26. [CrossRef]
28. Ma, C.; Jiang, L.; Wang, Y.; Gang, F.; Xu, N.; Li, T.; Liu, Z.; Chi, Y.; Wang, X.; Zhao, L.; et al. 3D Printing of Conductive Tissue Engineering Scaffolds Containing Polypyrrole Nanoparticles with Different Morphologies and Concentrations. *Materials* **2019**, *12*, 2491. [CrossRef]
29. Bellet, P.; Gasparotto, M.; Pressi, S.; Fortunato, A.; Scapin, G.; Mba, M.; Menna, E.; Filippini, F. Graphene-Based Scaffolds for Regenerative Medicine. *Nanomaterials* **2021**, *11*, 404. [CrossRef]
30. Shin, S.R.; Li, Y.C.; Jang, H.L.; Khoshakhlagh, P.; Akbari, M.; Nasajpour, A.; Zhang, Y.S.; Tamayol, A.; Khademhosseini, A. Graphene-Based Materials for Tissue Engineering. *Adv. Drug Deliv. Rev.* **2016**, *105*, 255–274. [CrossRef]
31. Hermenean, A.; Codreanu, A.; Herman, H.; Balta, C.; Rosu, M.; Mihali, C.V.; Ivan, A.; Dinescu, S.; Ionita, M.; Costache, M. Chitosan-Graphene Oxide 3D Scaffolds as Promising Tools for Bone Regeneration in Critical-Size Mouse Calvarial Defects. *Sci. Rep.* **2017**, *7*, 16641. [CrossRef] [PubMed]

32. Aparicio-Collado, J.L.; García-San-Martín, N.; Molina-Mateo, J.; Torregrosa Cabanilles, C.; Donderis Quiles, V.; Serrano-Aroca, A.; Sabater i Serra, R. Electroactive Calcium-Alginate/Polycaprolactone/Reduced Graphene Oxide Nanohybrid Hydrogels for Skeletal Muscle Tissue Engineering. *Colloids Surf. B Biointerfaces* **2022**, *214*, 112455. [CrossRef] [PubMed]
33. Qian, Y.; Zhao, X.; Han, Q.; Chen, W.; Li, H.; Yuan, W. An Integrated Multi-Layer 3D-Fabrication of PDA/RGD Coated Graphene Loaded PCL Nanoscaffold for Peripheral Nerve Restoration. *Nat. Commun.* **2018**, *9*, 323. [CrossRef] [PubMed]
34. Wang, X.; Su, M.; Liu, C.; Shen, C.; Liu, X. Poly (Vinyl Alcohol)/Graphene Nanocomposite Hydrogel Scaffolds for Control of Cell Adhesion. *J. Renew. Mater.* **2020**, *8*, 89–99. [CrossRef]
35. Das, B.; Eswar Prasad, K.; Ramamurty, U.; Rao, C.N.R. Nano-Indentation Studies on Polymer Matrix Composites Reinforced by Few-Layer Graphene. *Nanotechnology* **2009**, *20*, 125705. [CrossRef]
36. Sridhar, V.; Lee, I.; Chun, H.H.; Park, H. Graphene Reinforced Biodegradable Poly(3-Hydroxybutyrate-Co-4-Hydroxybutyrate) Nano-Composites. *Express Polym. Lett.* **2013**, *7*, 320–328. [CrossRef]
37. Aparicio-Collado, J.L.; Molina-Mateo, J.; Cabanilles, C.T.; Vidaurre, A.; Salesa, B.; Serrano-Aroca, Á.; Sabater i Serra, R. Pro-Myogenic Environment Promoted by the Synergistic Effect of Conductive Polymer Nanocomposites Combined with Extracellular Zinc Ions. *Biology* **2022**, *11*, 1706. [CrossRef]
38. Sabater i Serra, R.; Serrano-Aroca, Á.; Molina Mateo, J.; Aparicio Collado, J.L. Biomaterial. Biodegradable. Patent ES 2812048 B2, 5 May 2022.
39. Sgarminato, V.; Tonda-Turo, C.; Ciardelli, G. Reviewing Recently Developed Technologies to Direct Cell Activity through the Control of Pore Size: From the Macro- to the Nanoscale. *J. Biomed. Mater. Res. B Appl. Biomater.* **2020**, *108*, 1176–1185. [CrossRef]
40. Mansur, H.S.; Sadahira, C.M.; Souza, A.N.; Mansur, A.A.P. FTIR Spectroscopy Characterization of Poly (Vinyl Alcohol) Hydrogel with Different Hydrolysis Degree and Chemically Crosslinked with Glutaraldehyde. *Mater. Sci. Eng. C* **2008**, *28*, 539–548. [CrossRef]
41. Mansur, H.S.; Oréfice, R.L.; Mansur, A.A.P. Characterization of Poly(Vinyl Alcohol)/Poly(Ethylene Glycol) Hydrogels and PVA-Derived Hybrids by Small-Angle X-Ray Scattering and FTIR Spectroscopy. *Polymer* **2004**, *45*, 7193–7202. [CrossRef]
42. Paşcu, E.I.; Stokes, J.; McGuinness, G.B. Electrospun Composites of PHBV, Silk Fibroin and Nano-Hydroxyapatite for Bone Tissue Engineering. *Mater. Sci. Eng. C* **2013**, *33*, 4905–4916. [CrossRef]
43. Ţucureanu, V.; Matei, A.; Avram, A.M. FTIR Spectroscopy for Carbon Family Study. *Crit. Rev. Anal. Chem.* **2016**, *46*, 502–520. [CrossRef]
44. Singh, S.; Mohanty, A.K. Wood Fiber Reinforced Bacterial Bioplastic Composites: Fabrication and Performance Evaluation. *Compos. Sci. Technol.* **2007**, *67*, 1753–1763. [CrossRef]
45. Voronova, M.I.; Surov, O.V.; Guseinov, S.S.; Barannikov, V.P.; Zakharov, A.G. Thermal Stability of Polyvinyl Alcohol/Nanocrystalline Cellulose Composites. *Carbohydr. Polym.* **2015**, *130*, 440–447. [CrossRef]
46. Lewandowska, K. Miscibility and Thermal Stability of Poly(Vinyl Alcohol)/Chitosan Mixtures. *Thermochim. Acta* **2009**, *493*, 42–48. [CrossRef]
47. Wang, L.; Liu, F.; Jin, C.; Zhang, T.; Yin, Q. Preparation of Polypyrrole/Graphene Nanosheets Composites with Enhanced Thermoelectric Properties. *RSC Adv.* **2014**, *4*, 46187–46193. [CrossRef]
48. Zeinali, K.; Khorasani, M.T.; Rashidi, A.; Daliri Joupari, M. Preparation and Characterization of Graphene Oxide Aerogel/Gelatin as a Hybrid Scaffold for Application in Nerve Tissue Engineering. *Int. J. Polym. Mater. Polym. Biomater.* **2021**, *70*, 674–683. [CrossRef]
49. Jitphuthi, P.; Tangtrakulwanich, B.; Meesane, J. Hierarchical Porous Formation, Collagen and Mineralized Collagen Modification of Polylactic Acid to Design Mimicked Scaffolds for Maxillofacial Bone Surgery. *Mater. Today Commun.* **2017**, *13*, 46–52. [CrossRef]
50. Silva, M.; Pinho, I.; Gonçalves, H.; Vale, A.C.; Paiva, M.C.; Alves, N.M.; Covas, J.A. Engineering Ligament Scaffolds Based on PLA/Graphite Nanoplatelet Composites by 3D Printing or Braiding. *J. Compos. Sci.* **2023**, *7*, 104. [CrossRef]
51. Le, M.T.; Huang, S.C. Thermal and Mechanical Behavior of Hybrid Polymer Nanocomposite Reinforced with Graphene Nanoplatelets. *Materials* **2015**, *8*, 5526–5536. [CrossRef] [PubMed]
52. Wang, H.; Xie, G.; Fang, M.; Ying, Z.; Tong, Y.; Zeng, Y. Mechanical Reinforcement of Graphene/Poly(Vinyl Chloride) Composites Prepared by Combining the in-Situ Suspension Polymerization and Melt-Mixing Methods. *Compos. B Eng.* **2017**, *113*, 278–284. [CrossRef]
53. Kaur, T.; Thirugnanam, A.; Pramanik, K. Effect of Carboxylated Graphene Nanoplatelets on Mechanical and In-Vitro Biological Properties of Polyvinyl Alcohol Nanocomposite Scaffolds for Bone Tissue Engineering. *Mater. Today Commun.* **2017**, *12*, 34–42. [CrossRef]
54. Yang, D.; Yang, S.; Jiang, Z.; Yu, S.; Zhang, J.; Pan, F.; Cao, X.; Wang, B.; Yang, J. Polydimethyl Siloxane-Graphene Nanosheets Hybrid Membranes with Enhanced Pervaporative Desulfurization Performance. *J. Memb. Sci.* **2015**, *487*, 152–161. [CrossRef]
55. Carnes, M.E.; Pins, G.D. Skeletal Muscle Tissue Engineering: Biomaterials-Based Strategies for the Treatment of Volumetric Muscle Loss. *Bioengineering* **2020**, *7*, 85. [CrossRef]
56. Zhou, J.; Vijayavenkataraman, S. 3D-Printable Conductive Materials for Tissue Engineering and Biomedical Applications. *Bioprinting* **2021**, *24*, e00166. [CrossRef]
57. Pien, N.; Krzyslak, H.; Shastry Kallaje, S.; Van Meerssche, J.; Mantovani, D.; De Schauwer, C.; Dubruel, P.; Van Vlierberghe, S.; Pennisi, C.P. Tissue Engineering of Skeletal Muscle, Tendons and Nerves: A Review of Manufacturing Strategies to Meet Structural and Functional Requirements. *Appl. Mater. Today* **2023**, *31*, 101737. [CrossRef]

58. Yu, C.; Yao, F.; Li, J. Rational Design of Injectable Conducting Polymer-Based Hydrogels for Tissue Engineering. *Acta Biomater.* **2022**, *139*, 4–21. [CrossRef] [PubMed]
59. Du, Y.; Ge, J.; Li, Y.; Ma, P.X.; Lei, B. Biomimetic Elastomeric, Conductive and Biodegradable Polycitrate-Based Nanocomposites for Guiding Myogenic Differentiation and Skeletal Muscle Regeneration. *Biomaterials* **2018**, *157*, 40–50. [CrossRef]
60. Hosseinzadeh, S.; Mahmoudifard, M.; Mohamadyar-Toupkanlou, F.; Dodel, M.; Hajarizadeh, A.; Adabi, M.; Soleimani, M. The Nanofibrous PAN-PANi Scaffold as an Efficient Substrate for Skeletal Muscle Differentiation Using Satellite Cells. *Bioprocess Biosyst. Eng.* **2016**, *39*, 1163–1172. [CrossRef]
61. Deepthi, S.; Nivedhitha Sundaram, M.; Vijayan, P.; Nair, S.V.; Jayakumar, R. Engineering Poly(Hydroxy Butyrate-Co-Hydroxy Valerate) Based Vascular Scaffolds to Mimic Native Artery. *Int. J. Biol. Macromol.* **2018**, *109*, 85–98. [CrossRef]
62. Tebaldi, M.L.; Maia, A.L.C.; Poletto, F.; de Andrade, F.V.; Soares, D.C.F. Poly(-3-Hydroxybutyrate-Co-3-Hydroxyvalerate) (PHBV): Current Advances in Synthesis Methodologies, Antitumor Applications and Biocompatibility. *J. Drug Deliv. Sci. Technol.* **2019**, *51*, 115–126. [CrossRef]
63. Sreena, R.; Nathanael, A.J. Biodegradable Biopolymeric Nanoparticles for Biomedical Applications-Challenges and Future Outlook. *Materials* **2023**, *16*, 2364. [CrossRef] [PubMed]
64. Ye, R.; Song, W.; Feng, M.; Zhou, R. Potential Interference of Graphene Nanosheets in Immune Response: Via Disrupting the Recognition of HLA-Presented KK10 by TCR: A Molecular Dynamics Simulation Study. *Nanoscale* **2021**, *13*, 19255–19263. [CrossRef] [PubMed]
65. Zhang, Y.; Ali, S.F.; Dervishi, E.; Xu, Y.; Li, Z.; Casciano, D.; Biris, A.S. Cytotoxicity Effects of Graphene and Single-Wall Carbon Nanotubes in Neural Phaeochromocytoma-Derived Pc12 Cells. *ACS Nano* **2010**, *4*, 3181–3186. [CrossRef]
66. Gies, V.; Lopinski, G.; Augustine, J.; Cheung, T.; Kodra, O.; Zou, S. The Impact of Processing on the Cytotoxicity of Graphene Oxide. *Nanoscale Adv.* **2019**, *1*, 817–826. [CrossRef]
67. Fan, H.; Wang, L.; Zhao, K.; Li, N.; Shi, Z.; Ge, Z.; Jin, Z. Fabrication, Mechanical Properties, and Biocompatibility of Graphene-Reinforced Chitosan Composites. *Biomacromolecules* **2010**, *11*, 2345–2351. [CrossRef] [PubMed]
68. Nilawar, S.; BS, M.; Chatterjee, K. Nanoceria-Decorated Graphene Nanosheets Enhance Mechanical Properties and Bioactivity of a Degradable Polyurethane for Biomedical Applications. *J. Polym. Environ.* **2023**. [CrossRef]

materials

MDPI

Article

Comparison of the Influence of 45S5 and Cu-Containing 45S5 Bioactive Glass (BG) on the Biological Properties of Novel Polyhydroxyalkanoate (PHA)/BG Composites

Katharina Schuhladen [1], Barbara Lukasiewicz [2], Pooja Basnett [2], Ipsita Roy [3,*] and Aldo R. Boccaccini [1,*]

1 Department of Materials Science and Engineering, Institute of Biomaterials, University of Erlangen-Nuremberg, 91058 Erlangen, Germany; katharina.ks.schuhladen@fau.de

2 Applied Biotechnology Research Group, School of Life Sciences, College of Liberal Arts and Sciences, University of Westminster, London W1W 6UW, UK; barbara.lukasiewicz@gmail.com (B.L.); P.Basnett@westminster.ac.uk (P.B.)

3 Department of Material Science and Engineering, Faculty of Engineering, University of Sheffield, Mappin Street, Sheffield S1 3JD, UK

* Correspondence: I.Roy@sheffield.ac.uk (I.R.); aldo.boccaccini@ww.uni-erlangen.de (A.R.B.)

Received: 9 May 2020; Accepted: 5 June 2020; Published: 8 June 2020

Abstract: Polyhydroxyalkanoates (PHAs), due to their biodegradable and biocompatible nature and their ability to be formed in complex structures, are excellent candidates for fabricating scaffolds used in tissue engineering. By introducing inorganic compounds, such as bioactive glasses (BGs), the bioactive properties of PHAs can be further improved. In addition to their outstanding bioactivity, BGs can be additionally doped with biological ions, which in turn extend the functionality of the BG-PHA composite. Here, different PHAs were combined with 45S5 BG, which was additionally doped with copper in order to introduce antibacterial and angiogenic properties. The resulting composite was used to produce scaffolds by the salt leaching technique. By performing indirect cell biology tests using stromal cells, a dose-depending effect of the dissolution products released from the BG-PHA scaffolds could be found. In low concentrations, no toxic effect was found. Moreover, in higher concentrations, a minor reduction of cell viability combined with a major increase in VEGF release was measured. This result indicates that the fabricated composite scaffolds are suitable candidates for applications in soft and hard tissue engineering. However, more in-depth studies are necessary to fully understand the release kinetics and the resulting long-term effects of the BG-PHA composites.

Keywords: bioactive glass; copper; polyhydroxyalkanoates; tissue engineering; scaffolds

1. Introduction

Due to disease, injury and trauma, treatments to promote the repair, replacement or regeneration of damaged and degenerated tissues in the human body are necessary. These treatments typically involve living tissue and organs for transplantation and have been lifesaving [1–4]. However, due to donor limitations and organ rejection, tissue engineering (TE) as a suitable alternative is being increasingly investigated [5]. In the most common TE approach, cells grow on a scaffold made using suitable methods to provide a temporary support and a well-defined pore structure [6]. Furthermore, growth factors and other biomolecules can be incorporated within the scaffold in order to guide the regulation of cellular functions during tissue regeneration [7]. The first important step in the designing of TE scaffolds is to find suitable biomaterials for building a 3D structure, which would degrade appropriately, in a rate similar to the new tissue growth rate. Natural and synthetic polymers are widely used as biomaterials for TE scaffolds [8]. Natural polymers are extracted from animals or plants (e.g., collagen, chitosan), whereas synthetic polymers are synthesized chemically (e.g., poly-L-lactic acid, poly-ε-caprolactone) [9].

Polyhydroxyalkanoates (PHAs) are polyesters of 3-, 4-, 5- and 6-hydroxyalkanoic acid which are synthesized using microbial biotechnology approaches involving bacterial fermentation. Some bacteria store PHAs as an intracellular storage compounds for energy and carbon, normally under conditions of nutrient limitation (e.g., nitrogen, sulphur, oxygen, magnesium or phosphorus), with the excess of carbon (e.g., carbohydrates, lipids, fatty acids) [10–13]. The number of carbon atoms in a monomer unit is crucial to the properties of PHAs. Depending on this number, PHAs can be classified as short-chain length PHAs (scl-PHAs) and medium-chain length PHAs (mcl-PHAs). Scl-PHAs contain 3–5 carbon atoms, whereas mcl-PHAs contain 6–14 carbon atoms. Scl-PHAs are generally brittle and stiff, with a high melting point and crystallinity, except for poly(4-hydroxybutyrate), P(4HB). Typical examples for scl-PHAs include poly(3-hydroxybutyrate-co-3-hydroxyvalerate), P(3HB-*co*-3HV), and the most well studied one, poly(3-hydroxybutyrate), P(3HB) [14]. Mcl-PHAs, on the other hand, are semi-crystalline polymers, have a low melting point, and are extremely elastomeric. Poly(3-hydroxyhexanoate) and poly(3-hydroxyoctanoate) are typical examples for mcl-PHAs [10,15,16]. In general, PHAs are biodegradable, insoluble in water, nontoxic, biocompatible, piezo-electric (which stimulates bone growth and promotes wound healing), thermoplastic and/or elastomeric [15–17].

Polymers, such as PHAs, are highly biocompatible and can be easily formed in complex shapes and structures. However, in order to further enhance their biological properties and to tailor the properties of PHAs for different applications, these biomaterials can be mixed with inorganic components (e.g., hydroxyapatite, bioactive glasses) forming composites [9,18,19]. For example, such composites have been shown to have enhanced capability to form an apatite layer on the implant surface, which for instance is important in the regeneration of bone [20–22].

In 1969, Hench and co-workers invented the first bioactive glass (BG), showing a strong bone bonding ability of 45S5 BG [23]. This glass composition (45% SiO_2-24.5% Na_2O-24.5 CaO-6% P_2O_5 in wt.%) is highly reactive when in contact with an aqueous environment [23,24]. It has been reported that 45S5 BG has been used in more than a million patients to repair bone defects in the jaw and in orthopaedics [24]. The 45S5 BG is in general bioactive, biocompatible, biodegradable, osteoconductive, osteoinductive, angiogenic, nontoxic and noninflammatory because of its ability to form a hydroxycarbonate apatite surface layer in a biological medium and exploiting the release of ionic dissolution products that stimulate specific cellular pathways [24–29]. Since the release of ionic dissolution products, after exposure to a physiological environment, is believed to improve the bioactivity of materials, new approaches for enhancing BG bioactivity are being investigated by introducing therapeutic ions in BG compositions [30–32]. In this work, Cu-doped BG was used as an example of the use of inorganic ions to further enhance the bioactivity of PHAs. Copper has been shown to promote synergistically stimulating effects on angiogenesis by stabilizing the expression of hypoxia-inducible factors and promote the proliferation of human endothelial cells [33–38]. Moreover, Cu ions were shown to promote wound healing in rats, which has been linked to the up regulation of vascular endothelial growth factor (VEGF) by stimulated cells [6] and to the antimicrobial behavior of Cu. Indeed, Cu is able to artificially mimic hypoxia, which plays an important role in blood vessel formation as well as in the differentiation and recruitment of endothelial cells [37,39]. The angiogenic effect of Cu-doped 45S5 BG has been investigated and proven *in vitro* and *in vivo* using 3D BG scaffolds [40].

In the present work, 3D mcl/scl-PHA scaffolds containing 45S5 and Cu-doped 45S5 BGs were prepared using solvent casting/salt leaching technique in order to create an advanced PHAs based composite with Cu-ion delivery capability, and with tailored properties for use in different types of hard and soft tissue engineering. For the first time, the interaction of stromal (ST2) cells with the dissolution products of the neat and composite PHA scaffolds was studied by evaluating the cell viability, cell morphology and the release of VEGF from the cells cultivated in the presence of dissolution products from the scaffolds. The newly designed composites are considered to be suitable for soft and hard tissue engineering due to the enhanced angiogenic effect potentially due to the release of therapeutic ions, especially copper, from the BG particles embedded in the PHA matrix of the scaffold [10,16,23,24].

2. Materials and Methods

2.1. Materials

All the chemicals used for PHA production were purchased from VWR (Lutterworth, UK) or from Sigma-Aldrich (Dorset, UK). The two different BGs, the 45S5 composition (in wt.%: 45.0 SiO_2, 24.5 Na_2O, 24.5 CaO, 6 P_2O_5) and a Cu-doped 45S5 composition (in wt.%: 45.0 SiO_2, 24.5 Na_2O, 22.0 CaO, 6 P_2O_5, 2.5 CuO) used in this study were produced by melt-quenching, presented elsewhere [40]. Briefly, the BGs were produced by mixing SiO_2, Na_2CO_3, $CaCO_3$, $Ca_3(PO_4)_2$ and $CuCO_3 \cdot Cu(OH)_2$ (all analytical grade) and melting the raw materials in a platinum crucible at 1450 °C for 45 min. The produced BGs were milled using a jaw crusher and a planetary mill (both Retsch, Germany) to obtain a particle size of d_{50} = 5 μm.

2.2. Scl-PHA Production

A P(3HB) polymer was synthesized by using Gram-positive bacterium *Bacillus subtilis* OK2. The production with some modifications was carried out as described elsewhere [11], by using a modified seed culture medium and a modified Kannan and Rehacek medium. Briefly, a single colony of this strain was grown at 30 °C and 200 rpm for 16 h in a seed culture medium (containing (g/L): meat peptone, 4.3; casein peptone, 4.3; sodium chloride, 6.4). Afterwards, the inoculum was used to inoculate a fermenter containing the modified Kannan and Rehacek medium (containing (g/L): glucose, 35; yeast extract, 2.5; ammonium sulphate, 5.0, potassium chloride, 3.0). The culture was grown for 48 h with constant stirring (200 rpm) and an air flow (1 L of air/min/1 L of media) at 30 °C.

2.3. Mcl-PHA Production

P(3HO), P(3HO-*co*-3HD) and P(3HO-*co*-3HD-*co*-3HDD) polymers were produced using Gram-negative bacterium *Pseudomonas mendocina* CH50, obtained from the National Collection of the Industrial and Marine Bacteria (NCIMB 10541) and different carbon sources (Table 1) based on optimized protocols [41–43]. A single colony of this strain was grown under the same conditions and in the same seed culture medium as described for P(3HB). The remaining production steps for mcl-PHAs were described in Rai et al. [43]. Briefly, to inoculate the second stage, a modified mineral salt medium (containing (g/L): ammonium sulphate, 0.45; potassium phosphate monobasic, 2.38; Di-sodium hydrogen phosphate anhydrous, 3.42; magnesium sulphate heptahydrate, 0.8; trace element solution, 1 mL/L; suitable carbon sources, (Table 1)) and to inoculate the production stage, a second modified mineral salt medium (containing (g/L): ammonium sulphate, 0.5; potassium phosphate monobasic, 2.65; Di-sodium hydrogen phosphate anhydrous, 3.8; magnesium sulphate heptahydrate, 0.8; trace element solution, 1 mL/L; related carbon sources, (Table 1)) were used. In order to simplify the labelling of the different tested PHAs, the short versions of the mcl-PHAs were used according to Table 1.

Table 1. Produced mcl-PHAs with the related carbon sources.

Mcl-PHA	Abbreviations	Carbon Source	Concentration
P(3HO)	P(3HO)	Sodium octanoate	3.36 g/L
P(3HO-3HD)	P(3HD)	Glucose	20 g/L
P(3HO-3HD-3HDD)	P(3HDD)	Coconut oil	20 g/L

2.4. PHA Extraction

For the extraction of PHAs, the cells were harvested by centrifuging the cultures at 4600 rpm, then homogenized and lyophilized. In order to extract the polymer, Soxhlet extraction was used firstly with methanol and secondly with chloroform. The extracted solution was concentrated by evaporation, followed by the precipitation of the polymer, using chilled methanol with continuous stirring [43].

2.5. PHA and BG-PHA Scaffold Preparation

The PHA and BG-PHA scaffolds were prepared by using the salt-leaching technique. For composite scaffolds, 1 wt.% (in case of scl-PHAs) or 2 wt.% (in case of mcl-PHAs) of 45S5 BG/45S5-2.5Cu powder was added to the dissolved PHAs (to obtain a final BG content of 20 wt.%) in chloroform (5 wt.% of scl-PHAs/10 wt.% of mcl-PHAs) and dispersed by sonication. Sodium chloride (sieved, diameter, 355 μm, Sigma Aldrich, Dorset, UK) was then added in the 10:1 ratio to the PHAs, as a porogen. The solution was cast layer by layer in a mould (3.5 cm × 1.2 cm). After drying, the scaffolds were immersed in distilled water for 24 h in order to dissolve the sodium chloride, and thus to form pores [44].

2.6. Scanning Electron Microscopy (SEM)

Characterization of the scaffolds was carried out using a scanning electron microscope (LEO 435 VP; LEO Electron Microscopy Ltd., Cambridge, UK, and Ultra Plus; Zeiss, Jena, Germany). The samples were placed on the 8 mm diameter aluminum holder and images were then recorded at different magnifications. The microscope was operated at 1 kV and a working distance of 2 mm. The pore size range was calculated based on at least 5 different SEM images per sample (not shown here) using ImageJ (NIH, Bethesda, MD, USA) [45].

2.7. Cell Culture

Stromal ST2 cells (Deutsche Sammlung von Mikroorganismen and Zellkulturen GmbH, Braunschweig, Germany) derived from the mouse bone marrow of BC8 mice were used to study the biocompatibility of the produced scaffolds. ST2 cells were chosen due to their potential to differentiate in osteoblasts, adipocytes and hematopoietic cells [46]. The cells were grown, harvested and counted as described by Balasubramanian et al. [47]. For the biocompatibility test, 100,000 ST2 cells per mL of cell culture medium (CCM, RPMI 1640 medium (Gibco, Schwerte, Germany) containing 10 vol.% fetal bovine serum (Sigma-Aldrich, Darmstadt, Germany) and 1 vol.% penicillin/streptomycin (Gibco, Germany)) were transferred per well in a 48-well plate (VWR, Darmstadt, Germany) and incubated for 24 h. Scaffolds were cut in 0.1 ± 0.01 g cubes and disinfected by using UV for a period of 30 min on each side. The scaffolds were immersed in 1 mL of CCM and incubated for 24 h under the same conditions as the cells. After 24 h, 1 mL CCM containing ionic dissolution products (IDPs) of 0.1 ± 0.01 g of scaffolds was removed and named 10%-CCM in accordance to a previous publication [45] (although the 0.1 g of scaffold did not completely dissolve). By further diluting this 10%-CCM, dilutions were produced and named as 1%-, 0.1%- and 0.01%-CCM in order to simplify the labelling. After removing the CCM from the cells incubated for 24 h, the now attached ST2 cells were incubated with the different dilutions of the CCM containing IDPs for 48 h under the same conditions as described above. Cells grown in CCM containing no IDPs were taken as a reference. Every sample type was investigated as replicates of three. After 48 h, the CCM was collected in Eppendorf tubes for VEGF release measurement studies. The cell viability using a WST-8 assay (Sigma Aldrich) and the release of VEGF was tested as described by Balasubramanian et al. [47]. Briefly, the VEGF release was measured by using a RayBio mouse VEGF enzyme-linked immunosorbent assay (ELISA) kit. The cell viability, as well as the release of VEGF, was determined by following the manufacturer's protocol and then spectrometrically analyzed using a microplate reader (PHOmo, Anthos Mikrosysteme, Krefeld, Germany) at 450 nm. Further, hematoxylin and eosin (H&E) staining was used to investigate the morphology. Prior to staining, cells were washed with PBS and fixed using Fluoro-fix. After another washing step with distilled water, the cells were stained using Hematoxylin for 10 min. Subsequently afterwards, Hematoxylin was removed by first washing with tap water, followed by "Scott's tap water" and distilled water for 1–5 min. Then, cells stained with an Eosin solution (0.1% Eosin in 90% ethanol and 5% acetic acid) for 5 min and finally washed with 100% ethanol. Stained cells were observed using a light microscope (Primo Vert, Zeiss, Oberkochen, Germany).

3. Results

In Figure 1, SEM images of neat and composite P(3HB) and P(3HO) scaffolds can be seen. The pores were of irregular shape and of varied sizes (10–250 μm) proven by the SEM images in Figure 1A,B. Moreover, good pore interconnectivity was indicated by the SEM images, which needs to be proven in future work (e.g., by μCT). The fabrication technique used to prepare these samples leads to results similar to the ones achieved for other polymers using this method [48]. No obvious difference could be observed between the neat scl-PHAs and mcl-PHAs scaffolds. Furthermore, it was confirmed that the microstructures of the composite foams were similar to that of the neat ones. In Figure 1, SEM images of neat P(3HB) (C) and a composite P(3HB) (D) scaffold at high magnification are shown as examples. No BG particles could be found on the surface of the composite P(3HB) scaffolds, and the same was observed for all composite mcl-PHAs scaffolds. This observation confirmed that the BG particles were fully embedded in the polymer matrix. Figure 1E shows an exemplary digital image of a P(3HB)/BG and a P(3HO)/BG scaffold, indicating that the scaffolds maintained the shape of the used molds.

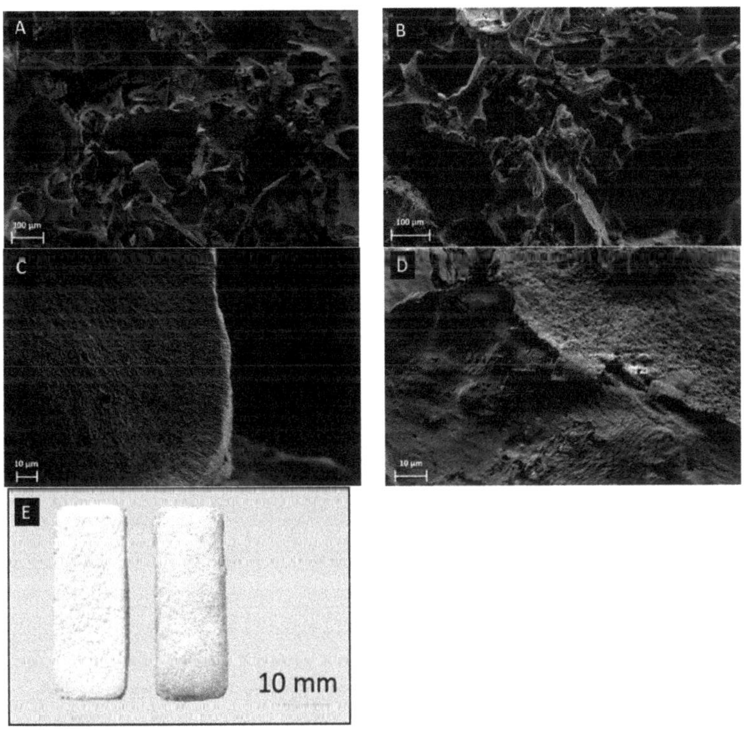

Figure 1. SEM images of neat P(3HB) scaffold (**A**), neat P(3HO) scaffolds (**B**), P(3HB)/Cu-doped bioactive glass (BG) composite (**C**), and P(3HO)/Cu-doped BG composite (**D**). (**E**) Digital camera images of P(3HB)/BG (left) and P(3HO)/BG (right) scaffolds.

There were no differences between the viability of cells grown in 0.01%-CCM and 0.1%-CCM of all the different scaffolds compared to the cells grown in CCM without any IDPs, and Figure 2 shows the results for the cells grown in 1%- and 10%-CCM. As seen in Figure 2a, a 60–80% decrease in cell viability was observed when 1%-CCM was added as compared to the reference for both the neat and composite PHA samples. No significant difference was found between the different PHAs. Furthermore, the addition of BG did not seem to have any influence. Figure 2b shows that cells grown in 10%-CCM exhibited remarkably less cell viability. Here too, no significant influence of adding BG into the polymer scaffold was detected.

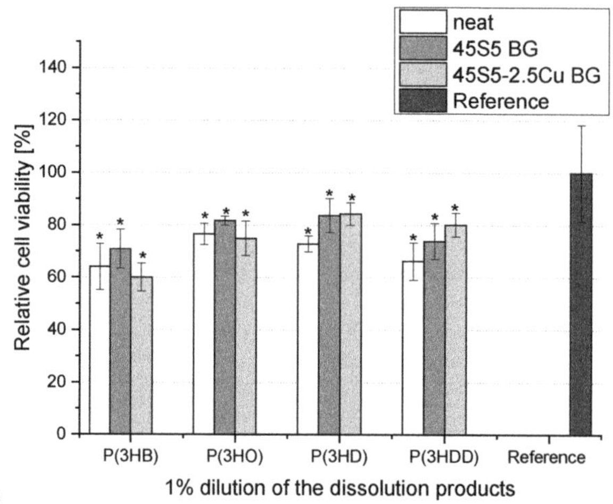

(**a**)

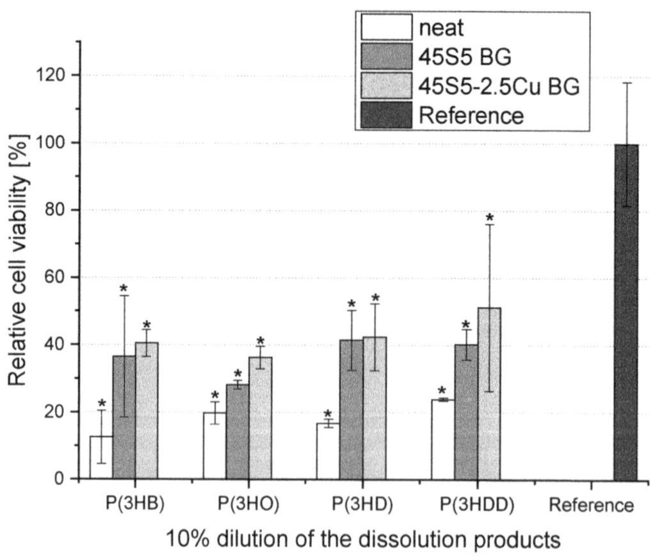

(**b**)

Figure 2. Relative viability of ST2 cells cultured in media containing 1% (**a**) and 10% (**b**) dilution of ionic dissolution products (IDPs) from different BG-PHA and neat PHA scaffolds. One-way ANOVA statistical analysis denotes significant differences compared to the reference (* $p < 0.05$). As reference pure CCM was used (Bonferroni's post hoc test was used).

Light microscopy images of H&E-stained ST2 cells cultured with 1%-CCM and 10%-CCM of neat PHA scaffolds and PHA scaffolds containing 45S5 BG are shown in Figure 3. Cells cultured in 1%-CCM of BG-PHA and neat PHA scaffolds exhibited their typical phenotypic cell morphology, and showed adhesion to the well plate. Moreover, the density of cells grown on 1%-CCM was seen to be almost the same as the density observed in the reference sample (control). This indicates that ST2 cells grown in contact with the dissolution products of the PHA scaffolds containing additional BG in concentrations lower than 10% do not have any adverse effect on the cell morphology. However, cells cultured in 10%-CCM of all scaffolds showed relatively poor cell proliferation and adhesion. These results confirm the data from the cell viability test. According to Figure 3, in all CCM containing the same concentration of IDPs, only minor differences were observed between the different PHAs as well as between the neat and BG-containing scaffolds.

The VEGF release from the ST2 cells cultured in media with different dilutions (10, 1, 0.1 and 0.01%) of dissolution products of different neat and composite PHA scaffolds was measured. Since no difference between the VEGF release from cells cultured in media containing 0.1%-CCM and 0.01%-CCM of all BG-PHA and neat PHA scaffolds compared to the reference could be observed, these results are not shown here. According to Figure 4a, it was observed that the VEGF release increased, up to 150–220%, for all the investigated scaffolds. In contrast to cells grown in 1%-CCM, VEGF released by cells cultured in the 10%-CCM of all other neat and composite scaffolds exhibited a major reduction. In summary, the cells cultured in 1%-CCM showed a minor reduction of cell viability, but the remaining cells released higher concentrations of VEGF, whereas cells grown in 10%-CCM were relatively less viable and able to release less VEGF as compared to the cells cultured in 1%-CCM.

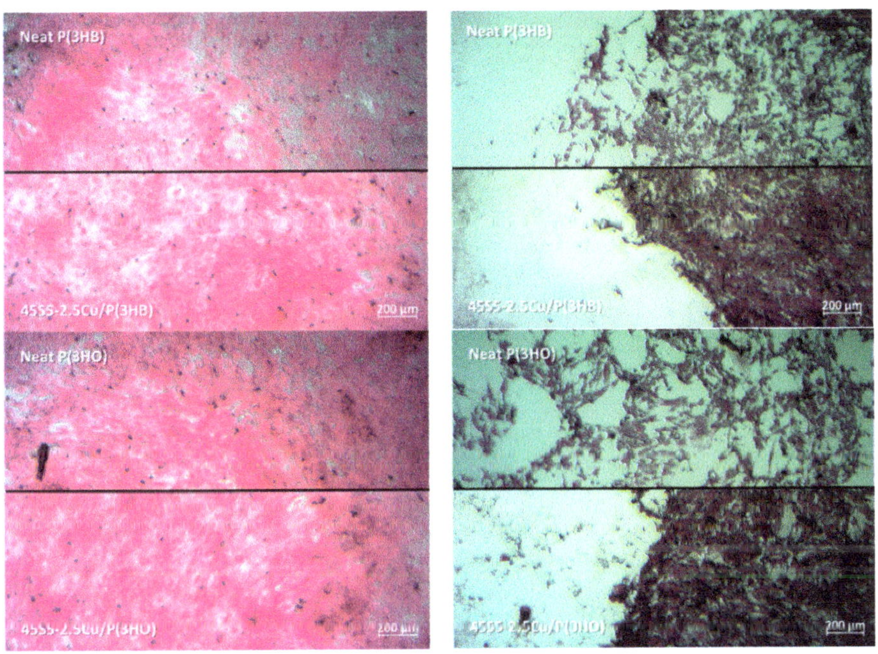

Figure 3. *Cont.*

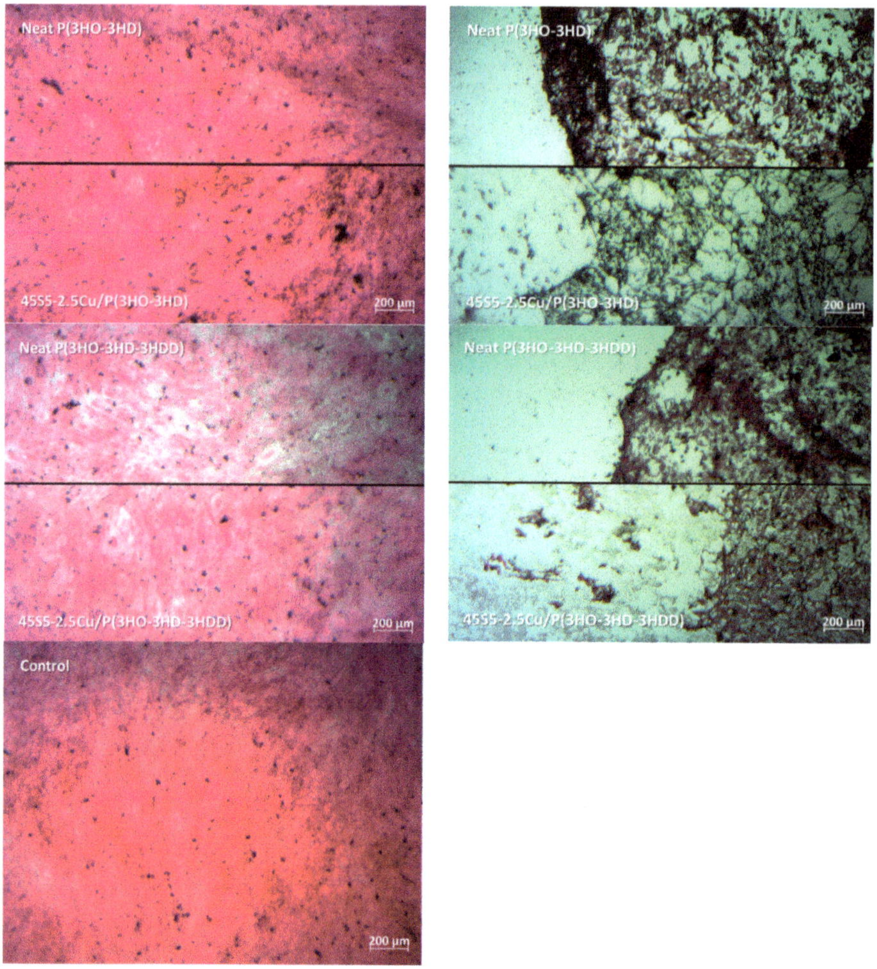

Figure 3. Images of hematoxylin and eosin (H&E)-stained cells cultured in 1% (left) and in 10% (right) dilution of the dissolution products of the different neat and composite PHA scaffolds. Cells cultured in CCM containing no dissolution products were used as control.

Overall, the promotion of the bioactivity by adding bioactive glass in the case of bioactive glass/PHA scaffolds could only be found in the case of 1%-CCM. Here, the addition of bioactive glass led to a major increase in VEGF release and to a minor increase in cell viability. Several studies in the past have already proven the biocompatibility of different PHAs [16]. Additionally this study indicates that the polymer can promote VEGF secretion. Angiogenesis plays a crucial role in tissue regeneration and therefore is essential for the success of scaffolds in tissue engineering [28]. VEGF is known to be involved in the formation of blood vessels, and therefore the ability of PHAs to increase VEGF secretion makes them interesting for applications in tissue engineering. However, in the case of 10% dilution, the dissolution products of all scaffolds showed a major reduction of cell viability and VEGF release, proven by the cell morphology. A possible reason could be that the used salt to produce the scaffolds was not completely leached out and could therefore have a negative impact on cell viability and VEGF release. Additionally, it is also possible that the lower concentration of the 1% dilution leads to the measured increase in VEGF release as observed in a previous study [49].

Moreover, the 0.01% and 0.1% dilutions showed no influence on cell viability and VEGF release compared to the reference. Therefore, the concentration of the CCM seems to play a crucial role and further studies, especially under dynamic conditions, are necessary to understand the dose-dependent effect on VEGF release. A significant effect of the addition of BGs in the polymer matrix could not be measured. It is well known that the degradation and therefore the dissolution of ions start from the surface of bioactive glass particles [6], hence it is important that bioactive glass particles are exposed to the fluid environment at various stages during the *in vitro* degradation of the composite. Here, most of the BG was not present on the surfaces of the scaffolds, which was proven by SEM. Instead, BG particles were embedded in the polymer matrix and therefore prevented from direct contact with the CCM. Therefore, it is suggested that an immersion time of 24 h is not enough to measure any significant release of BG dissolution products and their influence on cell viability and VEGF secretion. Moreover, future experiments should include direct cell biology studies in order to test the impact that the stiffness and the topography of the BG-PHA scaffolds have oncell viability and VEGF secretion.

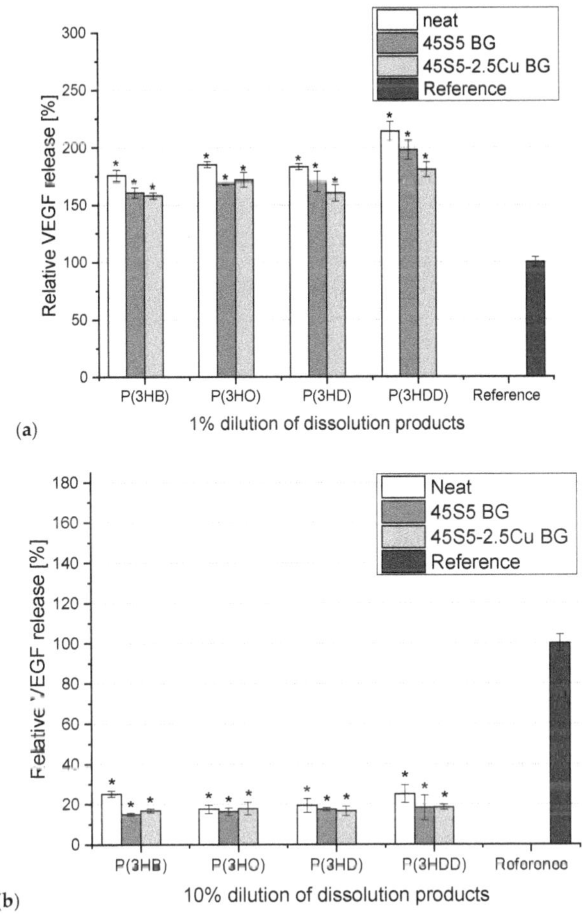

Figure 4. Relative vascular endothelial growth factor (VEGF) release from ST2 cells cultured in media containing 1% (**a**) and 10% (**b**) dilution of dissolution products of BG-PHA and neat PHA scaffolds. One-way ANOVA statistical analysis denotes significant difference (* $p < 0.05$) compared with the reference. As reference pure CCM was used (Bonferroni's post hoc test was used).

4. Conclusions

In this study, composite BG-PHA scaffolds were successfully produced using the salt leaching technique. To evaluate the biocompatibility and the effect of the addition of BG, cell studies on the fabricated scaffolds were performed. The results indicated good biocompatibility and thus the present composites are interesting for applications in tissue engineering. However, long-term cell studies need to be carried out to better examine the effect of dissolution products of BGs, which should be followed by *in vivo* studies.

Author Contributions: K.S., I.R. and A.R.B. conceived and designed the experiments. K.S., B.L. and P.B. performed the experiments and analyzed the data. K.S. wrote the manuscript. I.R. and A.R.B. revised and edited the manuscript. All authors have read and agreed to the published version of the manuscript.

Funding: This study was supported by the Erasmus+ program of the European Union.

Acknowledgments: The authors thank Rainer Detsch and Alina Grünewald for helping with the cell culture.

Conflicts of Interest: There are no conflicts to declare.

References

1. Murphy, C.; O'Brien, F.J.; Little, D.; Schindeler, A. Cell-scaffold interactions in the bone tissue engineering triad. *Eur. Cells Mater.* **2013**, *32*, 120–132. [CrossRef] [PubMed]
2. Damien, C.J.; Parsons, J.R. Bone graft and bone graft substitutes: A review of current technology and applications. *J. Appl. Biomater.* **1991**, *2*, 187–208. [CrossRef] [PubMed]
3. Giannoudis, P.V.; Dinopoulos, H.; Tsiridis, E. Bone substitutes: An update. *Injury* **2005**, *36*, S20–S27. [CrossRef] [PubMed]
4. Khan, S.N.; Cammisa, F.P.; Sandhu, H.S.; Diwan, A.; Girardi, F.P.; Lane, J.M. The Biology of Bone Grafting. *J. Am. Acad. Orthop. Surg.* **2005**, *13*, 77–86. [CrossRef] [PubMed]
5. Williams, S.F.; Martin, D.P.; Horowitz, D.M.; Peoples, O.P. PHA applications: Addressing the price performance issue: I. Tissue engineering. *Int. J. Biol. Macromol.* **1999**, *25*, 111–121. [CrossRef]
6. Rahaman, M.N.; Day, D.E.; Bal, B.S.; Fu, Q.; Jung, S.B.; Bonewald, L.F.; Tomsia, A.P. Bioactive glass in tissue engineering. *Acta Biomater.* **2011**, *7*, 2355–2373. [CrossRef]
7. Tayalia, P.; Mooney, D.J. Controlled Growth Factor Delivery for Tissue Engineering. *Adv. Mater.* **2009**, *21*, 3269–3285. [CrossRef]
8. O'Brien, F.J. Biomaterials & scaffolds for tissue engineering. *Mater. Today* **2011**, *14*, 88–95. [CrossRef]
9. Leong, K.F.; Chua, C.; Sudarmadji, N.; Yeong, W.Y. Engineering functionally graded tissue engineering scaffolds. *J. Mech. Behav. Biomed. Mater.* **2008**, *1*, 140–152. [CrossRef]
10. Nigmatullin, R.; Thomas, P.; Lukasiewicz, B.; Puthussery, H.; Roy, I. Polyhydroxyalkanoates, a family of natural polymers, and their applications in drug delivery. *J. Chem. Technol. Biotechnol.* **2015**, *90*, 1209–1221. [CrossRef]
11. Valappil, S.; Misra, S.; Boccaccini, A.; Keshavarz, T.; Bücke, C.; Roy, I. Large-scale production and efficient recovery of PHB with desirable material properties, from the newly characterised Bacillus cereus SPV. *J. Biotechnol.* **2007**, *132*, 251–258. [CrossRef]
12. Kaur, G. Strategies for Large-scale Production of Polyhydroxyalkanoates. *Chem. Biochem. Eng. Q.* **2015**, *29*, 157–172. [CrossRef]
13. Grigore, M.E.; Grigorescu, R.M.; Iancu, L.; Ion, R.-M.; Zaharia, C.; Andrei, E.R. Methods of synthesis, properties and biomedical applications of polyhydroxyalkanoates: A review. *J. Biomater. Sci. Polym. Ed.* **2019**, *30*, 695–712. [CrossRef]
14. Utsunomia, C.; Ren, Q.; Zinn, M. Poly(4-Hydroxybutyrate): Current State and Perspectives. *Front. Bioeng. Biotechnol.* **2020**, *8*, 257. [CrossRef] [PubMed]
15. Rai, R.; Keshavarz, T.; Roether, J.; Boccaccini, A.; Roy, I. Medium chain length polyhydroxyalkanoates, promising new biomedical materials for the future. *Mater. Sci. Eng. R Rep.* **2011**, *72*, 29–47. [CrossRef]
16. Valappil, S.P.; Misra, S.K.; Boccaccini, A.R.; Roy, I. Biomedical applications of polyhydroxyalkanoates, an overview of animal testing andin vivoresponses. *Expert Rev. Med. Devices* **2006**, *3*, 853–868. [CrossRef] [PubMed]

17. Dwivedi, R.; Pandey, R.; Kumar, S.; Mehrotra, D. Poly hydroxyalkanoates (PHA): Role in bone scaffolds. *J. Oral Boil. Craniofac. Res.* **2020**, *10*, 389–392. [CrossRef] [PubMed]
18. Chen, G.-Q.; Wu, Q. The application of polyhydroxyalkanoates as tissue engineering materials. *Biomaterials* **2005**, *26*, 6565–6578. [CrossRef] [PubMed]
19. Luo, Z.; Wu, Y.-L.; Li, Z.; Loh, X.J. Recent Progress in Polyhydroxyalkanoates-Based Copolymers for Biomedical Applications. *Biotechnol. J.* **2019**, *14*, e1900283. [CrossRef] [PubMed]
20. Kokubo, T.; Kim, H.-M.; Kawashita, M. Novel bioactive materials with different mechanical properties. *Biomaterials* **2003**, *24*, 2161–2175. [CrossRef]
21. Wang, M. Developing bioactive composite materials for tissue replacement. *Biomaterials* **2003**, *24*, 2133–2151. [CrossRef]
22. Hench, L.L. Bioceramics. *Stress Int. J. Biol. Stress* **1998**, *28*, 1705–1728. [CrossRef]
23. Hench, L.L. The story of Bioglass®. *J. Mater. Sci. Mater. Electron.* **2006**, *17*, 967–978. [CrossRef] [PubMed]
24. Jones, J. Reprint of: Review of bioactive glass: From Hench to hybrids. *Acta Biomater.* **2015**, *23*, S53–S82. [CrossRef]
25. Hoppe, A.; Güldal, N.S.; Boccaccini, A.R. A review of the biological response to ionic dissolution products from bioactive glasses and glass-ceramics. *Biomaterials* **2011**, *32*, 2757–2774. [CrossRef] [PubMed]
26. Jones, J.R.; Brauer, D.S.; Hupa, L.; Greenspan, D.C. Bioglass and Bioactive Glasses and Their Impact on Healthcare Introduction and Scope. *Int. J. Appl. Glass Sci.* **2016**, *7*, 423–434. [CrossRef]
27. Baino, F.; Novajra, G.; Miguez-Pacheco, V.; Boccaccini, A.R.; Vitale-Brovarone, C. Bioactive glasses: Special applications outside the skeletal system. *J. Non-Cryst. Solids* **2016**, *432*, 15–30. [CrossRef]
28. Miguez-Pacheco, V.; Hench, L.L.; Boccaccini, A. Bioactive glasses beyond bone and teeth: Emerging applications in contact with soft tissues. *Acta Biomater.* **2015**, *13*, 1–15. [CrossRef]
29. Gorustovich, A.; Roether, J.A.; Boccaccini, A.R. Effect of Bioactive Glasses on Angiogenesis: A Review of In Vitro and In Vivo Evidences. *Tissue Eng. Part B Rev.* **2010**, *16*, 199–207. [CrossRef]
30. Hoppe, A.; Mouriño, V.; Boccaccini, A. Therapeutic inorganic ions in bioactive glasses to enhance bone formation and beyond. *Biomater. Sci.* **2013**, *1*, 254–256. [CrossRef]
31. Jodati, H.; Güner, B.; Evis, Z.; Keskin, D.; Tezcaner, A. Synthesis and characterization of magnesium-lanthanum dual doped bioactive glasses. *Ceram. Int.* **2020**, *46*, 10503–10511. [CrossRef]
32. Malavasi, G.; Salvatori, R.; Zambon, A.; Lusvardi, G.; Rigamonti, L.; Chiarini, L.; Anesi, A. Cytocompatibility of Potential Bioactive Cerium-Doped Glasses based on 45S5. *Materials* **2019**, *12*, 594. [CrossRef] [PubMed]
33. Finney, L.; Vogt, S.; Fukai, T.; Glesne, D. COPPER AND ANGIOGENESIS: UNRAVELLING A RELATIONSHIP KEY TO CANCER PROGRESSION. *Clin. Exp. Pharmacol. Physiol.* **2009**, *36*, 88–94. [CrossRef]
34. Hu, G.-F. Copper stimulates proliferation of human endothelial cells under culture. *J. Cell. Biochem.* **1998**, *69*, 326–335. [CrossRef]
35. Gérard, C.; Bordeleau, L.-J.; Barralet, J.E.; Doillon, C.J. The stimulation of angiogenesis and collagen deposition by copper. *Biomaterials* **2010**, *31*, 824–831. [CrossRef] [PubMed]
36. Rodríguez, J.P.; Ríos, S.; González, M. Modulation of the proliferation and differentiation of human mesenchymal stem cells by copper. *J. Cell. Biochem.* **2002**, *85*, 92–100. [CrossRef]
37. Li, J.; Zhai, D.; Lv, F.; Yu, Q.; Ma, H.; Yin, J.; Yi, Z.; Liu, M.; Chang, J.; Wu, C. Preparation of copper-containing bioactive glass/eggshell membrane nanocomposites for improving angiogenesis, antibacterial activity and wound healing. *Acta Biomater.* **2016**, *36*, 254–266. [CrossRef]
38. Wang, C.; Jin, K.; He, J.; Wang, J.; Yang, X.; Yao, C.; Dai, X.; Gaoa, C.; Gou, Z.; Ye, J. Synergistic Effect of Copper-Containing Mesoporous Bioactive Glass Coating on Stimulating Vascularization of Porous Hydroxyapatite Orbital Implants in Rabbits. *J. Biomed. Nanotechnol.* **2018**, *14*, 688–697. [CrossRef]
39. Wang, X.; Molino, B.Z.; Pitkänen, S.; Ojansivu, M.; Xu, C.; Hannula, M.; Hyttinen, J.; Miettinen, S.; Hupa, L.; Wallace, G.G. 3D Scaffolds of Polycaprolactone/Copper-Doped Bioactive Glass: Architecture Engineering with Additive Manufacturing and Cellular Assessments in a Coculture of Bone Marrow Stem Cells and Endothelial Cells. *ACS Biomater. Sci. Eng.* **2019**, *5*, 4496–4510. [CrossRef]
40. Hoppe, A.; Mészáros, R.; Stähli, C.; Romeis, S.; Schmidt, J.; Peukert, W.; Marelli, B.; Nazhat, S.N.; Wondraczek, L.; Lao, J.; et al. In vitro reactivity of Cu doped 45S5 Bioglass® derived scaffolds for bone tissue engineering. *J. Mater. Chem. B* **2013**, *1*, 5659–5674. [CrossRef]
41. Ortiz, R.; Basnett, P.; Roy, I.; Quintana, I. Picosecond Laser Ablation of Polyhydroxyalkanoates (PHAs): Comparative Study of Neat and Blended Material Response. *Polymers* **2020**, *12*, 127. [CrossRef] [PubMed]

42. Basnett, P.; Marcello, E.; Lukasiewicz, B.; Panchal, B.; Nigmatullin, R.; Knowles, J.C.; Roy, I. Biosynthesis and characterization of a novel, biocompatible medium chain length polyhydroxyalkanoate by Pseudomonas mendocina CH50 using coconut oil as the carbon source. *J. Mater. Sci. Mater. Electron.* **2018**, *29*, 179. [CrossRef] [PubMed]
43. Rai, R.; Yunos, D.M.; Boccaccini, A.R.; Knowles, J.C.; Barker, I.A.; Howdle, S.M.; Tredwell, G.; Keshavarz, T.; Roy, I. Poly-3-hydroxyoctanoate P(3HO), a Medium Chain Length Polyhydroxyalkanoate Homopolymer from Pseudomonas mendocina. *Biomacromolecules* **2011**, *12*, 2126–2136. [CrossRef] [PubMed]
44. Lu, L.; Mikos, A.G. The Importance of New Processing Techniques in Tissue Engineering. *MRS Bull.* **1996**, *21*, 28–32. [CrossRef] [PubMed]
45. Schneider, C.A.; Rasband, W.S.; Eliceiri, K.W. NIH Image to ImageJ: 25 years of image analysis. *Nat. Methods* **2012**, *9*, 671–675. [CrossRef]
46. Otsuka, E.; Yamaguchi, A.; Hirose, S.; Hagiwara, H. Characterization of osteoblastic differentiation of stromal cell line ST2 that is induced by ascorbic acid. *Am. J. Physiol. Content* **1999**, *277*, C132–C138. [CrossRef]
47. Balasubramanian, P.; Hupa, L.; Jokic, B.; Detsch, R.; Grünewald, A.; Boccaccini, A. Angiogenic potential of boron-containing bioactive glasses: In vitro study. *J. Mater. Sci.* **2016**, *52*, 8785–8792. [CrossRef]
48. Boccaccini, A.R.; Blaker, J.J. Bioactive composite materials for tissue engineering scaffolds. *Expert Rev. Med. Devices* **2005**, *2*, 303–317. [CrossRef]
49. Amara, S.; Tiriveedhi, V. Inflammatory role of high salt level in tumor microenvironment (Review). *Int. J. Oncol.* **2017**, *50*, 1477–1481. [CrossRef]

materials

MDPI

Article

Accelerated Endothelialization of Nanofibrous Scaffolds for Biomimetic Cardiovascular Implants

Claudia Matschegewski [1], Stefanie Kohse [2], Jana Markhoff [2], Michael Teske [2], Katharina Wulf [2], Niels Grabow [2], Klaus-Peter Schmitz [1,2] and Sabine Illner [2,*]

[1] Institute for Implant Technology and Biomaterials e.V., Friedrich-Barnewitz-Straße 4, 18119 Rostock, Germany; claudia.matschegewski@uni-rostock.de (C.M.); schmitz@iib-ev.de (K.-P.S.)

[2] Institute for Biomedical Engineering, Rostock University Medical Center, Friedrich-Barnewitz-Straße 4, 18119 Rostock, Germany; stefaniekohse@web.de (S.K.); jana.markhoff@uni-rostock.de (J.M.); michael.teske@uni-rostock.de (M.T.); katharina.wulf@uni-rostock.de (K.W.); niels.grabow@uni-rostock.de (N.G.)

* Correspondence: sabine.illner@uni-rostock.de; Tel.: +49-381-54345-521

Abstract: Nanofiber nonwovens are highly promising to serve as biomimetic scaffolds for pioneering cardiac implants such as drug-eluting stent systems or heart valve prosthetics. For successful implant integration, rapid and homogeneous endothelialization is of utmost importance as it forms a hemocompatible surface. This study aims at physicochemical and biological evaluation of various electrospun polymer scaffolds, made of FDA approved medical-grade plastics. Human endothelial cells (EA.hy926) were examined for cell attachment, morphology, viability, as well as actin and PECAM 1 expression. The appraisal of the untreated poly-L-lactide (PLLA L210), poly-ε-caprolactone (PCL) and polyamide-6 (PA-6) nonwovens shows that the hydrophilicity (water contact angle > 80°) and surface free energy (<60 mN/m) is mostly insufficient for rapid cell colonization. Therefore, modification of the surface tension of nonpolar polymer scaffolds by plasma energy was initiated, leading to more than 60% increased wettability and improved colonization. Additionally, NH_3-plasma surface functionalization resulted in a more physiological localization of cell–cell contact markers, promoting endothelialization on all polymeric surfaces, while fiber diameter remained unaltered. Our data indicates that hydrophobic nonwovens are often insufficient to mimic the native extracellular matrix but also that they can be easily adapted by targeted post-processing steps such as plasma treatment. The results achieved increase the understanding of cell–implant interactions of nanostructured polymer-based biomaterial surfaces in blood contact while also advocating for plasma technology to increase the surface energy of nonpolar biostable, as well as biodegradable polymer scaffolds. Thus, we highlight the potential of plasma-activated electrospun polymer scaffolds for the development of advanced cardiac implants.

Keywords: nonwoven; cardiovascular stent; human endothelial cells; biocompatibility; CD31

Citation: Matschegewski, C.; Kohse, S.; Markhoff, J.; Teske, M.; Wulf, K.; Grabow, N.; Schmitz, K.-P.; Illner, S. Accelerated Endothelialization of Nanofibrous Scaffolds for Biomimetic Cardiovascular Implants. *Materials* 2022, 15, 2014. https://doi.org/10.3390/ma15062014

Academic Editors: Roser Sabater i Serra and Ángel Serrano-Aroca

Received: 4 February 2022
Accepted: 7 March 2022
Published: 9 March 2022

1. Introduction

Structural and valvular heart diseases are of increasing incidence and a leading cause of mortality worldwide [1–3]. Minimally invasive surgery for coronary revascularization by percutaneous coronary intervention (PCI) based on stent implantation has become the standard procedure of care to relieve the stenosis and to restore the dysfunctional stenotic vessel. However, stent surgery is accompanied with local vessel injury, comprising disruption of the intimal smooth muscle cell layer and the luminal endothelial cell layer. The most important clinical complications following PCI are stent restenosis and stent thrombosis, associated with hyperplasia, delayed endothelialization, and acute and chronic inflammation events [4]. Since stent restenosis is mainly related to bare metal stents, drug-eluting stents (DES) have been shown to effectively reduce hyperplasia by comprising a polymer coating with incorporated antiproliferative drugs that inhibit smooth muscle cell proliferation following implantation [5]. However, clinical follow-up studies reported higher rates of late stent thrombosis attributed to DES compared to bare metal stents [6]. Although those DES-related complications are relatively rare, their severity demands further improvements in cardiac implant technology [7].

Contemporary efforts in stent technology focus on polymer-based bioresorbable coronary scaffolds or nanofibrous covered stent systems such as the commercial Papyrus (Biotronik), Jostent Graft-Master (Abbott) or Bioweb (Zeus), in order to decrease incidence of late stent thrombosis [3,8–10]. Artificial scaffolds can not only be applied in DES systems but could also be adopted for the development of a variety of other minimally invasive cardiac implants, such as biomimetic heart valves or occluder devices [11]. At the moment, these prosthetics are most commonly xenografts of porcine aortic valves or calf pericardium, with the disadvantages of structural valve deterioration often being accompanied by subsequent valve thickening and calcification, as well as their limited durability [12].

Polymer-based nano- and microfibrous scaffolds hold great potential for overcoming some of the major disadvantages of current polymer stent and cardiac implant designs. Polymeric fibrous scaffolds can be produced in layers or as a composite, in the form of single, core-sheath, blended or co-polymer film or fiber coating systems. Within a comprehensive search using the search engine Web of science, PubMed and Scopus, the increasing interests in electrospun nanofibers over the last 20 years are clearly visible (see Supplementary Materials Figures S1–S4). In addition to synthetic polymers, natural biopolymers can also be used for scaffold fabrication. Intensive research is being conducted by mixing synthetic biopolymers with, e.g., hyaluronic acid derivatives, silk fibroin, cellulose, chitosan, shellac, gelatin, alginates or heparin [13–16]. The advantages of natural biopolymers can include biostatic properties and biomimetic properties due to extraction from natural sources such as bones, plants, skins, cocoons or algae. However, their source can also be a disadvantage because of their batch and origin dependence. It is difficult to produce reproducible and high-quality implants for medical technology if there is a possibility that (i) allergens are present in the material or (ii) not all components are known or (iii) complex matrices can only be analyzed with great effort.

In contrast, synthetic scaffolds are often limited due to suboptimal biomechanical properties, low biocompatibility or their biodegradation behavior [11,13–15]. Since scaffold structure directly influences these properties, the introduction of innovative nano- and microfabrication techniques, such as micropatterning, electrospinning or 3D-bioprinting, will support the development of novel scaffold designs for cardiac implants and similarly will enable the fine-tuning of the desired mechanical requirements as well as biological properties by controlling fiber structure. In particular, electrospinning enables the fabrication of biomaterial scaffolds with tunable parameters, including fiber diameter and alignment, porosity, pore size, inter-connectivity and mesh thickness, which makes them suitable for a range of applications [16–19]. In addition, electrospinning allows for the fabrication of fused fiber biomaterial scaffolds in the nano- to micrometer range, with fiber diameters starting at 5 nm [20], including the range of feature sizes known to facilitate cellular contact guidance and directed cellular response [21]. Moreover, electrospun nonwovens can offer small pore sizes, up to 90% porosity and pronounced interfibrous integrity [16] with controlled degradation kinetics, if required. Since implants holding the capability of inducing directed cellular response are in high demand, nonwovens are favored, due to their biomimetic surface, which allows for cell attachment, while additionally being highly porous for nutrient transport and cell migration events [19]. Electrospun polymeric scaffolds have already been reported to possess superior capacity in shaping cell morphology, guiding cell migration and affecting cell differentiation in vivo and in vitro [22–26]. The influence of nanofibrous polymer matrices on cell physiology and endothelialization is determined by the specific surface topography and by the physicochemical properties [21,27]. The attachment of cells to the biomaterial surface and the initiated subsequent cell–biomaterial interaction triggers a signaling cascade, which substantially regulates diverse cell functions, including viability, proliferation and the expression of cell-specific proteins [28]. Thereby, alterations in cellular function are suggested to coincide with morphological changes and vice versa.

Alterations in surface topography, e.g., porosity, surface roughness or fiber diameter of electrospun scaffolds have been shown to enhance endothelialization of biomaterials [17,18,29,30]. Additionally, modifications of physicochemical properties such as surface functionalization with plasma, accompanied by changes in hydrophilic properties, have also been demonstrated to influence cell physiology [25].

The first use of plasma-generated amino groups on polymer surfaces to improve blood compatibility was reported in 1969 [31]. Since then, several studies have been performed that have achieved biocompatible surfaces based on the controlled surface properties of the materials, in particular, wettability, chemistry, morphology, crystallinity or surface charge. Plasma modifications, especially cold plasmas, only affect the surface and do not affect bulk properties such as the mechanical and optical properties of the material. Cold plasma is very gentle and therefore suitable for a diverse range of polymers, especially temperature-sensitive materials. Screenings have to be performed for each material and plasma system to optimize parameters to the needs of the desired application. Additionally, a wide range of surface modifications are possible based on the choice of gas or gas mixture. The method can be performed quickly and is suitable for complex geometric samples. In addition to cold low vacuum methods, cold atmospheric plasma systems were developed leading to new applications in plasma medicine such as plasma pens [16,31].

The desired application of the scaffold material, and thus the cell type chosen, is crucial as the same surfaces can show different effects, depending on the cell type, or improving permeation of cell nutrients through nonwovens [32]. For example, enhanced osteogenic differentiation was observed for human mesenchymal stem cells on ammonia plasma treated titanium [33]. Additionally, reduced antimicrobial activity was observed for wound dressings [16,33] or increased C_2C_{12} proliferation for PLA surfaces [34]. Several studies have been devoted to enhancing the hydrophilicity of polymer surfaces to promote protein and cell attachment [35–38]. Moreover, plasma surface treatment was shown to directly affect biocompatibility since surface functionalization with NH_3- or O_2-plasma has been observed to improve cell growth [39–45]. NH_3-plasma generates amino groups on the surface, and it has been demonstrated to be more effective than O_2-plasma treatments regarding improved cell growth patterns on polymer surfaces or scaffolds [39,42,44,46–49]. In this context, plasma treatment has been extensively studied and applied to thin films and bulk polymer scaffolds but has not yet been sufficiently investigated for electrospun hydrophobic fiber scaffolds [50]. We selected ammonia plasma based on previous studies of the materials in the field of cardiovascular application [42,49,51]. Thus, this study examined the potential of NH_3-plasma functionalization of electrospun fibrous biomaterials for promoting cell attachment and physiological growth patterns of human endothelial cells in vitro for the development of biocompatible and biomimetic cardiac scaffolds.

Therefore, we aimed to combine the possibility of electrospinning and low-pressure cold plasma, to meet the needs of medical applications based on the combination of (i) the morphology of nonwoven fibers, with their structure and mechanical properties; and (ii) the surface energy and chemistry of plasma treatment without affecting the fiber morphology. Ammonia plasma treatment was tested on diverse nonwovens of FDA-approved and clinically tested polymers, namely, polylactide (PLA), polycaprolactone (PCL) and polyamide (PA), which represent both biodegradable and biostable biocompatible materials (see Figure 1).

Untreated and plasma-activated polymeric nonwovens were analyzed regarding their influence on endothelial cell characteristics, including viability, spreading and PECAM-1 expression, for judging endothelial activation and phenotype maintenance to evaluate their potential as innovative biomaterials (see Figure 2). These results are a decisive step in the development of novel, advanced scaffolds for cardiac regeneration and a better understanding of cell–biomaterial interactions, not least because endothelialization is a prerequisite for the successful integration of cardiovascular devices, as it forms a natural hemocompatible surface that prevents inflammation and thrombosis events, thus ensuring implant integrity.

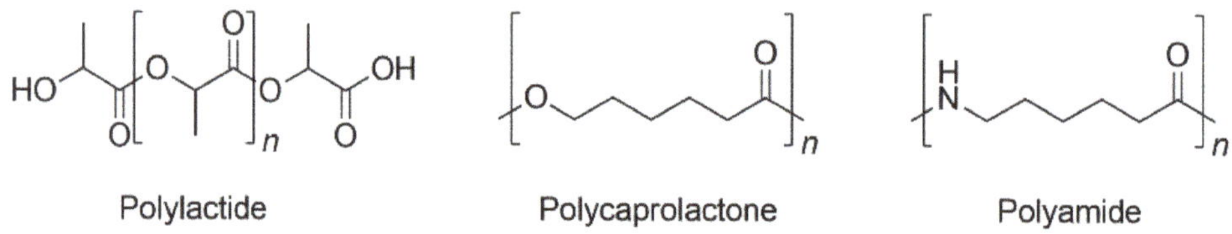

Figure 1. Chemical structure of the well-established polymer types for biomedical application used in this study.

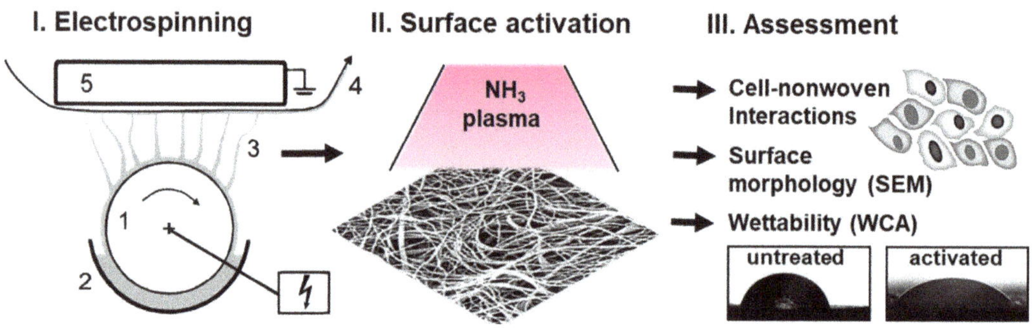

Figure 2. Systematic illustration of the study design and methodological background. (**I**). Preparation of the different used nonwovens using Elmarco nanospider setup, including (1) high-voltage source at the rotating emitter; (2) a bath filled with a polymer solution; (3) fiber formation under solvent evaporation; and (4) a nonwoven collecting unit, (5) which is electrical driven. (**II**). Surface plasma activation of a nonwoven illustrated by a SEM image. (**III**). Characterization of the prepared polymer-based nonwovens via comprehensive biological studies, SEM and contact angle measurements, before and after plasma treatment.

2. Materials and Methods

2.1. Fabrication of Polymeric Nanofiber Nonwovens by Electrospinning

Nanofibrous nonwovens were fabricated out of poly-(L-lactide) (PLLA L210), polyamide (PA-6) and poly-ε-caprolactone (PCL) by electrospinning as following: clear and homogenous polymer solutions of 11 to 12 wt% poly-ε-caprolactone (Capa 6800, Perstop UK Limited) and of 12 wt% polyamide (Ultramid B24N03, BASF, Ludwigshafen, Germany) were obtained by dissolving the polymer in a solvent mixture of formic acid and acetic acid (ratio 1:2, v/v) at 37 °C. A polymer solution of 2 wt% poly-(L-lactide) (RESOMER® L210, Mw ~400,000 g/mol, Evonik, Essen, Germany) with addition of 1vol% surfactant Triton X-100 (Sigma-Aldrich, Darmstadt, Germany) was obtained by dissolving in a solvent mixture of chloroform and methanol (ratio 4:1, v/v) at 37 °C.

Fibrous nonwovens were fabricated from these different polymer solutions by free-surface, needleless electrospinning via the Nanospider Lab 200 (ELMARCO, Liberec, Czech Republic) using a rotating wire emitter in a high-volume spinning tube, and a static collector. Emitter-to-collector distances of 18 cm at 16 rpm, 17 cm at 12 rpm or 16.5 cm at 13 rpm were used accordingly for PLLA L210, PCL and PA-6, each resulting in nonwoven samples with randomized fibers. The applied high voltages were 49 to 58 kV, 60 to 80 kV and 72 to 76 kV for PLLA L210, PCL and PA-6, respectively, each under ambient conditions of 23 °C and humidity of 35%. The generated polymeric nonwoven mats were dried for 12 h at 40 °C using the vacuum oven VO 200 (Memmert GmbH and Co., Schwabach, Germany, 40 mbar).

2.2. Nonwoven Characterization by Scanning Electron Microscopy

Fiber morphology of the PLLA L210, PCL and PA-6 nonwovens was examined by scanning electron microscopy (SEM) using a QUANTA FEG 250 (FEI Company, Dreieich, Germany) with an Everhart–Thornley secondary electron detector (ETD) at an acceleration voltage of 10 kV; a working distance of around 10 mm, at a high vacuum of 3.5×10^{-6} mbar; and a spot size of 3.0. The samples were fixed onto aluminum trays with conductive tape and sputter coated with gold by Agar Sputter Coater (Agar Scientific Ltd., Essex, UK). Therefore, the samples were put under vacuum of 0.2 mbar and exposed to gold flow 2 times for 120 s. SEM images were taken at magnification 500×, 1000×, 5000× and 8000×. For quality assurance of the produced polymeric nonwovens and for determining the average fiber diameter, SEM analysis was performed at different areas of the nonwovens. Fiber diameters were calculated from SEM images by using EDAX Genesis software, measuring 50 random fibers from five micrographs for each nonwoven at high magnification.

2.3. Surface Plasma Modification of Nonwovens

Plasma-chemical surface modification was conducted by plasma etching (PE) of nonwoven samples in an ammonia (NH_3) plasma, generating radical species and amino groups. The short plasma activation process was performed for 1 min and 60% generator output in an ammonia radio frequency (RF) plasma generator (frequency 13.56 MHz, power 100 W, Diener electronic GmbH and Co. KG, Ebhausen, Germany) at a low pressure of 0.3 mbar based on previously studies [42,49,51]. The screening data were not presented because the energy density of the plasma in the chamber depends on a wide range of factors such as chamber material, design, sample positioning and mounting, electrode spacing, size, shape and material, and the method of excitation. Generalization or the transfer of parameters from one system to another is not possible, so screening of suitable parameters depending on the objective must always be performed.

2.4. Water Contact Angle and Surface Free Energy

Water contact angle measurements were performed by the sessile drop method (water) on the nonwoven polymer surface using a goniometer (OCA 20, Dataphysics Instruments GmbH, Filderstadt, Germany) equipped with SPSS software 15.0. Nonwovens of PLLA L210 were washed three times for 10 min each with pure water to remove Triton X-100. Nonwovens were attached to glass slides, and water contact angles were determined by the sessile drop method with water droplets of 5 µL. A time-resolved measurement over 60 s was performed, whereby the smallest standard deviation was obtained after 10 s. Mean values and standard deviations were calculated from five independent samples with $n = 4$ measurements per sample.

To calculate the surface free energy (SFE) of untreated nonwovens according to Owens–Wendt–Rabel–Kaelble (OWRK) [52], further measurements were performed with a mobile surface analyzer (MSA, KRÜSS GmbH, Hamburg, Germany) with ADVANCE 1.9.2 software. The initial contact angles of two liquids, water and diiodomethane, were determined against air, whereby drops with a volume of 1 or 2 µL were deposited and measured within a few seconds ($n = 3$).

2.5. Cell Culture

Human vascular endothelial cells EA.hy926 (ATCC®, CRL-2922™, Manassas, VA, USA) were cultured in Dulbecco's Modified Eagle Medium (DMEM), including 4.5 g/L glucose and 3.7 g/L $NaHCO_3$ with 10% fetal calf serum (FCS) (both: PAN Biotech, Aidenbach, Germany) and penicillin (100 units/mL), streptomycin (100 ng/mL) at 37 °C, and 5% CO_2 under humidified atmosphere. For the experiments, cells were seeded at a concentration of 2×10^4 cells/cm^2 on the nanofibrous polymer scaffolds and incubated for 48 h at 37 °C and 5% CO_2 under humidified atmosphere.

2.6. Cell Viability Assay

Cell viability of human endothelial EA.hy926 cells was determined using the CellQuanti-Blue™ assay (BioAssaySystems, Hayward, CA, USA) according to the manufacturer's instructions. Briefly, cell viability was assessed via quantification of cellular metabolic activity by the reduction of the substrate resazurin to resorufin by cellular reductases. Therefore, cells were incubated with CellQuanti-Blue™ as 10% of the culture medium volume for 2 h after a 46-h cell cultivation period on the polymeric surfaces. The resulting fluorescence of resorufin was measured at an emission wavelength of 590 nm with an excitation wavelength of 544 nm using a microplate reader (FLUOstar OPTIMA, BMG Labtech, Offenburg, Germany). For each polymer, six independent biological replicates were measured. Data were normalized to viability of EA.hy926 cells grown on planar non-cytotoxic tissue culture polystyrene (TCPS) as control surface (NC) [40,42].

2.7. Cell Morphology Analysis by Scanning Electron Microscopy

Cell morphology of human endothelial EA.hy926 cells grown for 48 h on either polymeric nonwovens or tissue culture polystyrene control surface (NC) was observed by scanning electron microscopy. After the incubation on the polymeric surfaces, cells were fixed with 2.5% glutaraldehyde and 0.2 M sodium cacodylate, in PBS for 30 min. Samples were then washed with sodium phosphate buffer, dehydrated in a graded series of ethanol (50%, 75%, 90% and 100%) and dried with CO_2 in a critical point dryer (CPD 7501, Quorum Technologies Ltd., Laughton, Lewes, East Sussex, UK). Samples were sputter-coated with gold by Agar Sputter Coater (Canemco Inc., QC, Canada), and image acquisition was performed with the scanning electron microscope Quanta™ FEG 250 (FEI Company, Hillsboro, OR, USA) at 10 kV under high vacuum conditions by using the Everhart–Thornley secondary electron detector (ETD).

2.8. Endothelialization Analysis by Cell Spreading and Cell Shape Index

Endothelialization potential of polymeric nonwovens was assessed by quantification of cellular spreading of EA.hy926 endothelial cells and analysis of phenotype maintenance, evaluated by cell shape index after a 48 h cultivation period on all surfaces. This quantification was done based on SEM images of fixed cells. For determination of cell spreading, cell areas of 40 cells per specimen of three independent replicates were measured using the area measurement function of ImageJ software. Cell shape index (CSI) analysis was performed with ImageJ software by using the formula $CSI = 4\pi \times area/(perimeter)^2$ [53]. Calculated CSI defines cellular morphological shape ranging from 0 to 1, corresponding to a circular shape (CSI = 1) or a straight line, i.e., maximum elongated shape (CSI = 0).

2.9. Immunofluorescence of PECAM-1 (CD31) and Actin Cytoskeleton

For immunostaining, endothelial EA.hy926 cells were examined after incubation for 48 h on the polymeric nonwovens as well as on the control surface (NC) at 37 °C and 5% CO_2 under humidified atmosphere. Cells were fixed in 4% paraformaldehyde (PFA) for 30 min at room temperature (RT), rinsed in PBS (pH = 7.4) and permeabilized with Triton X-100 (Sigma-Aldrich, Darmstadt, Germany) for 30 min at RT. Cells were then incubated with primary murine monoclonal anti-human PECAM-1 (CD31) antibody (1:20, DAKO, Agilent, Santa Clara, CA, USA) overnight. Afterwards, cells were rinsed in PBS and incubated with secondary donkey anti-mouse antibody conjugated with Alexa Fluor 488 (Life Technologies GmbH, Darmstadt, Germany) for 1 h at RT. For actin staining, TRITC-conjugated phalloidin (500 μg/mL, Sigma-Aldrich, Taufkirchen, Germany) was used by incubating the cells in the staining solution for 1 h at RT. Cell nuclei were stained with Hoechst 33,342 (1:500, Sigma-Aldrich, Taufkirchen, Germany) for 1 h at RT. Cells were mounted in VectaShield mounting medium (Vector Laboratories, Burlingame, CA, USA) and examined by confocal laser scanning microscopy (FluoView FV1000, Olympus, Hamburg, Germany). Quantification of mean fluorescence intensity of the respective markers in cell images was evaluated by CellProfiler software.

2.10. Statistical Analysis

Data were reported as mean value with standard deviation and analyzed by one-way ANOVA carried out with GraphPad©Prism 5 software (La Jolla, CA, USA). Statistical significance was defined as ns = not significant *** $p < 0.001$.

3. Results

3.1. Morphology of Polymeric Nonwovens

Different polymeric nonwovens were produced via electrospinning and then compared in terms of fiber diameter. Representative images of the three polymer classes are shown in Figure 3 (additional magnifications of the SEM images are presented in Figure S5). Morphological differences between the individual polymer classes are evident but not after NH_3-plasma treatment. The PLLA L210 fibers are about twice as thick as the PCL fibers and four times thicker than PA-6 fibers.

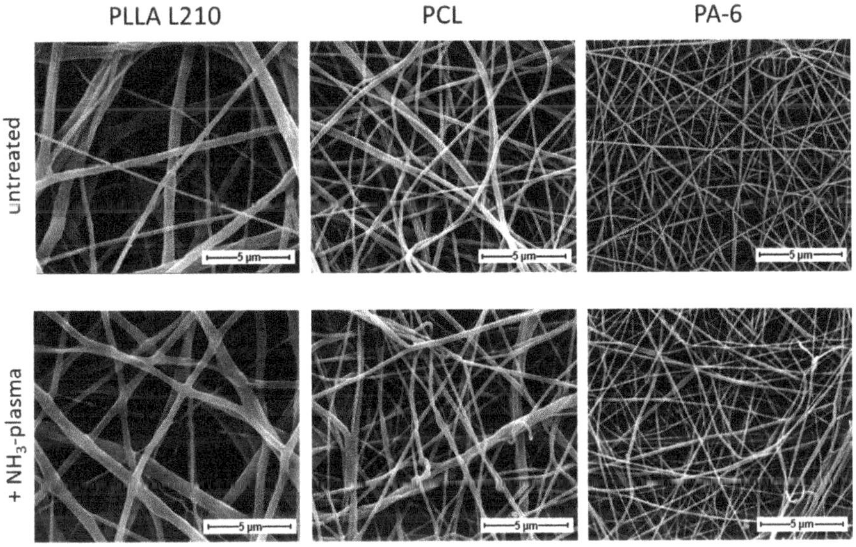

Figure 3. High-resolution scanning electron micrographs of morphology of untreated (**first row**) and NH_3-plasma treated (**second row**) nanofibrous PLLA L210, PCL and PA-6 nonwovens (magnification ×8000, bar = 5 µm).

The fiber diameters of the different polymeric nonwovens were determined from 50 measurements each and are shown in absolute terms in Figure 4. Whereas the diversity of the fibers varies the most within PLLA nonwovens, the fiber diameter of untreated and plasma functionalized PLLA L210, PCL and PA-6 nonwovens differ only marginally. This implies that the fiber structure is not changed by the plasma treatment. Additionally, the frequency of the respective fiber diameters for each investigated polymer is presented in Figure S6.

3.2. Wettability Analysis and Surface Free Energy of Polymeric Nonwovens

First, the wettability of the nonwovens was determined in a time-resolved manner (Figure 5A). A second measurement, performed instantaneously after only a few seconds, provides a direct comparison of the initial wettability of untreated and plasma-activated nonwovens (Figure 5B).

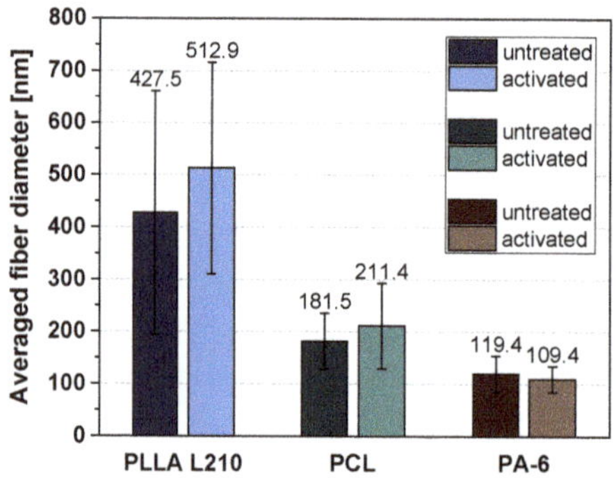

Figure 4. Fiber diameter of untreated and NH_3-plasma functionalized PLLA L210, PCL and PA-6 nonwovens based on individual measurement points. For each polymer, 10 fibers were measured manually in 5 different SEM images ($n = 50$).

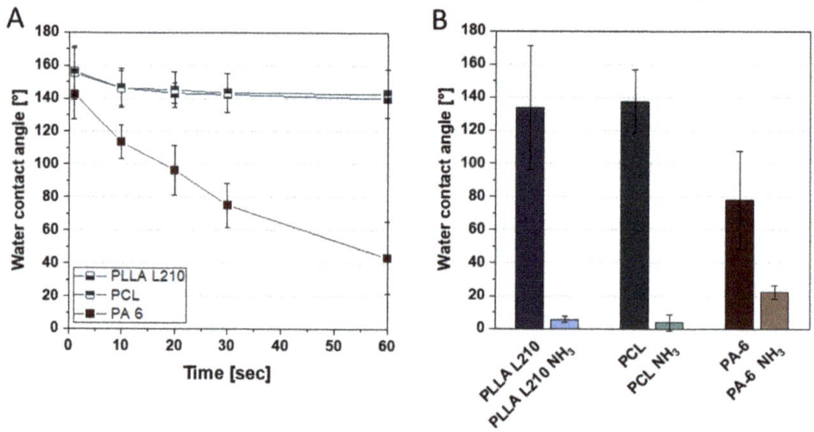

Figure 5. Water contact angle of (**A**) time-resolved measurement over 60 s using untreated nonwoven material and (**B**) measurement after a few seconds using untreated and NH_3-plasma functionalized PLLA L210, PCL and PA-6 nonwovens.

It can be seen that the water contact angle (WCA) of PA-6 nonwovens does not remain constant and decreases rapidly over 60 s (Figure 5A). In contrast, the WCA of the PLLA L210 and PCL nonwovens is very high at 130 to 140° as expected, which is due to the special surface morphology of nonwoven structures. As opposed to film or foil materials, nanofibrous nonwovens have pores that are filled with air. The wetting properties of nonwovens are influenced by their weight, layer thickness or (more precisely) their pore structure. Such a surface, where air is trapped between the liquid and the solid and thus can only be wetted incompletely, is called a composite interface. [54] The initial WCA before and after plasma treatment is shown in Figure 5B. It can be clearly seen that the WCA for all investigated nonwovens can be greatly reduced by one-minute NH_3-plasma treatment, indicating that plasma treatment is very efficient at improving the wettability of nonwoven fibers.

Further results on surface free energy and its division into polar and disperse fractions are summarized in Table 1. For their calculation by means of OWRK, the diiodomethane contact angle was additionally determined. The surface free energies of untreated PLLA L210 and PCL nonwovens differ only marginally. The SFE of PA-6 nonwoven is slightly lower, but in this case the measurement is affected by a high standard deviation. After plasma activation, the SFE is noticeably improved for PCL and PA-6, whereby the disperse fraction is reduced and the polar fraction strongly increased for all polymers.

Table 1. Summary of wettability and surface free energy of untreated and NH_3-plasma functionalized PLLA L210, PCL and PA-6 nonwovens. MSA measurement after 2 s, $n = 3$.

	Untreated Nonwovens			NH_3-Plasma-Activated Nonwovens		
	PLLA L210 *	**PCL**	**PA-6**	**PLLA L210 ***	**PCL**	**PA-6**
	Mean ± SD	**Mean ± SD**	**Mean ± SD**	**Mean ± SD**	**Mean ± SD**	**Mean ± SD**
Water [°]	133.6 ± 5.6	137.6 ± 3.8	78.0 ± 22.2	37.8 ± 1.7	19.3 ± 4.8	29.8 ± 4.1
Diiodomethane [°]	18.1 ± 2.4	21.6 ± 11.1	26.8 ± 7.6	64.6 ± 0.3	25.9 ± 1.0	31.6 ± 3.3
Surface free energy [mN/m]	57.1 ± 2.8	57.3 ± 5.3	48.5 ± 9.6	59.5 ± 1.3	75.9 ± 1.9	70.7 ± 3.5
Disperse fraction [mN/m]	48.3 ± 0.7	47.3 ± 3.5	45.5 ± 2.9	25.9 ± 0.2	45.8 ± 0.4	43.6 ± 1.4
Polar fraction [mN/m]	8.8 ± 2.1	10.0 ± 1.8	3.1 ± 6.8	33.6 ± 1.0	30.1 ± 1.6	27.1 ± 2.0

* washed.

3.3. Biocompatibility of Polymeric Nonwovens Assessed by Cell Viability

Cell viability of human endothelial EA.hy926 cells grown for 48 h on untreated and NH_3-plasma modified polymeric nanofiber nonwovens of PLLA L210, PCL and PA-6 is shown in Figure 6. Values were normalized to values from the control surface (NC, TCPS), which was set to 100%. For the untreated nonwovens, relative cell viability was highest for PLLA L210 (57.1%), followed by nonwovens of PA-6 (54.1%) and PCL (49.6%); however, these differences were not statistically significant.

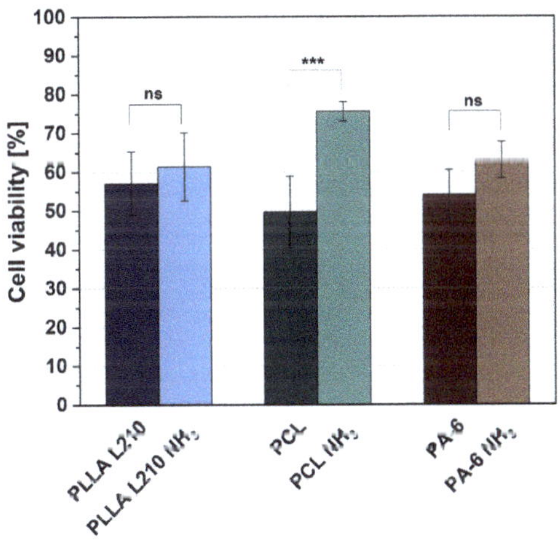

Figure 6. Relative viability of human endothelial EA.hy926 cells on unmodified and NH_3-plasma functionalized polymeric nonwovens after 48 h (mean + SD, $n = 6$, one-way ANOVA, ns = not significant, and *** $p < 0.001$).

In contrast, NH_3-plasma modification has indeed shown to increase the relative cell viability of endothelial EA.hy926 cells on all polymeric nonwovens. This effect was most prominent for PCL nonwoven, where NH_3-plasma modification yielded the highest and most statistically significant ($p < 0.001$) increase in EA.hy926 cell viability of 75.6%, corresponding to an enhancement of 26.0% compared to untreated PCL nonwovens. Additionally, for PLLA L210 and PA-6 nonwovens, a positive effect of NH_3-plasma treatment on cell viability was observed, although it was not significant and slightly lower than what was detected for PCL. NH_3-surface treatment of PLLA L210 and PA-6 resulted in an increased viability of endothelial EA.hy926 cells up to 61.3% for PLLA L210 and 63.0% for PA-6.

3.4. Influence of Nanofiber Scaffolds on Cell Attachment and Morphology

Cell morphology analysis using scanning electron microscopy revealed phenotypic differences of human endothelial EA.hy926 cells grown on untreated nonwovens of PLLA L210, PCL and PA-6 compared to those grown on NH_3-plasma-modified polymeric nonwovens (Figure 7). Only moderate cell attachment and spreading of EA.hy926 cells could be observed on the untreated nanofiber nonwovens of PLLA L210, PCL and PA-6. In particular, cells that were grown on electrospun PLLA L210 and PCL scaffolds were more likely to exhibit spherical phenotypes than cells that were grown on PA-6 nonwovens where they appeared more flattened. Especially on PLLA L210 nonwovens, EA.hy926 cells were shown to grow around the scaffold nanofibers to some degree. After NH_3-plasma treatment, all of the three types of polymeric nonwovens were shown to facilitate better cell attachment of human endothelial cells and showed increased cell spreading of EA.hy926 cells when compared to the corresponding untreated polymer surfaces. In general, endothelial EA.hy926 cells grown on all NH_3-plasma treated surfaces were larger and exhibited a flattened and more elongated phenotype with more filopods than on the unmodified nonwovens. Furthermore, this altered effect on cell spreading and the phenotype change was most obvious for cells that were grown on NH_3-plasma treated PLLA L210 nonwovens. It was also apparent that on all nanofibrous scaffolds, human endothelial cells were able to spread between individual fibers of the mats, but none of the cells were observed to have grown deeper into the pores of the analyzed nonwovens.

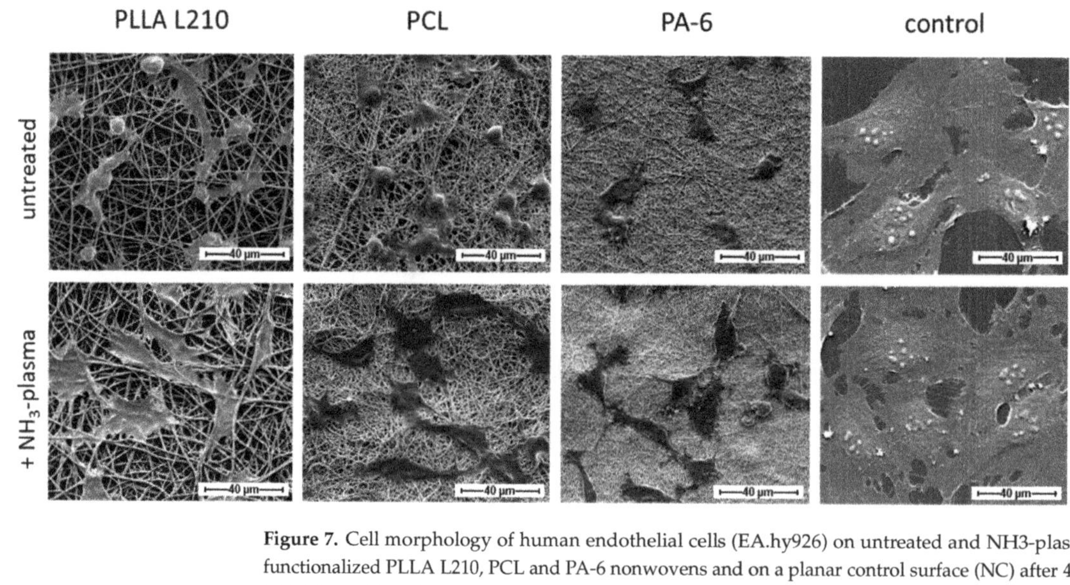

Figure 7. Cell morphology of human endothelial cells (EA.hy926) on untreated and NH3-plasma functionalized PLLA L210, PCL and PA-6 nonwovens and on a planar control surface (NC) after 48 h (SEM, bar = 40 µm).

Regarding phenotype maintenance, the morphology of human EA.hy926 cells grown on the NH_3-plasma modified polymeric nonwovens was comparable to those on the control surface (NC), although on the control surface, cells generally exhibited a much larger cell area compared to those on all of the polymeric nonwovens.

3.5. *Quantification of Endothelialization*

To evaluate the impact of NH_3-plasma treatment on endothelialization potential as a substantial biological requisite for successful cardiac scaffolds, cell spreading of endothelial EA.hy926 cells was quantified by measuring cell areas (Figure 8). Measurements were therefore conducted on cells either grown on NH_3-modified or untreated polymeric nonwovens of PLLA L210, PCL and PA-6.

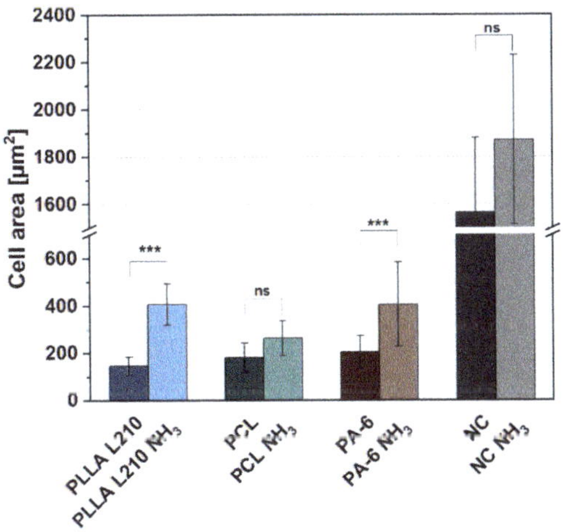

Figure 8. Spreading of human endothelial EA.hy926 cells on untreated and NH_3-plasma functionalized polymeric nonwovens after 48 h (mean + SD, $n = 40$, one-way ANOVA, ns = not significant, and *** $p < 0.001$).

Mean cell area of human endothelial EA.hy926 cells on untreated nonwovens ranged from 144.9 μm^2 to 203.2 μm^2 for PLLA L210 and PA-6, respectively. Modification with NH_3-plasma led to a remarkable increase of the cell area of human endothelial cells for all three types of investigated polymeric nonwovens. The effect was most obvious for PLLA L210 and PA-6 nonwovens, where the cell area was significantly increased and even doubled after NH_3-plasma treatment, ranging from 404.5 μm^2 to 406.0 μm^2, respectively. The cell area also tended to increase on PCL nonwovens after NH_3-plasma modification, although the difference was not statistically significant compared to untreated PCL mats. However, cell area of human EA.hy926 cells was highest on the control surface (NC), ranging from 1565.0 μm^2 for NC and 1873.0 μm^2 for NH_3-plasma treated NC.

In order to judge the quality of endothelialization based on maintenance of the endothelial phenotype, cell shape of human endothelial cells was analyzed by calculating cell circularity by cell shape index (CSI). The index, ranging from 0 to 1, is either corresponding to a circular shape (1) or a straight line (0 for maximum elongated shape). CSI of human endothelial cells on untreated polymeric nonwovens ranged between 0.64 for PCL nonwoven and 0.73 for PLLA nonwoven, compared to 0.61 for the control surface (NC) (Figure 9). After NH_3-plasma modification, CSI was diminished on all polymeric surfaces, with lower circularity measurements averaged to 0.67, 0.65 and 0.58 for PLLA L210, PCL and PA-6 nonwovens, respectively, thus representing a less circular shape and more elongated phenotype of human EA.hy926 cells on the NH_3-plasma modified nonwovens. In particular, CSI was lowest on plasma-modified PA-6 nonwoven and therefore expressed the smallest difference compared to control surface with respective CSI of 0.54. Nonetheless, cells were not detected to be aligned in one specific direction.

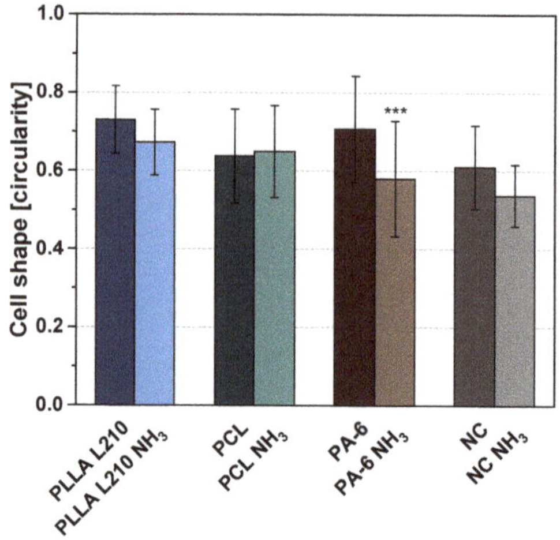

Figure 9. Cell shape described by cell circularity of human endothelial EA.hy926 cells on untreated and NH_3-plasma functionalized polymeric nonwovens after 48 h (mean + SD, n = 40, one-way ANOVA, *** $p < 0.001$).

3.6. Endothelial Actin Cytoskeleton Formation on Polymeric Nonwovens

The formation of the actin cytoskeleton of human endothelial EA.hy926 grown on specific polymer nonwovens is shown in Figure 10 (see also Figure S7) with regard to NH_3-activation status and is compared to the respective control surface (NC).

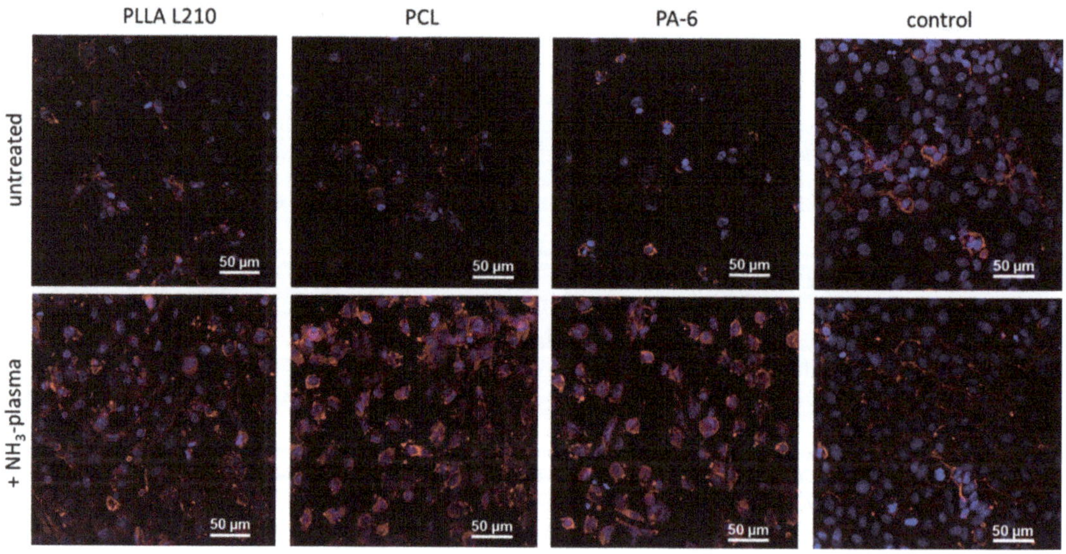

Figure 10. Fluorescent staining of F-actin in human endothelial EA.hy926 cells grown for 48 h on untreated and NH_3-plasma functionalized polymeric nonwovens (red: phalloidin-TRITC for F-actin, blue: Hoechst-staining indicating cell nuclei, confocal microscopy, and bar = 50 µm).

On all of the unmodified polymeric nonwovens comprising PLLA L210, PCL and PA-6, EA.hy926 cells was shown to exhibit a much reduced actin formation. This reduction in actin formation was evident due to the merely faint background staining of F-actin, with a concentrated F-actin ring at the outer cell membrane and a lack of any any visible intracellular actin fibers. In contrast, intracellular actin fibers could be observed spanning the entire cell body on the control surface.

NH_3-plasma treatment also showed a striking positive effect on the intracellular cytoskeleton formation regardless of the observed polymeric nonwovens. In particular, intracellular F-actin expression was enhanced on PLLA L210, PCL and PA-6 nonwovens when compared to the respective untreated counterparts. However, distinct stress fiber formation, as it is seen in EA.hy926 cells grown on the control surface, still remained relatively poor on all of the plasma-activated polymer nonwovens.

3.7. Expression of Endothelial Cell-Specific PECAM-1 Marker

Immunofluorescent staining was examined to determine the expression of endothelial cell specific cell–cell contact PECAM-1 (CD31) marker in human EA.hy926 endothelial cells growing on polymeric nonwovens of PLLA L210, PCL and PA-6, with or without NH_3-plasma surface modification and control surface (NC). Analysis of CD31 expression was used as a positive indicator of successful formation of cell–cell junctions in human endothelial EA.hy926 cells. Results indicate that on all untreated polymeric nonwovens, only a sparse CD31 expression in EA.hy926 cells was observed compared to the control surface (NC) (Figures 11 and S8).

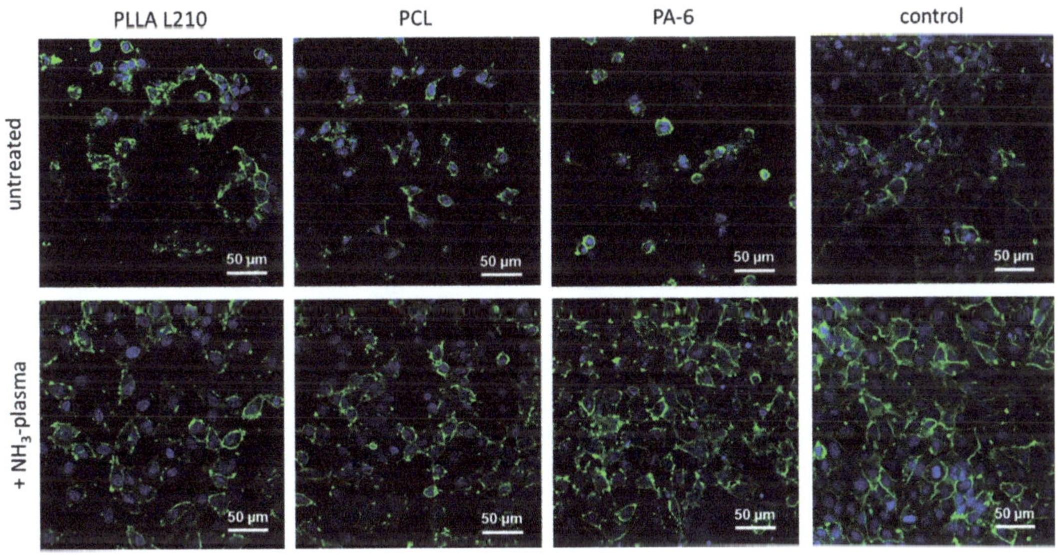

Figure 11. Immunostainings of CD31 in human endothelial EA.hy926 cells on unmodified and NH_3 plasma functionalized polymeric nonwovens after 48 h (green: CD31, blue: Hoechst-staining indicating cell nuclei, confocal microscopy, and bar = 50 µm).

On the contrary, EA.hy926 cells that were grown on NH_3-plasma-modified polymeric nonwovens demonstrated an increased CD31 expression that was concentrated in the cell-to-cell contact sites. Thus, CD31 expression patterns on NH_3-plasma-modified nonwovens were more similar to general CD31 expression on the respective control surface. While comparing the different NH_3-plasma-treated nonwovens, CD31 expression was highest and most homogeneous for EA.hy926 cells grown on NH_3-plasma treated nonwovens of PA-6. Here, it showed a faint intracellular constitutive background expression with intense cell–cell contact sites and thus was most comparable to the control (NC).

4. Discussion

Advancements in micro- and nanotechnology provide the technical platform to fabricate innovative biomaterials for tissue engineering in order to restore, maintain or even improve biological function. Electrospun polymeric scaffolds possess similar structural and morphological properties to extracellular matrix that might make them excellent biomimetic scaffolds for several cardiac interventions [18,26,55–58]. This includes drug-eluting systems, artificial heart valve prosthetics, or occluder systems. Although contemporary efforts have been made in stent technology regarding efficacy and safety, polymeric scaffolds are still facing challenges such as inflammation or late thrombosis events [9]. Delayed endothelialization is believed to play a key role in the occurrence of late stent thrombosis, and therefore research has attempted to improve stent designs to accelerate endothelialization. Since the endothelium represents an inherently antithrombogenic surface, fast formation of an intact endothelium at the implant site can prevent thrombus formation, or inflammation, events.

The present study is focused on the fabrication of nanofibrous polymer matrices as biomimetic scaffolds and the characterization of their physicochemical properties and biological performance in order to evaluate their suitability for the development of innovative artificial grafts for structural and valvular heart diseases [11,15,57]. Additionally, the effect of plasma functionalization of polymeric nonwovens was evaluated in attempt to improve endothelialization and thus biocompatibility of polymeric nanofibrous matrices [42,48,59]. Critical cellular parameters of human endothelial cells were examined to determine distinct growth patterns on polymeric nonwovens and investigate whether plasma functionalization with NH_3 affects biocompatibility and the maintenance of the endothelial cell phenotype.

4.1. *Surface Characterization and Chemical Modification by Plasma Treatment*

This study demonstrated that nanofibrous polymer scaffolds of biodegradable PLLA L210 and PCL as well as of biostable PA-6 were successfully fabricated by needleless electrospinning. Polymeric nonwovens exhibit uniform meshes with randomly distributed fibers with mean fiber diameters of 450 $\pm$ 250 nm for PLLA L210 and 100 $\pm$ 50 nm for PA-6 and 200 $\pm$ 50 nm for PCL while forming interconnected pores. Additionally, fiber morphology was shown to lack the formation of beads and junctions, which indicates sufficient evaporation of organic solvent mixtures during the electrospinning process.

The physicochemical surface analysis of polymeric nonwovens demonstrated clear differences in wettability and surface free energy among the fabricated polymeric nonwovens (see Table 1). Water contact angles of unmodified polymeric nonwovens ranged from hydrophobic to highly hydrophobic, which likely indicatives that, despite the chemical composition, the hydrophobicity is a consequence of the nanostructured surface topography itself [60–63]. Contact angle measurements confirmed that surface treatment with NH_3-plasma increases hydrophilicity of all types of investigated polymeric nonwovens. Moreover, the NH_3-plasma led to a higher polar fraction of the surface free energy for all examined polymeric nonwovens. Thus, NH_3-plasma functionalization was observed to exhibit a significant decrease of water contact angles (less than 40°) after NH_3-plasma deposition for all polymeric nonwovens, indicating a more hydrophilic surface. Among the tested polymer scaffolds, the very high contact angles of untreated PLLA L210 and PCL nonwovens, compared to the nearly zero apparent contact angles after NH_3-plasma-treatment, might be induced by capillary effects and the highly porous fibrous structure. That effect was observed to be highest for PLLA L210 nonwovens, while PCL- and PA-6 nonwovens exhibited less than half of the mean fiber diameters of PLLA L210.

4.2. *Evaluation of Cell Physiology and Cell Phenotype Maintenance on Nanofibrous Polymer Scaffolds*

Generally, cells are known to be able to sense aspects of their environment, including distinct surface topographical and chemical features, and adopt cellular physiology in response to those physicochemical properties of the biomaterial [28]. Thus, the results of the present study show that the nanofibrous topography of the fabricated electrospun polymeric matrices alone influences cell physiology and phenotype maintenance of human endothelial EA.hy926 cells, while distinct chemical characteristics, due to NH_3-plasma treatment, have a significant influence on these properties. In this study, human endothelial EA.hy926 cells were successfully proven to grow on fabricated nanofibrous nonwovens of PLLA L210, PCL and PA-6. However, for all of those untreated nonwovens, only a moderate biocompatibility, assessed by cell viability analysis (ranging from 49.6 to 57.1%), could be observed. Correspondingly, compared to the planar control surface, cells grown on the polymeric nonwovens also demonstrated less cell adhesion, as well as reduced cell viability nearly independent of the type of polymer. This could be a consequence of the nanofibrousness of the polymeric scaffolds, causing initially altered cell physiology affecting cell attachment, cell morphology and cell-functional parameters. Beyond the effect of the nanofibrous surface, in general, the nonwoven scaffolds also provide a different growing environment since the nonwovens are a more three-dimensional scaffold compared to the planar negative control (NC), which is simply two-dimensional with an additionally cell culture treated surface [40]. In consequence, the different dimensions between nonwoven and the planar control surface might be the reason for the observed differences in cell attachment and vitality of human endothelial EA.hy926 cells. In this context, Del Gaudio et al. [56] also reported lower cell vitality for primary human endothelial HUVEC cells seeded on PCL nonwovens in comparison to a planar TCPS surface, which is in agreement with the results on cell viability and attachment in the present and prior studies [42].

Regarding the evaluation of distinct cell–nanofiber interactions, Ahmed et al. [55] also described alterations in cell shape and function of primary human endothelial HUVEC cells grown on nano- and microfibrous poly(lactic-co-glycolic acid) PLGA meshes in a fiber diameter dependent manner. In particular, HUVECs became more spherical with smaller fiber diameters in the nanometer range (up to 500 nm), indicated by higher cell shape index, when compared to cells grown on micrometer fibers. Similarly, the results in the present study also demonstrate high CSI values of human endothelial cells on the tested nanofiber polymeric nonwovens.

In addition to the nanofibrous surface topography that might be responsible for distinct growth patterns of endothelial cells on polymeric nonwovens, the observed differences in wettability and surface free energy of the polymeric meshes could also contribute to changes in cell morphology. Limited cell attachment and viability can be attributed to the high hydrophobicity of all nanofibrous polymer surfaces that are known to negatively influence cell growth patterns. According to other studies, material surfaces that possess moderate hydrophilicity, i.e., water contact angle ranging between 40° and 80°, are shown to be favored by several cell types regarding cell attachment and vitality [64,65]. Therefore, the results of this study suggest that enhanced cell adhesion and viability of human endothelial EA.hy926 cells grown on PLLA L210 and PA-6 nonwovens might be attributed to the higher hydrophilicity than of PCL nonwovens. However, it is worth noting that the hydrophilic surface of PLLA L210 (less than 20° before washing, results not shown) was artificially induced by the addition of a surfactant and could not be maintained after the introduction of washing processes. Consequently, after release of Triton X-100, the contact angle of PLLA L210 was highly hydrophobic and similar to that of PCL nonwoven (see Table 1). As a general conclusion, it is essential to consider all additives used during production to simplify the electrospinning process and for a correct comparison of different final polymer fiber surfaces.

While distinct fiber geometries and surface wettability among the investigated untreated polymeric nonwovens were exhibited, differences in cell viability and cell shape were only slight and not significant. This might also be based on only slight differences in fiber diameters within the three polymer scaffolds, which only ranged within 100 to 450 nm for PCL, PA-6 and PLLA L210 nonwovens. In consequence, nanofiber-dependent differences in cell morphology and physiology patterns might become more obvious when comparing a higher range of fiber diameters. This might be evidenced by other studies that reported altered cell physiology according to differences in fiber diameters of polymeric nonwovens. Here, Ko et al. [58] showed increased proliferation and growth behavior of endothelial cells with increased fiber diameters of PLGA matrices by comparing nonwovens produced in nanofiber diameters of 200 nm and 600 nm.

Moreover, since no significant differences regarding cell viability and growth patterns among the investigated polymer nonwovens could be demonstrated, it seems that surface chemistry might be less influential than surface topography because endothelial EA.hy926 cells exhibited similar growth patterns on all polymeric nonwovens, regardless of their distinct chemical composition, i.e., PLLA L210 vs. PCL vs. PA-6. In regard to this differentiation between the relative influence of topographical and chemical effects, Cousin et al. [66] showed that changes in the morphology of fibroblasts could be predominantly attributed to the surface topography of nanoparticulated coatings, irrespective of the surface chemistry, which is consistent with the results of the present study.

4.3. Plasma Functionalization of Nanofibrous Polymer Scaffolds towards Improved Biocompatibility and Cellular Growth Patterns

Surface functionalization with NH_3-plasma was applied in order to improve biocompatibility and endothelialization of the polymeric nonwovens and thus their biological functionality and integrity regarding their usage as potential artificial cardiac or cardiovascular grafts.

Indeed, NH_3-plasma functionalization resulted in remarkable elevation of biocompatibility and phenotype maintenance of human endothelial cells on all of the investigated nanofibrous polymer scaffolds. In particular, NH_3-functionalization increased cell viability by up to 26% above untreated nanofibrous nonwovens as observed for PCL scaffolds. Additionally, phenotypic traits of endothelial cells were improved after NH_3-treatment, seen in enhanced cell spreading along with more elongated cell shapes (represented by lower CSI) as well as improved formation of actin cytoskeleton and more physiological CD31-expression. In consequence, surface functionalization with NH_3-plasma was proven to exert positive effects on endothelialization, which could potentially improve biocompatibility and hemocompatibility of PLLA L210, PCL and PA-6 nanofiber matrices. Those results are in accordance with other studies that also have shown a positive effect of NH_3-plasma of polymer films on cell viability and expression of endothelial markers such as CD31 in human endothelial HUVEC and HCAEC cells [42]. The positive effects seen in this study of NH_3-plasma treatment of the polymeric nonwovens of PLLA L210, PCL and PA-6 could be attributed to the incorporation of functional NH_2-groups that caused an increase in hydrophilicity of the polymeric nonwovens, subsequently leading to improved cell viability, which is also consistent with previous studies [67]. Because the surface morphology characterization did not reveal any significant effect of plasma treatment on the fiber diameters, it seems plausible that the enhancement of cell spreading and viability on the plasma-treated surfaces was induced solely by the increased hydrophilicity and the incorporation of amino functionalities, i.e., NH_2-groups. In particular, NH_2-group surface coupling might facilitate the adsorption of serum proteins to the surface that elevates cell attachment, subsequently leading to improved spreading and improved physiological patterns of human endothelial cells. Additionally, surface-adsorbed serum proteins might possess favorable configurations that are advantageous for cell attachment, spreading and growth patterns, as previous studies already reported [68]. Previous studies that demonstrated increased cell spreading of endothelial cells on NH_3-treated fibrous PLLA scaffolds postulate that the long-term beneficial effect of NH_3 plasma-treatment of polymer surfaces is a result of NH_2-group coupling to the surfaces [42,45].

4.4. Limitations and Potential of Plasma-Activated Nonwovens for Cardiovascular Applications

Within this study, we were able to gain basic knowledge about the biological interaction between cell-nonwoven surfaces under the influence of NH_3 plasma activation. However, so far we have only been able to make limited statements about the colonization of the fiber networks with human endothelial cells [58,69,70]. Both untreated and plasma-treated nonwovens were only studied within in vitro cell culture for short periods of time (48 h). In literature, a doubling time of approximately 25 h of the used endothelial cells is described [71,72]. Therefore, longer cell culture experiments in the future could yield more information. As already mentioned before, no general conclusions about polymer- or nonwoven-specific plasma treatment can be predicted. In general, parameters such as polymer type and electrospinning processing or plasma treatment time need to be adjusted to the desired applications. In addition, insights into biological response such as cell colonization and inflammatory behavior are limited in vitro experiments. In preparation for in vivo experiments, new methods such as cell behavior under flow conditions have to be established to quantify effects of shear stress on a confluent endothelial cell layer and to examine more complex situations such as biofilm formation or thrombotic events.

According to literature [11,69,73], the following important current questions cannot yet be conclusively or adequately addressed: (i) how long does complete endothealization of nonwoven materials take in vivo (with blood circulation)? (ii) do specific cells penetrate through the fiber network and does biodegradation influence the functionality? (iii) which polymer type should be preferred for which implant region—biodegradable vs. biostable?

In addition, Table 2 provides a brief summary of further considerations regarding plasma surface treatment of electrospun polymer scaffolds.

Table 2. Exemplary potentials and limitations of plasma surface treatment of electrospun polymer scaffolds.

	Control of	Selected Parameters	Selected Limitations
Plasma ↔ Electrospinning	Fiber topography	Adjustable fiber diameters ranging from nano- to micrometers	No electrospinning of chemical inert polymers, complex 3D scaffolds
	Biological response	Biofilm inhibition and endothelial cell enhancement	Polymer specific nonwovens, long-term stability of plasma activation
	Tribological properties	Surface roughness, sliding vs. adhesive	Specific hydrophilicity
	Local- or side-specific modifications	Layer structure, composites, and top vs. bottom	Possible material defects, e.g., delamination, etching or radiation
	Patient individuality	Easy and fast plasma modification, e.g., during interventions	Transferability, e.g., no defined electrospinning and plasma parameters
	Surface chemistry	Specific functional groups for graftings, coatings or drugs	Material and plasma dependence

5. Conclusions

In this study, nanofibrous scaffolds of PLLA L210, PCL and PA-6 were successfully fabricated by electrospinning for enabling their usage as biomimetic matrices for endothelialization. Results showed surface-topography-dependent alterations in cell physiology and endothelialization potential of human EA.hy926 endothelial cells. Untreated polymer meshes showed moderate biocompatibility and low cell spreading, which might complicate their use as implants. In order to improve the biocompatibility and biological integrity of the nanofibrous polymeric scaffolds, surface modification by NH_3-plasma functionalization was performed as it is described as a promising method to optimize biomaterials towards better cell compatibility and implant integrity. Chemical surface modification by NH_3-plasma treatment was demonstrated to alter hydrophilicity of all investigated polymer nonwovens while not affecting fiber morphology or structure integrity. Short functionalization with NH_3-plasma was demonstrated to effectively promote cell attachment and cell growth patterns of human endothelial cells on polymer nanofiber scaffolds. The positive effect was proven by enhanced cell attachment and spreading as well as increased cell viability. Thus, surface functionalization by NH_2-plasma could further promote endothelialization and hemocompatibility of polymeric nonwovens of PLLA L210, PCL and PA-6, making them suitable as advanced biomaterials for several cardiac and vascular interventions.

Further studies should more closely examine the influence of fiber thickness and pore size of nanofiber nonwovens using one specific polymer type. We know that, for instance, flat human endothelial cells forming a uniform monolayer with cell clusters of 200–400 μm are probably too large for the pores of our nanofiber meshes [74]. However, complex biodegradable PLLA or PCL matrices, where fiber degradation and cavity generation over time plays a role, need to be considered in vitro. To study degradation or ingrowth behavior under in vivo conditions for longer time periods, fast-degrading medically approved polymers could also become interesting as model fiber systems, if the influence of the polymer type on cell colonization is incidental. In addition, growth-promoting substances can be released via fiber degradation [74].

Controlling the fabrication parameters of the electrospinning process to optimize fiber diameter, pore structure, and mesh density and thickness is an important task to ensure adequate cell functionality and biological integrity. Together with the finely tunable fabrication properties of the electrospinning technique, plasma-treated polymer nonwovens are promising biomaterials for specifically directing cellular response. Thereby, systematic modulation of biomaterial's physicochemical properties will support the elucidation of key cell–biomaterial interactions to further improve implant technology.

Supplementary Materials: The following supporting information can be downloaded at: https://www.mdpi.com/article/10.3390/ma15062014/s1, Figure S1: The annual number of publications for the last 20 years with topic "engineer* cardi* scaffold*" provided by the search engine of Scopus, PubMed and Web of Science before the 15 February 2022; Figure S2: The annual number of publications for the last 20 years with topic "engineer* cardi* scaffold*" associated with supplementary search terms: "polymer*", "hydrogel*", "micropattern*", "nanofib*", "bioprint*" or "decellular*", provided by the search engine of Web of Science before 15 February 2022; Figure S3: The annual number of publications for the last 20 years with topic "engineer* cardi* scaffold*" associated with supplementary search terms: "polymer*", "hydrogel*", "micropattern*", "nanofib*", "bioprint*" or "decellular*", provided by the search engine of PubMed before before 15 February 2022; Figure S4: The annual number of publications for the last 20 years with topic "engineer* cardi* scaffold*" associated with supplementary search terms: "polymer*", "hydrogel*", "micropattern*", "nanofib*", "bioprint*" or "decellular*", provided by the search engine of Scopus before 15 February 2022; Figure S5: Representative SEM images of PLLA L210, PCL and PA-6 nonwovens at magnification 1000× and 5000×; Figure S6: Frequency of the respective fiber diameters within the individually defined increments for untreated and activated (A) PLLA L210, (B) PCL and (C) PA-6 nonwovens based on 50 individual measurement points (where for each polymer 10 fibers were measured manually in 5 SEM images); Figure S7: Fluorescence intensity of F-actin formation (phalloidin-TRITC for F-actin) in human endothelial EA.hy926 cells grown for 48 h on untreated and NH3-plasma functionalized polymeric nonwovens and polystyrene control surface (n = 1); Figure S8: Mean fluorescence intensity of CD 31 expression by human endothelial EA.hy926 cellsgrown for 48 h on untreated and NH3-plasma functionalized polymeric nonwovens (n = 1).

Author Contributions: Conceptualization, C.M. and S.K.; methodology, S.K., C.M., M.T. and S.I.; software, C.M. and S.I.; data curation, S.I.; writing—original draft preparation, C.M., S.I., S.K.; writing—review and editing, all; visualization, C.M., S.I. and K.W.; supervision, K.-P.S. and N.G.; project administration, S.I. and N.G.; funding acquisition, K.-P.S. and N.G. All authors have read and agreed to the published version of the manuscript.

Funding: Partial financial support by the European Regional Development Fund (ERDF) and the European Social Fund (ESF) within the excellence research program Card-ii-Omics and the collaborative research between economy and science of the state Mecklenburg-Vorpommern and by the Federal Ministry of Education and Research (BMBF) within RESPONSE "Partnership for Innovation in Implant Technology" is gratefully acknowledged.

Data Availability Statement: The data presented in this study are available upon request from the corresponding author.

Acknowledgments: The authors acknowledge technical assistance of Gabriele Karsten, Martina Nerger and Babette Hummel.

Conflicts of Interest: The authors declare no conflict of interest.

References

1. Mensah, G.A.; Roth, G.A.; Fuster, V. The Global Burden of Cardiovascular Diseases and Risk Factors: 2020 and Beyond. *J. Am. Coll. Cardiol.* **2019**, *74*, 2529–2532. [CrossRef] [PubMed]
2. Fioretta, E.S.; Motta, S.E.; Lintas, V.; Loerakker, S.; Parker, K.K.; Baaijens, F.P.T.; Falk, V.; Hoerstrup, S.P.; Emmert, M.Y. Next-generation tissue-engineered heart valves with repair, remodelling and regeneration capacity. *Nat. Rev. Cardiol.* **2021**, *18*, 92–116. [CrossRef] [PubMed]
3. Roshandel, M.; Dorkoosh, F. Cardiac tissue engineering, biomaterial scaffolds, and their fabrication techniques. *Polym. Adv. Technol.* **2021**, *32*, 2290–2305. [CrossRef]
4. Finn, A.V.; Joner, M.; Nakazawa, G.; Kolodgie, F.; Newell, J.; John, M.C.; Gold, H.K.; Virmani, R. Pathological correlates of late drug-eluting stent thrombosis: Strut coverage as a marker of endothelialization. *Circulation* **2007**, *115*, 2435–2441. [CrossRef] [PubMed]
5. Wessely, R. New drug-eluting stent concepts. *Nat. Rev. Cardiol.* **2010**, *7*, 194–203. [CrossRef]
6. Finn, A.V.; Nakazawa, G.; Joner, M.; Kolodgie, F.D.; Mont, E.K.; Gold, H.K.; Virmani, R. Vascular responses to drug eluting stents: Importance of delayed healing. *Arterioscler. Thromb. Vasc. Biol.* **2007**, *27*, 1500–1510. [CrossRef] [PubMed]
7. Dangas, G.D.; Claessen, B.E.; Caixeta, A.; Sanidas, E.A.; Mintz, G.S.; Mehran, R. In-stent restenosis in the drug-eluting stent era. *J. Am. Coll. Cardiol.* **2010**, *56*, 1897–1907. [CrossRef]
8. Hytönen, J.P.; Taavitsainen, J.; Tarvainen, S.; Ylä-Herttuala, S. Biodegradable coronary scaffolds: Their future and clinical and technological challenges. *Cardiovasc. Res.* **2018**, *114*, 1063–1072. [CrossRef]
9. Kilic, I.D.; Fabris, E.; Serdoz, R.; Caiazzo, G.; Foin, N.; Abou-Sherif, S.; Di Mario, C. Coronary covered stents. *EuroIntervention* **2016**, *12*, 1288–1295. [CrossRef]
10. Montoya, Y.; Cardenas, J.; Bustamante, J.; Valencia, R. Effect of sequential electrospinning and co-electrospinning on morphological and fluid mechanical wall properties of polycaprolactone and bovine gelatin scaffolds, for potential use in small diameter vascular grafts. *Biomater. Res.* **2021**, *25*, 38. [CrossRef]
11. Capulli, A.K.; MacQueen, L.A.; Sheehy, S.P.; Parker, K.K. Fibrous scaffolds for building hearts and heart parts. *Adv. Drug Deliv. Rev.* **2016**, *96*, 83–102. [CrossRef] [PubMed]
12. Carpentier, A. Hemodynamic factors affecting the fate of valvular bioprosthesis. *Circulation* **2010**, *121*, 2083–2084. [CrossRef] [PubMed]
13. Hoerstrup, S.P.; Sodian, R.; Daebritz, S.; Wang, J.; Bacha, E.A.; Martin, D.P.; Moran, A.M.; Guleserian, K.J.; Sperling, J.S.; Kaushal, S.; et al. Functional living trileaflet heart valves grown in vitro. *Circulation* **2000**, *102*, III44-9. [CrossRef] [PubMed]
14. Stock, A.M.; Robinson, V.L.; Goudreau, P.N. Two-component signal transduction. *Annu. Rev. Biochem.* **2000**, *69*, 183–215. [CrossRef] [PubMed]
15. Davoudi, P.; Assadpour, S.; Derakhshan, M.A.; Ai, J.; Solouk, A.; Ghanbari, H. Biomimetic modification of polyurethane-based nanofibrous vascular grafts: A promising approach towards stable endothelial lining. *Mater. Sci. Eng. C Mater. Biol. Appl.* **2017**, *80*, 213–221. [CrossRef]
16. Emmert, S.; Pantermehl, S.; Foth, A.; Waletzko-Hellwig, J.; Hellwig, G.; Bader, R.; Illner, S.; Grabow, N.; Bekeschus, S.; Weltmann, K.-D.; et al. Combining Biocompatible and Biodegradable Scaffolds and Cold Atmospheric Plasma for Chronic Wound Regeneration. *Int. J. Mol. Sci.* **2021**, *22*, 9199. [CrossRef]
17. Ramakrishna, S. *An introduction to Electrospinning and Nanofibers*; World Scientific: Singapore; Hackensack, NJ, USA, 2005; ISBN 9789812567611.
18. Xu, C.; Inai, R.; Kotaki, M.; Ramakrishna, S. Electrospun nanofiber fabrication as synthetic extracellular matrix and its potential for vascular tissue engineering. *Tissue Eng.* **2004**, *10*, 1160–1168. [CrossRef]
19. Khorshidi, S.; Solouk, A.; Mirzadeh, H.; Mazinani, S.; Lagaron, J.M.; Sharifi, S.; Ramakrishna, S. A review of key challenges of electrospun scaffolds for tissue-engineering applications. *J. Tissue Eng. Regen. Med.* **2016**, *10*, 715–738. [CrossRef]
20. Murugan, R.; Ramakrishna, S. Nano-featured scaffolds for tissue engineering: A review of spinning methodologies. *Tissue Eng.* **2006**, *12*, 435–447. [CrossRef]
21. Singhvi, R.; Stephanopoulos, G.; Wang, D.I. Effects of substratum morphology on cell physiology. *Biotechnol. Bioeng.* **1994**, *43*, 764–771. [CrossRef]
22. Hashi, C.K.; Zhu, Y.; Yang, G.-Y.; Young, W.L.; Hsiao, B.S.; Wang, K.; Chu, B.; Li, S. Antithrombogenic property of bone marrow mesenchymal stem cells in nanofibrous vascular grafts. *Proc. Natl. Acad. Sci. USA* **2007**, *104*, 11915–11920. [CrossRef] [PubMed]
23. Schnell, E.; Klinkhammer, K.; Balzer, S.; Brook, G.; Klee, D.; Dalton, P.; Mey, J. Guidance of glial cell migration and axonal growth on electrospun nanofibers of poly-epsilon-caprolactone and a collagen/poly-epsilon-caprolactone blend. *Biomaterials* **2007**, *28*, 3012–3025. [CrossRef] [PubMed]

24. Xin, X.; Hussain, M.; Mao, J.J. Continuing differentiation of human mesenchymal stem cells and induced chondrogenic and osteogenic lineages in electrospun PLGA nanofiber scaffold. *Biomaterials* **2007**, *28*, 316–325. [CrossRef]
25. Ren, X.; Feng, Y.; Guo, J.; Wang, H.; Li, Q.; Yang, J.; Hao, X.; Lv, J.; Ma, N.; Li, W. Surface modification and endothelialization of biomaterials as potential scaffolds for vascular tissue engineering applications. *Chem. Soc. Rev.* **2015**, *44*, 5680–5742. [CrossRef] [PubMed]
26. Badami, A.S.; Kreke, M.R.; Thompson, M.S.; Riffle, J.S.; Goldstein, A.S. Effect of fiber diameter on spreading, proliferation, and differentiation of osteoblastic cells on electrospun poly(lactic acid) substrates. *Biomaterials* **2006**, *27*, 596–606. [CrossRef] [PubMed]
27. Anselme, K.; Ploux, L.; Ponche, A. Cell/Material Interfaces: Influence of Surface Chemistry and Surface Topography on Cell Adhesion. *J. Adhes. Sci. Technol.* **2010**, *24*, 831–852. [CrossRef]
28. Matschegewski, C.; Staehlke, S.; Loeffler, R.; Lange, R.; Chai, F.; Kern, D.P.; Beck, U.; Nebe, B.J. Cell architecture–cell function dependencies on titanium arrays with regular geometry. *Biomaterials* **2010**, *31*, 5729–5740. [CrossRef]
29. Xu, H.; Nguyen, K.T.; Brilakis, E.S.; Yang, J.; Fuh, E.; Banerjee, S. Enhanced endothelialization of a new stent polymer through surface enhancement and incorporation of growth factor-delivering microparticles. *J. Cardiovasc. Transl. Res.* **2012**, *5*, 519–527. [CrossRef]
30. Chung, T.-W.; Lu, Y.-F.; Wang, H.-Y.; Chen, W.-P.; Wang, S.-S.; Lin, Y.-S.; Chu, S.-H. Growth of human endothelial cells on different concentrations of Gly-Arg-Gly-Asp grafted chitosan surface. *Artif. Organs* **2003**, *27*, 155–161. [CrossRef]
31. Tabares, F.L.; Junkar, I. Cold Plasma Systems and their Application in Surface Treatments for Medicine. *Molecules* **2021**, *26*, 1903. [CrossRef]
32. Sultana, A.; Zare, M.; Luo, H.; Ramakrishna, S. Surface Engineering Strategies to Enhance the In Situ Performance of Medical Devices Including Atomic Scale Engineering. *Int. J. Mol. Sci.* **2021**, *22*, 11788. [CrossRef] [PubMed]
33. Kwon, J.-S.; Choi, S.-H.; Choi, E.H.; Kim, K.-M.; Chu, P.K. Enhanced Osteogenic Differentiation of Human Mesenchymal Stem Cells on Amine-Functionalized Titanium Using Humidified Ammonia Supplied Nonthermal Atmospheric Pressure Plasma. *Int. J. Mol. Sci.* **2020**, *21*, 6085. [CrossRef] [PubMed]
34. Yang, Y.-W.; Wu, J.-Y.; Liu, C.-T.; Liao, G.-C.; Huang, H.-Y.; Hsu, R.-Q.; Chiang, M.-H.; Wu, J.-S. Fast incorporation of primary amine group into polylactide surface for improving C_2C_{12} cell proliferation using nitrogen-based atmospheric-pressure plasma jets. *J. Biomed. Mater. Res. A* **2014**, *102*, 160–169. [CrossRef] [PubMed]
35. Ehtesabi, H.; Massah, F. Improvement of hydrophilicity and cell attachment of polycaprolactone scaffolds using green synthesized carbon dots. *Mater. Today Sustain.* **2021**, *13*, 100075. [CrossRef]
36. Wang, W.; Caetano, G.; Ambler, W.S.; Blaker, J.J.; Frade, M.A.; Mandal, P.; Diver, C.; Bártolo, P. Enhancing the Hydrophilicity and Cell Attachment of 3D Printed PCL/Graphene Scaffolds for Bone Tissue Engineering. *Materials* **2016**, *9*, 992. [CrossRef]
37. Cho, S.J.; Jung, S.M.; Kang, M.; Shin, H.S.; Youk, J.H. Preparation of hydrophilic PCL nanofiber scaffolds via electrospinning of PCL/PVP-b-PCL block copolymers for enhanced cell biocompatibility. *Polymer* **2015**, *69*, 95–102. [CrossRef]
38. Baek, H.S.; Park, Y.H.; Ki, C.S.; Park, J.-C.; Rah, D.K. Enhanced chondrogenic responses of articular chondrocytes onto porous silk fibroin scaffolds treated with microwave-induced argon plasma. *Surf. Coat. Technol.* **2008**, *202*, 5794–5797. [CrossRef]
39. Park, H.; Lee, K.Y.; Lee, S.J.; Park, K.E.; Park, W.H. Plasma-treated poly(lactic-co-glycolic acid) nanofibers for tissue engineering. *Macromol. Res.* **2007**, *15*, 238–243. [CrossRef]
40. Martins, A.; Pinho, E.D.; Faria, S.; Pashkuleva, I.; Marques, A.P.; Reis, R.L.; Neves, N.M. Surface modification of electrospun polycaprolactone nanofiber meshes by plasma treatment to enhance biological performance. *Small* **2009**, *5*, 1195–1206. [CrossRef]
41. Prabhakaran, M.P.; Venugopal, J.; Chan, C.K.; Ramakrishna, S. Surface modified electrospun nanofibrous scaffolds for nerve tissue engineering. *Nanotechnology* **2008**, *19*, 455102. [CrossRef]
42. Rudolph, A.; Teske, M.; Illner, S.; Kiefel, V.; Sternberg, K.; Grabow, N.; Wree, A.; Hovakimyan, M. Surface Modification of Biodegradable Polymers towards Better Biocompatibility and Lower Thrombogenicity. *PLoS ONE* **2015**, *10*, e0142075. [CrossRef] [PubMed]
43. Yoo, H.S.; Kim, T.G.; Park, T.G. Surface-functionalized electrospun nanofibers for tissue engineering and drug delivery. *Adv. Drug Deliv. Rev.* **2009**, *61*, 1033–1042. [CrossRef] [PubMed]
44. Barry, J.J.A.; Silva, M.M.C.G.; Shakesheff, K.M.; Howdle, S.M.; Alexander, M.R. Using Plasma Deposits to Promote Cell Population of the Porous Interior of Three-Dimensional Poly(D,L-Lactic Acid) Tissue-Engineering Scaffolds. *Adv. Funct. Mater.* **2005**, *15*, 1134–1140. [CrossRef]
45. Cheng, Q.; Lee, B.L.-P.; Komvopoulos, K.; Yan, Z.; Li, S. Plasma surface chemical treatment of electrospun poly(L-lactide) microfibrous scaffolds for enhanced cell adhesion, growth, and infiltration. *Tissue Eng. Part A* **2013**, *19*, 1188–1198. [CrossRef]
46. Park, K.; Ju, Y.M.; Son, J.S.; Ahn, K.-D.; Han, D.K. Surface modification of biodegradable electrospun nanofiber scaffolds and their interaction with fibroblasts. *J. Biomater. Sci. Polym. Ed.* **2007**, *18*, 369–382. [CrossRef]
47. Gugala, Z.; Gogolewski, S. Attachment, growth, and activity of rat osteoblasts on polylactide membranes treated with various low-temperature radiofrequency plasmas. *J. Biomed. Mater. Res. A* **2006**, *76*, 288–299. [CrossRef]
48. Slepička, P.; Malá, Z.; Rimpelová, S.; Slepičková Kasálková, N.; Švorčík, V. Plasma treatment of the surface of poly(hydroxybutyrate) foil and non-woven fabric and assessment of the biological properties. *React. Funct. Polym.* **2015**, *95*, 71–79. [CrossRef]
49. Wulf, K.; Teske, M.; Löbler, M.; Luderer, F.; Schmitz, K.-P.; Sternberg, K. Surface functionalization of poly(ε-caprolactone) improves its biocompatibility as scaffold material for bioartificial vessel prostheses. *J. Biomed. Mater. Res. Part B Appl. Biomater.* **2011**, *98*, 89–100. [CrossRef]

50. Guex, A.G.; Frobert, A.; Valentin, J.; Fortunato, G.; Hegemann, D.; Cook, S.; Carrel, T.P.; Tevaearai, H.T.; Giraud, M.N. Plasma-functionalized electrospun matrix for biograft development and cardiac function stabilization. *Acta Biomater.* **2014**, *10*, 2996–3006. [CrossRef]
51. Teske, M.; Sternberg, K. Surface functionalization of poly(ε-caprolactone) and poly(3-hydroxybutyrate) with VEGF. *BioNanoMaterials* **2017**, *18*. [CrossRef]
52. Owens, D.K.; Wendt, R.C. Estimation of the surface free energy of polymers. *J. Appl. Polym. Sci.* **1969**, *13*, 1741–1747. [CrossRef]
53. Malek, A.M.; Izumo, S. Mechanism of endothelial cell shape change and cytoskeletal remodeling in response to fluid shear stress. *J. Cell Sci.* **1996**, *109 Pt 4*, 713–726. [CrossRef] [PubMed]
54. Song, K.; Lee, J.; Choi, S.-O.; Kim, J. Interaction of Surface Energy Components between Solid and Liquid on Wettability, and Its Application to Textile Anti-Wetting Finish. *Polymers* **2019**, *11*, 498. [CrossRef] [PubMed]
55. Ahmed, M.; Ramos, T.; Wieringa, P.; van Blitterswijk, C.; de Boer, J.; Moroni, L. Geometric constraints of endothelial cell migration on electrospun fibres. *Sci. Rep.* **2018**, *8*, 6386. [CrossRef]
56. Del Gaudio, C.; Bianco, A.; Folin, M.; Baiguera, S.; Grigioni, M. Structural characterization and cell response evaluation of electrospun PCL membranes: Micrometric versus submicrometric fibers. *J. Biomed. Mater. Res. A* **2009**, *89*, 1028–1039. [CrossRef]
57. Hasan, A.; Memic, A.; Annabi, N.; Hossain, M.; Paul, A.; Dokmeci, M.R.; Dehghani, F.; Khademhosseini, A. Electrospun scaffolds for tissue engineering of vascular grafts. *Acta Biomater.* **2014**, *10*, 11–25. [CrossRef]
58. Ko, Y.-G.; Park, J.H.; Lee, J.B.; Oh, H.H.; Park, W.H.; Cho, D.; Kwon, O.H. Growth behavior of endothelial cells according to electrospun poly(D,L-lactic-co-glycolic acid) fiber diameter as a tissue engineering scaffold. *Tissue Eng. Regen. Med.* **2016**, *13*, 343–351. [CrossRef]
59. Masaeli, E.; Morshed, M.; Tavanai, H. Study of the wettability properties of polypropylene nonwoven mats by low-pressure oxygen plasma treatment. *Surf. Interface Anal.* **2007**, *39*, 770–774. [CrossRef]
60. Bico, J.; Tordeux, C.; Quéré, D. Rough wetting. *Europhys. Lett.* **2001**, *55*, 214–220. [CrossRef]
61. El-Saber, M.; Abou-Gabal, H.; Aloufy, A.; El Saghir, A. Enhancing the Wetting Properties of Polypropylene and Polyethylene Nonwoven Fabrics Using a Cost-Effective Surface Dielectric Barrier Discharge. *KEM* **2018**, *786*, 258–266. [CrossRef]
62. Palasantzas, G.; de Hosson, J. Wetting on rough surfaces. *Acta Mater.* **2001**, *49*, 3533–3538. [CrossRef]
63. Zhu, L.; Perwuelz, A.; Lewandowski, M.; Campagne, C. Wetting behavior of thermally bonded polyester nonwoven fabrics: The importance of porosity. *J. Appl. Polym. Sci.* **2006**, *102*, 387–394. [CrossRef]
64. Arima, Y.; Iwata, H. Effect of wettability and surface functional groups on protein adsorption and cell adhesion using well-defined mixed self-assembled monolayers. *Biomaterials* **2007**, *28*, 3074–3082. [CrossRef]
65. Harnett, E.M.; Alderman, J.; Wood, T. The surface energy of various biomaterials coated with adhesion molecules used in cell culture. *Colloids Surf. B Biointerfaces* **2007**, *55*, 90–97. [CrossRef]
66. Cousins, B.G.; Zekonyte, J.; Doherty, P.J.; Garvey, M.J.; Williams, R.L. Manufacturing a nanometre scale surface topography with varying surface chemistry to assess the combined effect on cell behaviour. *IJNBM* **2008**, *1*, 320. [CrossRef]
67. Mahmoodi, M.; Zamanifard, M.; Safarzadeh, M.; Bonakdar, S. In vitro evaluation of collagen immobilization on polytetrafluoroethylene through NH_3 plasma treatment to enhance endothelial cell adhesion and growth. *Biomed. Mater. Eng.* **2017**, *28*, 489–501. [CrossRef] [PubMed]
68. Lim, C.T.; Tan, E.P.S.; Ng, S.Y. Effects of crystalline morphology on the tensile properties of electrospun polymer nanofibers. *Appl. Phys. Lett.* **2008**, *92*, 141908. [CrossRef]
69. Abdal-Hay, A.; Abdelrazek Khalil, K.; Al-Jassir, F.F.; Gamal-Eldeen, A.M. Biocompatibility properties of polyamide 6/PCL blends composite textile scaffold using EA.hy926 human endothelial cells. *Biomed. Mater.* **2017**, *12*, 35002. [CrossRef]
70. Garipcan, B.; Maenz, S.; Pham, T.; Settmacher, U.; Jandt, K.D.; Zanow, J.; Bossert, J. Image Analysis of Endothelial Microstructure and Endothelial Cell Dimensions of Human Arteries—A Preliminary Study. *Adv. Eng. Mater.* **2011**, *13*, B54–B57. [CrossRef]
71. Lu, Z.J.; Ren, Y.Q.; Wang, G.P.; Song, Q.; Li, M.; Jiang, S.S.; Ning, T.; Guan, Y.S.; Yang, J.L.; Luo, F. Biological behaviors and proteomics analysis of hybrid cell line EAhy926 and its parent cell line A549. *J. Exp. Clin. Cancer Res.* **2009**, *28*, 16. [CrossRef]
72. Pinevich, A.A.; Vartanyan, N.L.; Terekhina, L.A.; Krutetskaya, I.Y.; Shashkova, O.A.; Smirnov, I.V.; Samoylovich, M.P. Endoglin Expression and Surface Renewal in Mesenchymal Stem Cells and Endothelial Cells. *Cell Tiss. Biol.* **2021**, *15*, 107–119. [CrossRef]
73. Zhao, J.; Feng, Y. Surface Engineering of Cardiovascular Devices for Improved Hemocompatibility and Rapid Endothelialization. *Adv. Healthc. Mater.* **2020**, *9*, e2000920. [CrossRef] [PubMed]
74. Gapizov, S.S.; Petrovskaya, L.E.; Shingarova, L.N.; Svirschevskaya, E.V.; Dolgikh, D.A.; Kirpichnikov, M.P. The effect of tnf and VEGF on the properties of EA. Hy926 endothelial cells in a model of multi-cellular spheroids. *Acta Nat.* **2018**, *10*, 34–42. [CrossRef]

materials

Article

Improvement of Drug Release and Compatibility between Hydrophilic Drugs and Hydrophobic Nanofibrous Composites

Hazim J. Haroosh [1], Yu Dong [1,*], Shaimaa Jasim [2] and Seeram Ramakrishna [3]

[1] School of Civil and Mechanical Engineering, Curtin University, Perth, WA 6845, Australia; hazim.haroosh@curtin.edu.au
[2] Department Biomedical Science, Murdoch University, Perth, WA 6150, Australia; 32862216@student.murdoch.edu.au
[3] Department of Mechanical Engineering, National University of Singapore, Singapore 117575, Singapore; seeram@nus.edu.sg
* Correspondence: y.dong@curtin.edu.au

Abstract: Electrospinning is a flexible polymer processing method to produce nanofibres, which can be applied in the biomedical field. The current study aims to develop new electrospun hybrid nanocomposite systems to benefit the sustained release of hydrophilic drugs with hydrophobic polymers. In particular, electrospun hybrid materials consisting of polylactic acid (PLA):poly(ε-caprolactone) (PCL) blends, as well as PLA:PCL/halloysite nanotubes-3-aminopropyltriethoxysilane (HNT-ASP) nanocomposites were developed in order to achieve sustained release of hydrophilic drug tetracycline hydrochloride (TCH) using hydrophobic PLA:PCL nanocomposite membranes as a drug carrier. The impact of interaction between two commonly used drugs, namely TCH and indomethacin (IMC) and PLA:PCL blends on the drug release was examined. The drug release kinetics by fitting the experimental release data with five mathematical models for drug delivery were clearly demonstrated. The average nanofiber diameters were found to be significantly reduced when increasing the TCH concentration due to increasing solution electrical conductivity in contrast to the presence of IMC. The addition of both TCH and IMC drugs to PLA:PCL blends reduced the crystallinity level, glass transition temperature (T_g) and melting temperature (T_m) of PCL within the blends. The decrease in drug release and the impairment elimination for the interaction between polymer blends and drugs was accomplished by mobilising TCH into HNT-ASP for their embedding effect into PLA:PCL nanofibres. The typical characteristic was clearly identified with excellent agreement between our experimental data obtained and Ritger–Peppas model and Zeng model in drug release kinetics. The biodegradation behaviour of nanofibre membranes indicated the effective incorporation of TCH onto HNT-ASP.

Keywords: electrospun nanofibres; nanocomposites; polylactic acid (PLA); poly(ε-caprolactone) (PCL); halloysite nanotubes (HNTs); drug release; release kinetic modelling

Citation: Haroosh, H.J.; Dong, Y.; Jasim, S.; Ramakrishna, S. Improvement of Drug Release and Compatibility between Hydrophilic Drugs and Hydrophobic Nanofibrous Composites. *Materials* **2021**, *14*, 5344. https://doi.org/10.3390/ma14185344

Academic Editors: Csaba Balázsi and Manuel Arruebo

Received: 29 July 2021
Accepted: 13 September 2021
Published: 16 September 2021

1. Introduction

Electrospinning, as a combination of two processing methods, namely electrospraying and spinning [1,2], is an efficient, convenient and versatile method for creating reliable ultrafine fibres from polymer solutions. It has been a hot subject with significant growing interest to scientific communities during recent years. It offers a promising technique to manufacture continuous fibres with the diameters ranging from nanometres to microns [3,4]. It is also deemed as an alternative approach to fabricate hybrid fibre membranes based on polymer nanocomposites when compared with commonly used manufacturing processes of twin-screw extrusion, high shear mixing and injection moulding. Polymer nanocomposites when incorporated with nanoparticles become very attractive advanced materials due to their well-tailored properties [5] and unique structures, which might not be possibly accomplished in conventional composites [6,7]. Halloysite nanotubes (HNTs) have also gained widespread attention as tubular clay-based nanoparticles with cost-effectiveness mainly used for the reinforcement purpose [8]. In general, poly(ε-caprolactone) (PCL) is a hydrophobic and semi-crystalline polymer [9]. It degrades in the body due to the existence of an aliphatic ester bond for hydrolysis [10], and the resulting materials can be metabolised via tricarboxylic acid cycle or eliminated directly by the renal secretion [11]. On the other hand, polylactic acid (PLA), as one of the most promising biodegradable polymers, has remarkable material merits including easy processability, ability to be dissolved in common solvents, good sustainability and reasonable biocompatibility [12,13]. Since PLA has a low degree of crystallinity, when compared with PCL, which can degrade to form nontoxic monomers, it is well utilised in biomedical applications such as medical devices, drug delivery, tissue scaffolding, etc. [14]. As a typical example, it submits the secession to monomeric units of lactic acid in the body, which naturally takes place in the carbohydrate metabolism [15,16]. Whereas, PCL is a slowly degraded biopolymer as opposed to many other counterparts such as PLA owing to its semi-crystalline nature [17,18] and a high degree of crystallinity. Moreover, PCL is not harmful to local tissues because it does not form an acidic environment like poly(lactide-co-glycolide) (PLGA) or PLA. In the past decades, the use of polymers, as a carrier and release controller for drug delivery systems, has been the major research focus due to the improvement of healing effectiveness and the reduction in toxic side-effect [19]. In addition, the interaction between drugs and their carriers is paramount to control a sustained drug release [20,21]. Hydrophobic drugs including clindamycin, ciprofloxacin, and cephalexin can be simply incorporated with, and thus interacted with hydrophobic polymers. However, hydrophilic drugs cannot be embedded into hydrophobic polymers due to their weak interactions while the drug release rate may not be well controlled [22]. This study aims to develop new electrospun hybrid nanocomposite systems to benefit the sustained release of hydrophilic drugs when encountered with hydrophobic polymers in order to ultimately overcome the weak interaction issue identified between the drugs and their carriers.

2. Materials and Methods

2.1. Materials

PLA 3051D (molecular weight (*MW*) = 93,500 g/mol) was supplied by Nature Works, Blair, Nebraska, USA. PCL (*MW* = 80,000 g/mol), tetracycline hydrochloride (TCH) ($C_{22}H_{24}N_2O_8 \cdot HCl$, *MW*= 480.9 g/mol) shown in Figure 1a, Indomethacin (IMC) ($C_{19}H_{16}ClNO_4$, MW = 357.79 g/mol) illustrated in Figure 1b, phosphate buffer solution (PBS), chloroform and methanol were all purchased from Sigma-Aldrich Ltd., Castle Hill, NSW, Australia, and these materials were used without any purification. Halloysite nanotubes (HNT) were obtained from Imerys Ceramics, Kaeo, Northland, New Zealand.

2.2. Electrospinning

TCH and IMC drugs with a fixed concentration of 5 wt% were initially mixed with blend solutions (i.e., PLA:PCL at a mix ratio of 1:1), in which TCH was mixed with 1wt%/v HNT-ASP and subsequently added to PLA:PCL solution. The solvent used in all cases was the mixture of chloroform and methanol (volume ratio: 2:1). In the following electrospinning process, different types of solutions were transferred to a 10 mL syringe pump (A Fusion 100 syringe pump, Chemyx Inc., Stafford, TX, USA) with a needle specification of 20 G (inner diameter: 0.584 mm). The flow rate of the solution was set to 2 mL/h, the applied voltage was in the range of 25–28 kV, and the needle-to-collector distance was fixed at 13 cm. The electrospinning process was carried out at 24 °C in the closed system and in ventilated fume cupboards with optimal conditions of humidity and temperature. The polymer solution was electrospun immediately after being prepared to reduce the effect of environmental conditions. The resulting fibres were collected on a ground collector covered by flat aluminium foil. A sharp blade was used to cut fibre mats and they were removed with fine forceps. The thicknesses of fibre mats produced were measured using a digital electronic micrometer, recorded in the range from 330 to 450 μm.

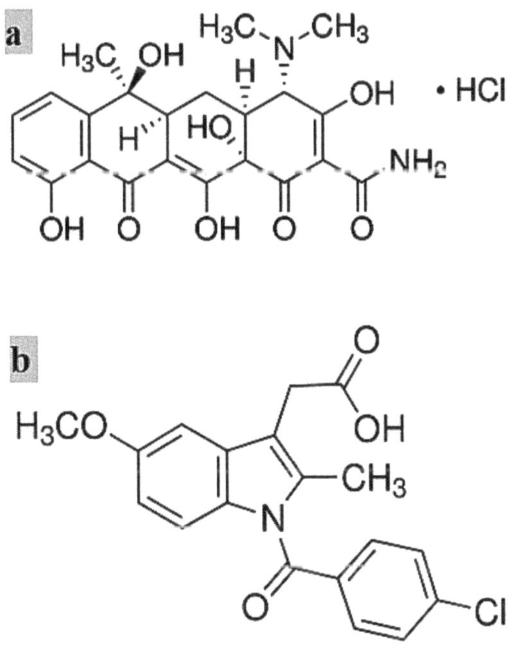

Figure 1. Chemical structure of two model drugs: (**a**) TCH [23] and (**b**) IMC [24].

2.3. In Vitro Drug Release Study

The drug-loaded fibre mat samples (sample size: 2 cm × 2 cm) were incubated in 20 mL PBS (pH = 7.4) using a rotary shaker at 37 °C. After the required incubation time for drug release, the samples were transferred to 20 mL fresh buffer solution and 36 separate incubations were conducted to obtain TCH and IMC release data for the three nanofibre mats (i.e., PLA:PCL/IMC, PLA:PCL/TCH and PLA:PCL/HNT-ASP/TCH) and subsequently, the released drug amount in the buffer solution was determined accordingly.

The percentage of the released drug was calculated from the primary weight of the drug (i.e., TCH and IMC) contained in the electrospun mats. The cumulative amount of drug released from nanofibre mats was calculated according to the following Equation:

$$\text{Cumulative amount of drug released } (\%) = \left(\frac{M_t}{M_\infty}\right) \times 100\% \quad (1)$$

where M_t is the number of drugs released up to time t and M_∞ is the initial amount of drug within electrospun fibres.

2.4. Mathematical Models for Drug Release Kinetics

The kinetics of TCH and IMC release from nanofibre mats were determined from the release curves against time t. The experimental drug release data were fitted to five typical mathematical models for drug release kinetics consisting of conventional zero-order model, first-order model, Higuchi model, Ritger–Peppas model, as well as Zeng model [20].

2.4.1. Zero-Order Model

A zero-order model is employed to a drug delivery system in which the drug release rate is independent of drug concentration. Its empirical equation [25] can be described as

$$\frac{M_t}{M_\infty} = K_0 t \quad (2)$$

where M_t, M_∞ and K_0 are the absolute cumulative amounts by mass of drug released at time t and infinite time, and zero-order release constant, respectively. The ratio M_t/M_∞ denotes the cumulative amount released in percentage [26].

2.4.2. First-Order Model

The first-order equation can be expressed in a derived mathematical form [27] for the released drug in an aqueous phase [28] given below

$$\frac{M_t}{M_\infty} = 1 - e^{K_1 t} \tag{3}$$

in which K_1 is the first-order release constant.

2.4.3. Higuchi Model

Higuchi [29] developed important mathematical models to study the release of water-soluble and low soluble drugs loaded in semi-solid and solid matrices. The equation can be simplified by

$$\frac{M_t}{M_\infty} = K_H t^{\frac{1}{2}} \tag{4}$$

where K_H is Higuchi kinetic constant that reflects design variables in a drug delivery system.

2.4.4. Ritger–Peppas Model

Ritger and Peppas [25] established a simple exponential relationship to investigate both Fickian and non-Fickian drug release conditions in swelling and non-swelling polymeric delivery systems. Such a mathematical equation is given by

$$\frac{M_t}{M_\infty} = K_R t^n \tag{5}$$

where K_R is Ritger–Peppas kinetic constant, which incorporates structural and geometric characteristics of a macromolecular network system and the drug. n is the diffusion exponent for the transport mechanism through the polymer.

2.4.5. Zeng Model

Zeng and co-workers [20] developed a three-parameter model with the close-form analytical solution to consider the reversible drug–carrier interaction and first-order drug release from liposomes. The equation is relatively complex as given by

$$\frac{M_t}{M_\infty} = \frac{K_{off}}{K_{on} + K_{off}}\left(1 - e^{-K_S t}\right) + \frac{K_{on}}{K_{on} + K_{off}}\left(1 - e^{-K_{off} t}\right) \tag{6}$$

and

$$\Delta G = -k_B T\left(\frac{K_{on}}{K_{off}}\right) \tag{7}$$

In Equation (6), K_{on} is the rate constant of association for non-dispersed drug molecules in the system to be disassociated from carriers prior to drug release. Conversely, K_{off} is the rate constant of disassociation accordingly and K_S is a constant proportional to the surface-to-volume ratio of the carriers in order to improve drug release. In addition, ΔG is the difference of free energy between the free and bound states, k_B is the Boltzmann's constant and T is the absolute temperature where T = 300 K in Equation (7). K_{off}, K_S and ΔG are three critical parameters in Zeng model to explain the effect of cumulative drug release.

2.5. In Vitro Biodegradation Study

With respect to in vitro degradation studies, nanofibre mats with the thickness ranging from 300 to 450 μm were cut into 2 cm × 2 cm testing samples. They were measured to determine the initial weight (m) and then were placed into an incubated rotary shaker at the rotor speed of 100 rpm and the temperature of 37 °C during the biodegradation study where 15 mL PBS (pH = 7.4) was utilised for the required incubation time.

Such nanofibre mats were removed at each designated incubation period and washed with deionised water. These mats were dried under vacuum at 37 °C until they reached a constant weight (m_1). Mass loss (%) was determined using the equation below

$$\text{mass loss } (\%) = \left(\frac{m - m_1}{m}\right) \times 100\% \tag{8}$$

3. Characterisation Techniques

The morphology of electrospun nanofibres was examined via an EVO 40XVP scanning electron microscope (SEM) (Carl Zeiss AG, Jena, Germany) at the accelerating voltage of 5 kV. Before SEM observation, the samples were sputter-coated with platinum. Fibre diameters were calculated from the SEM images by using an image analysis tool within Zeiss Smart SEM software. A minimum of 150 fibres was selected from multiple scanned SEM images for the measurements per sample based on a sampling rate of 15 fibres per image.

Solution viscosity was determined with the aid of a Visco 88 portable viscometer from Malvern Instruments, Malvern, UK. The electrical conductivity of the solution was measured by using a WP-81 Waterproof Conductivity Meter, TPS, Brendale, QLD, Australia).

XRD measurements of prepared samples were undertaken in a Bruker Discover 8 X-ray diffractometer, Bruker Corporation, Berlin, Germany. It operated at 40 kV and 40 mA using Cu-Kα radiation subjected to the monochromatisation with graphite sample monochromators in a 2θ range from 5° to 40° (scanning rate: 0.05°/s).

Thermal analysis was performed via a DSC6000 Perkin Elmer, Boston, MA, USA with a cryofill liquid nitrogen cooling system. Approximately 10 mg fibre mat was cut and sealed in an aluminium pan. The thermal behaviour was analysed during the first heating scan in a temperature range from −90 °C to 200 °C at the ramp rate of 10 °C/min.

Thermogravimetric analysis (TGA) was carried out by using a SeikoSII Exstar 6000 (TG/DTA 6200), Tokyo, Japan to evaluate the thermal decomposition effect of electrospun PLA:PCL fibres loaded with drugs. About 6–10 mg TGA samples were heated from 40 °C to 900 °C at a heating ramp rate of 10 °C/min under a nitrogen flow of 200 mL/min [30].

Fourier transform infrared spectroscopy (FTIR) was performed in a Spectrum 100 FTIR Spectrometer, Perkin Elmer, Kanagawa, Japan. Resulting spectra were recorded in a wavenumber range of 4000–550 cm^{-1} with the resolution of 4 cm^{-1} by using an attenuated total reflectance (ATR) technique [31].

The amount of TCH present in the release PBS was obtained by means of a UV-vis spectrophotometer, JascoV-67, Easton, MD, USA at two specific wavelengths of 360 nm [32] and 319 nm [33] for TCH and IMC drugs accordingly.

4. Results and Discussion

4.1. Drug Effect on Fiber Morphology

To investigate the drug effect, PLA:PCL polymer solution was blended with IMC and TCH at a fixed concentration of 5 wt%. As observed in Figures 2 and 3, loaded IMC gave rise to uniform composite fibrous structures in possession of much larger fibre diameters at approximately 794 nm as opposed to 623 nm for TCH counterparts. No remarkable variation in fibre diameter was identified with the addition of IMC when compared with PLA:PCL blend fibres. Similarly, PLA:PCL/HNT-ASP/TCH composites were also found to attain uniform fibrous structures despite a small fibre diameter of 716 nm relative to those PLA:PCL blends and PLA:PCL/HNT-ASP composites.

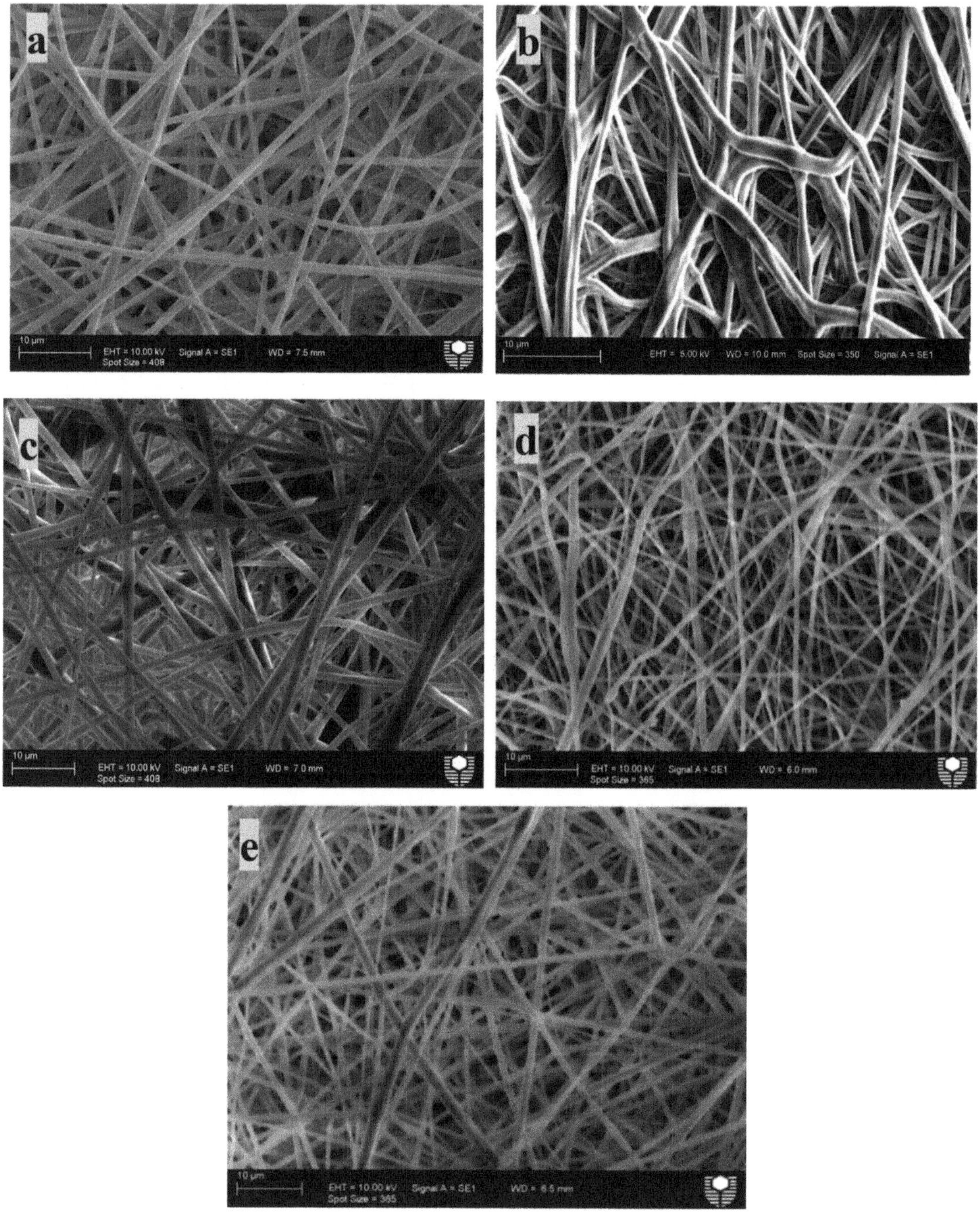

Figure 2. Scanning electron micrographs of electrospun fibre mats (**a**) PLA:PCL (**b**) PLA:PCL/HNT-ASP (**c**) PLA:PCL/IMC (**d**) PLA:PCL/TCH and (**e**) PLA:PCL/HNT-ASP /TCH. All the scale bars represent 10 μm in size.

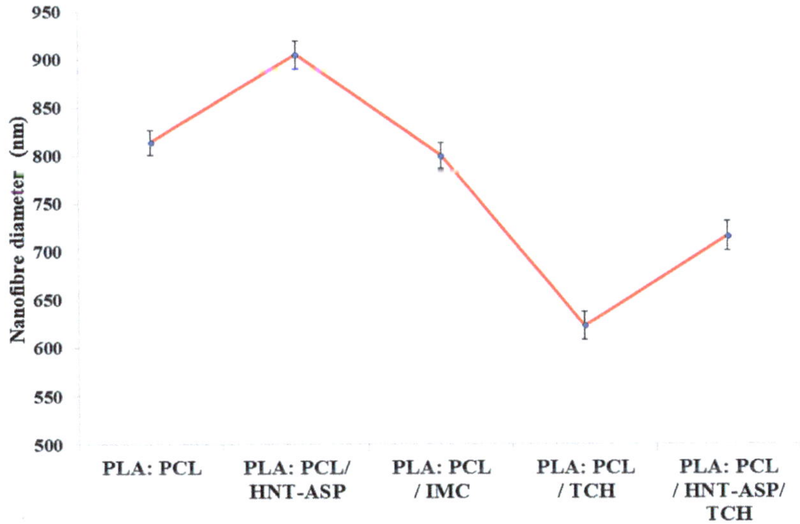

Figure 3. Effects of HNT-ASP, IMC and TCH on nanofibre diameter.

Figure 4 reveals that the addition of IMC to PLA:PCL solution did not appear to significantly affect the solution electrical conductivity in good accordance with previous work [34]. However, as expected, the use of cationic drug TCH further increases the solution conductivity. The additional amphoteric molecules of TCH, having several ionisable functional groups [35], further increases the solution conductivity as compared to that without TCH.

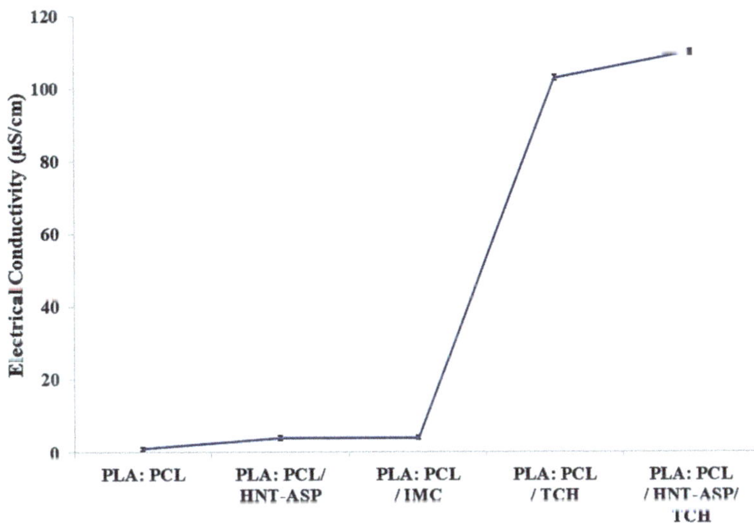

Figure 4. Effects of HNT-ASP, IMC and TCH on solution electrical conductivity.

4.2. Crystallinity Level

An XRD examination was carried out to investigate the impact of drug type on crystallinity level and fibrous structures of electrospun nanocomposites fibre mats. The associated crystalline peaks in XRD patterns were directly detected with the aid of a DICVOL program used in the FullProf software [36] and labelled with XRD reflection and crystal planes (*hkl*) accordingly. Figure 5 illustrates the XRD patterns with respect to fibre mats of PLA:PCL, PLA:PCL/TCH, PLA:PCL/IMC and PLA:PCL/HNT-ASP/TCH. Such patterns possess two distinct diffraction peaks at the angles of $2\theta = 20.1^\circ$ and 23.2° corresponding to the crystal planes (101) and (200), respectively. In particular, the characteristic XRD peak of PCL/HNT-ASP/TCH was detected at $2\theta = 12.4^\circ$, which is indicative of the HNT reflection plane (001). The basal reflection of HNTs results from their tubular morphological structures and small crystal size $L = 21.7$ nm obtained according to the Scherrer relation [8]. No significant difference in relation to peak position was manifested regardless of different material samples, indicating the minor effect arising from loaded TCH and IMC to alter crystalline structures of electrospun nanocomposite fibre mats. As such, it may be revealed that TCH and IMC are most likely to be dispersed in an amorphous state into PLA:PCL fibre mats. This may be attributed to rapid solvent evaporation, which did not have enough time to cause relaxation of chain orientation. A short period for drug recrystallisation and the formation of a preferred configuration in the amorphous state would take place accordingly [37].

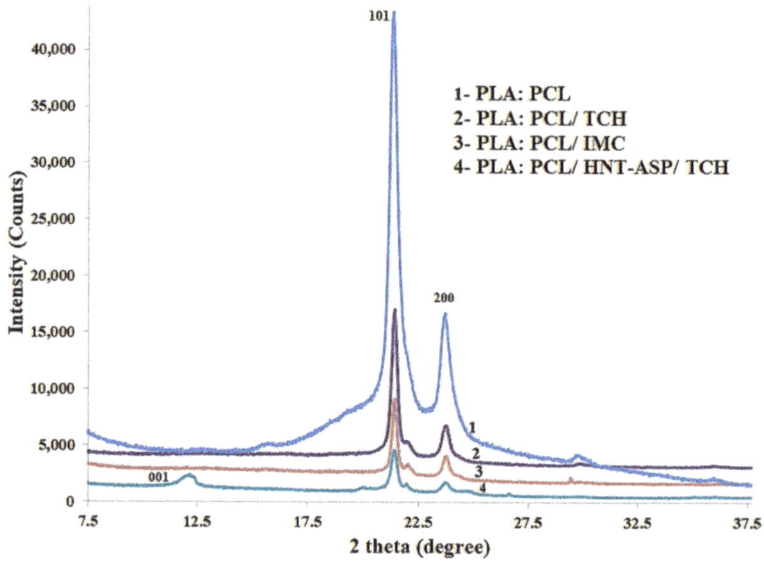

Figure 5. XRD patterns for selected material samples showing the relative positions of reflection peaks when loaded with TCH and IMC drugs.

Figure 6 suggests that the degree of crystallinity (X_c) of nanocomposite fibre mats is reduced more remarkably when TCH is concurrently added, as opposed to that of PLA: PCL counterparts. Additionally, the use of IMC loaded to PLA:PCL fibre mats was found to influence X_c more significantly in comparison with TCH loaded counterparts. It is evidently shown that the presence of TCH and IMC enables to greatly accelerate the nucleation effect, thus resulting in a shorter period than the required time for the disentanglement of molecular chains. This is because the resulting degree of crystallinity can be impacted by the restricted mobility of polymeric chains in order to prevent the more rapid growth of developed crystals. In addition, the crystallinity of nanocomposites was reduced when HNT-ASP was embedded into polymers due to the diminished mobility of polymeric chains.

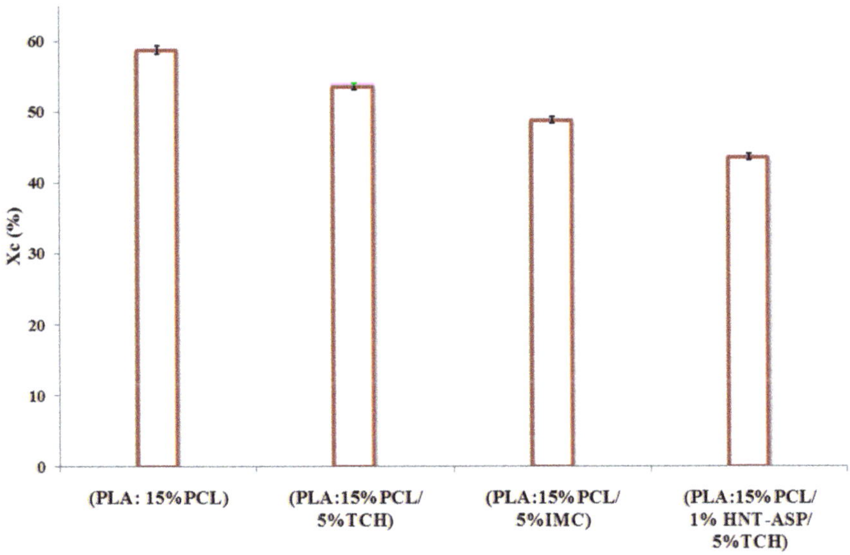

Figure 6. Degree of crystallinity (X_c) of electrospun PLA: PCL fibre mats loaded with TCH and IMC drugs.

4.3. Thermal Properties

The thermal properties of electrospun nanocomposite fibre mats loaded with IMC and TCH separately via differential scanning calorimetry (DSC) are shown in Table 1 and Figure 7. There is a considerable decreasing trend with respect to the T_g of PCL within nanocomposite fibre mats despite a slight decline for the T_m levels of both PLA and PCL when embedded with HNT-ASP and drug-loaded with 5 wt%/v TCH. The effect of decreased T_g can be associated with the low molecular weight of TCH. Short molecular chains of TCH are believed to yield the decline in the packing density of polymeric chains, thus facilitating the chain mobility as a result of lower T_g. On the contrary, the addition of TCH offers a significant increase in T_c with respect to electrospun nanocomposites.

Table 1. Thermal properties of PLA:PCL fibre mats and PLA:PCL nanocomposite fibre mats loaded with TCH and IMC.

Material Sample	T_g (°C) PCL	T_m (°C) PCL	T_C (°C) PLA	T_m (°C) PLA	X_c (%)
PLA:PCL	−52.19	63.75	83.97	152.54	58.84
PLA:PCL/IMC	−58.55	54.57	82.01	148.98	48.89
PLA:PCL/TCH	−58.48	59.64	105.77	151.93	53.69
PLA:PCL/HNT-ASP/TCH	−61.43	59.17	102.77	149.07	43.60

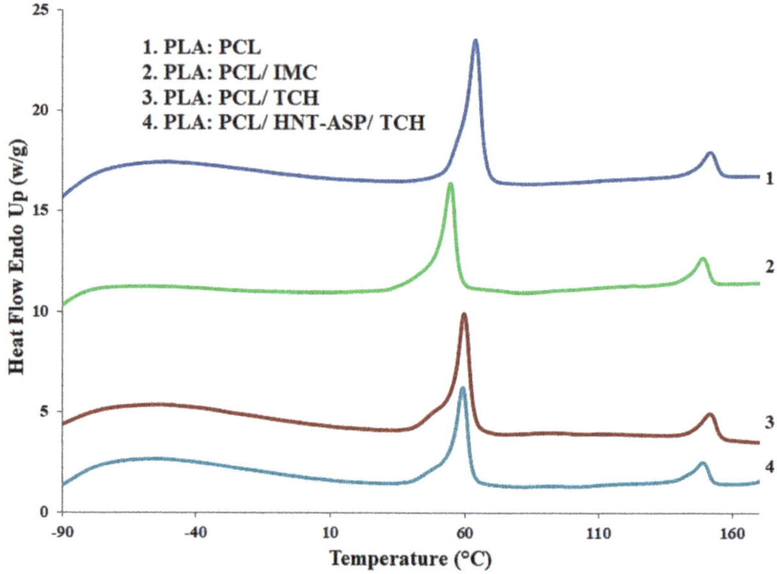

Figure 7. DSC thermograms for selected PLA:PCL material samples.

PLA:PCL/HNT-ASP/TCH fibre mats possess the crystallisation temperature (T_c) about 103 °C as opposed to 106°C for PLA:PCL/TCH and 84 °C for PLA in PLA:PCL counterparts. This result suggests that low-molecular-weight TCH enables to hinder the cold crystallisation process of PLA in an anti-nucleating agent role as opposed to HNT-ASP. On the other hand, loaded IMC results in an evident drop in T_m in contrast with a slight decline in T_c. This finding lies in a better interaction between PLA:PCL fibre mats when compared with the use of TCH. The T_g of PLA within the blends is hardly detected as it has overlapped the melting peak of PCL

4.4. TGA Analysis

TGA spectra of electrospun nanocomposite fibre mats loaded with IMC and TCH are exhibited in Figure 8. The addition of TCH, IMC or HNT-ASP does not appear to induce a significant alteration in thermal stability of PCL with a narrow degradation peak range from 398 to 399 °C. When TCH, IMC and HNT-ASP/TCH are incorporated, it is evidently demonstrated that a peak shift to higher temperatures takes place within nanocomposite fibre mats, thus retarding the thermal degradation. The temperature associated with the TGA peak, which is assigned to PLA at 334 °C within PLA:PCL fibre mats, shifts to 345 °C when loaded with IMC and further increases up to 350 °C with the addition of TCH and HNT-ASP/TCH, resulting in much better thermal stability. Furthermore, the residual masses were increasingly recorded to be 0.36%, 0.68%, 2.4% and 15.6% for PLA:PCL, PLA:PCL/IMC, PLA:PCL/TCH and PLA:PCL/HNT-ASP/TCH fibre mats, respectively. In particular, the higher residue mass of PLA:PCL/TCH when compared with that of PLA:PCL/IMC can be ascribed to different molecular structures of TCH (i.e., $C_{22}H_{25}ClN_2O_8$) with a larger number of carbon atoms relative to that of IMC (i.e., $C_{19}H_{16}ClNO_4$).

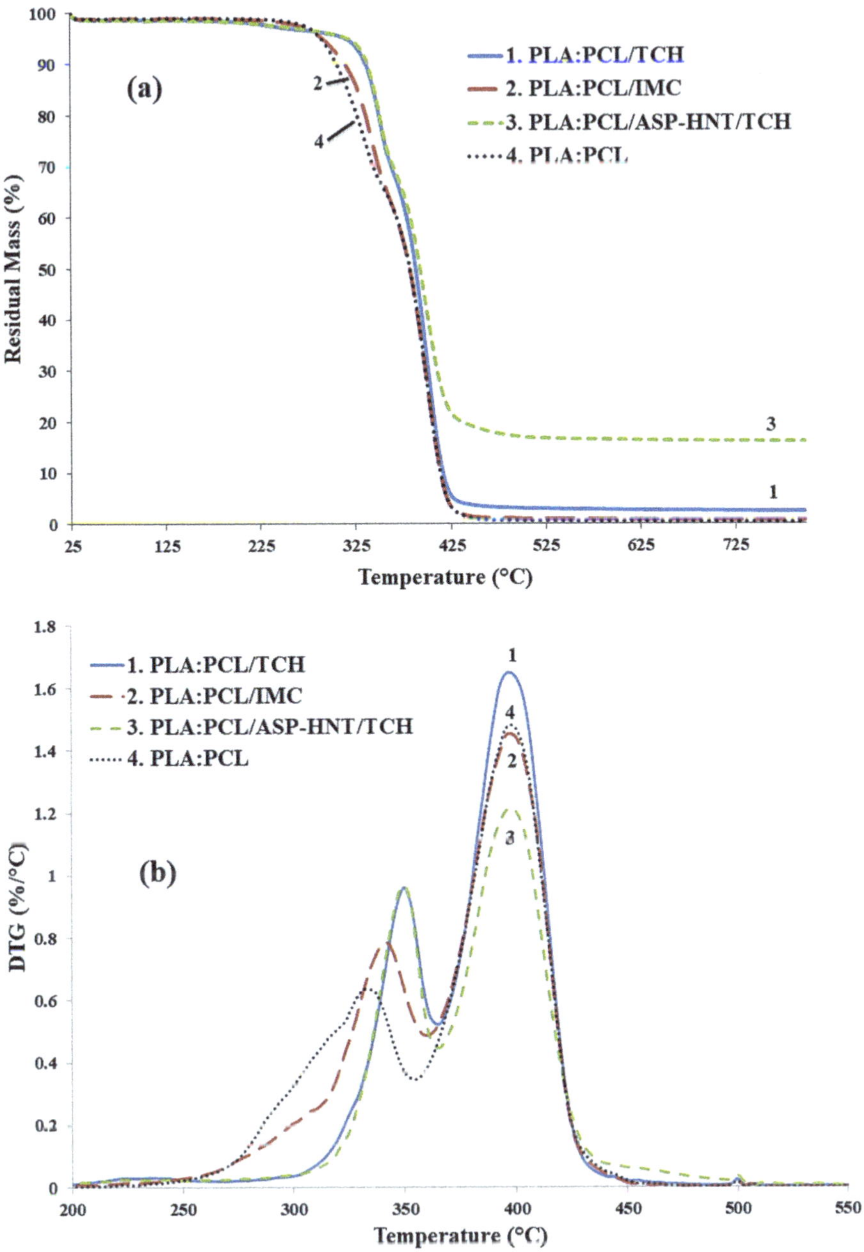

Figure 8. (**a**) TGA curves and (**b**) derivative thermogravimetric (DTGA) curves for typical PLA: PCL blend fibre mats.

4.5. FTIR Evaluation

Figure 9 depicts the FTIR spectra of PLA:PCL fibre mats loaded with or without TCH and IMC for comparison in terms of their chemical bonding effect. The spectra of TCH within PLA:PCL fibre mats and corresponding nanocomposite fibre mats do not appear to be easily assigned to the band shift. Nonetheless, corresponding TCH bands recorded at 1614 and 1581 cm^{-1} within nanocomposite fibre mats are assigned to C=O stretchings at ring A and ring C, respectively [38], signifying the successful encapsulation of TCH. In a similar manner, effective encapsulation of IMC is also manifested, as evidenced by the existing bond taking place at 1560 cm^{-1} associated with the ionisation of carboxyl groups [39] according to the FTIR spectra of PLA:PCL/IMC fibre mats.

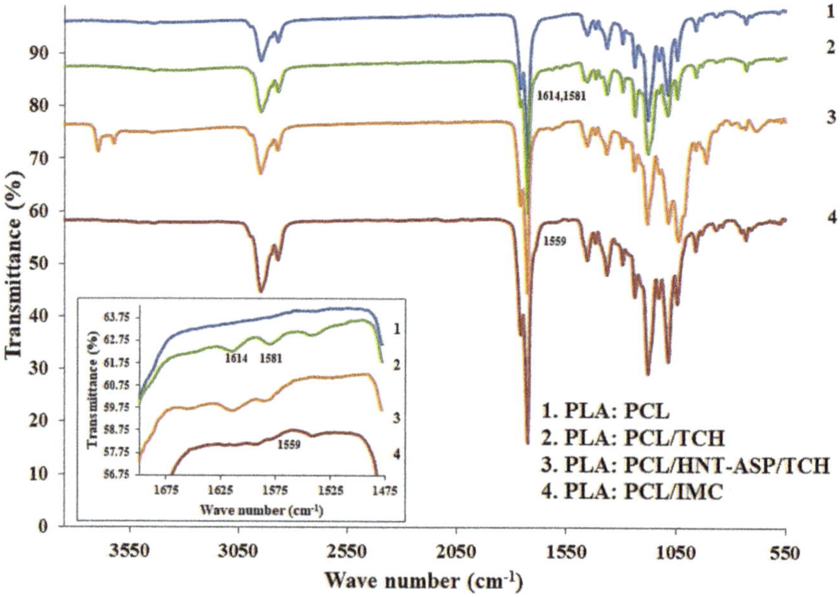

Figure 9. FTIR spectra for typical PLA:PCL based fibre mats showing drug effect on FTIR peaks.

4.6. *In Vitro Drug Release*

Drug particles within PLA:PCL fibre mats tend to remain on the fibre surfaces owing to the fast solvent evaporation arising from the corresponding blend solution in electrospinning along with high ionic interactions [40]. As such, it is quite convincing that a considerable burst release may happen at the initial drug-release stage. As a result of weak interaction, the burst release becomes more pronounced on the condition that it is incompatible between the drug and PLA:PCL solution. In comparison, the release rate of hydrophilic drug TCH, when interacting with hydrophobic PLA:PCL blends, becomes faster than that of hydrophobic drug IMC. As seen from Figure 10, drug release intends to be quite rapid during the first 5 h, which is especially the case when using TCH (i.e., 42% for PLA:PCL/TCH vs. 30% for PLA:PCL/IMC). This phenomenon is attributed to better interaction and more active compatibility taking place between IMC and PLA:PCL fibre mats, as opposed to the use of TCH. It is worth noting that chemical interaction between the drug and its carrier may impede drug crystallisation within its carrier leading to a sustained drug-release condition in a crystalline state [41]. The addition of 1 wt%/v HNT-ASP into PLA:PCL blends can reduce initial burst release to 30% after 5 h as opposed to initial 42% for PLA: PCL blends alone. In the meantime, a similar release trend was observed over the steady release period from 50 to 250 h, indicating a robust drug release control over both short and long evaluation periods, which appears to arise from the embedded HNT-ASP. It is well known that HNTs possess negative charges on the outer surfaces while positive charges on their inner surfaces [42]. Such distinct charges on HNT-ASP may give rise to the electrostatic interaction identified between TCH and outer surfaces, as well as the lumen structures of HNT-ASP. Accordingly, the drug release rate of TCH can be reduced when PLA:PCL/HNT-ASP nanocomposite fibre mats act as a drug carrier.

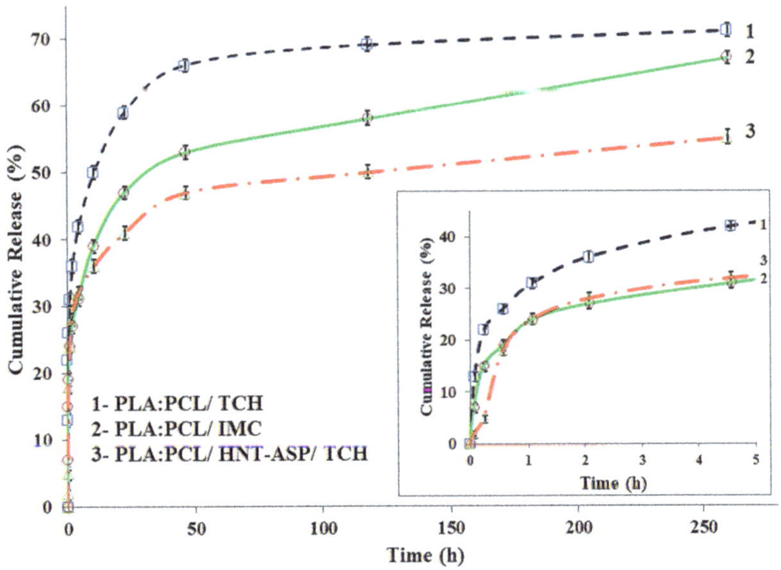

Figure 10. Drug release profiles for PLA:PCL fibre mats incorporated with IMC, TCH and HNT-ASP/TCH.

4.7. Release Kinetics

The mechanism of drug release kinetics is vital to investigate the efficacy of drug release with the carriers. Consequently, our drug release data obtained were fitted with five mathematical models whose equations and corresponding parameters are explicitly listed in Table 2. The conventional first-order model reveals that drug release rate can completely depend on drug concentration and its carrier feature. Whereas, Ritger–Pappas model demonstrates a particular drug release mechanism following a Fickian transport phenomenon [26] according to n values in a dominant diffusion process. On the other hand, Zeng model [20] reveals that drug release may be significantly influenced by the diffusion process and close interaction between TCH and IMC drugs and PLA:PCL blend fibre mats as an effective carrier. In particular, IMC has relatively low free energy variation between free and bound states, namely ΔG shown in Table 2.It is found that $\Delta G = -2.1 \times 10^{-21}$ for IMC when compared with $\Delta G = -2.0 \times 10^{-22}$ for TCH. PLA:PCL fibre mats are supposed to decrease the release rate of hydrophobic IMC though it may not work for hydrophilic TCH owing to their enhanced interaction. Additionally, the decrease in K_{off} values of IMC presented in Table 2 implies a strong drug-carrier interaction when compared with K_{off} of TCH ranging from 0.0033 to 0.0025 h^{-1}.

Table 2. Drug release parameters determined by fitting drug release data to five different mathematical models for drug release kinetics.

Mathematical Model	Zero-Order	First-Order	Higuchi	Ritger–Peppas	Zeng
Equation	$M_t/M_\infty = K_0 t$	$M_t/M_\infty = 1 - e^{-K_1 t}$	$M_t/M_\infty = K_H\, t^{1/2}$	$M_t/M_\infty = K_R t^n$	**
PLA:PCL/IMC	$K_0 = 0.002$ $R^2 = 0.564$	$K_1 = 0.436$ $R^2 = 0.822$	$K_H = 0.037$ $R^2 = 0.801$	$K_R = 1.618$ $n = 0.251$ $R^2 = 0.938$	$K_{on} = 0.0049\ h^{-1}$ $K_{off} = 0.0025\ h^{-1}$ $K_s = 1.006\ h^{-1}$ * $\Delta G = -2.1 \times 10^{-21} J$ $R^2 = 0.945$
PLA:PCL/TCH	$K_0 = 0.002$ $R^2 = 0.434$	$K_1 = 0.582$ $R^2 = 0.839$	$K_H = 0.038$ $R^2 = 0.691$	$K_R = 1.262$ $n = 0.204$ $R^2 = 0.927$	$K_{on} = 0.0034\ h^{-1}$ $K_{off} = 0.0033\ h^{-1}$ $K_s = 1.160\ h^{-1}$ * $\Delta G = -2.0 \times 10^{-22} J$ $R^2 = 0.920$
PLA:PCL/ HNT-ASP / TCH	$K_0 = 0.0012$ $R^2 = 0.314$	$K_1 = 0.857$ $R^2 = 0.961$	$K_H = 0.024$ $R^2 = 0.534$	$K_R = 2.025$ $n = 0.323$ $R^2 = 0.684$	$K_{on} = 0.0012\ h^{-1}$ $K_{off} = 0.00068\ h^{-1}$ $K_s = 1.025\ h^{-1}$ * $\Delta G = -2.2 \times 10^{-21} J$ $R^2 = 0.986$

** $M_t/M_\infty = (K_{off}/(K_{on} + K_{off}))(1 - e^{-K_S t}) + (K_{on}/(K_{on} + K_{off}))(1 - e^{-K_{off} t})$. * $\Delta G = -k_B T \ln(K_{on}/K_{off})$ where k_B is the Boltzmann's constant and T is the absolute temperature (300 K).

It is worth noting that with the incorporation of HNT-ASP into PLA:PCL fibre mats, very minor alteration is evidently observed in K_S though ΔG could decrease from -2.0×10^{-22} to -2.2×10^{-21} J. This finding confirms that embedded HNT-ASP may decrease the drug release rate of TCH, and further overcome the fast release issue induced by the poor interaction between loaded TCH and PLA:PCL fibre mats. Generally speaking, the close interaction between TCH molecules and HNT-ASP enables to yield hydrogen bonds with silanol groups mounted on HNT-ASP. Since our proposed models do not take erosion/biodegradation and dimensional alteration of drug carriers into consideration, TCH release data are hard to be completely fitted with those available models. On the flip side, the fitting results based on Zeng model is seemingly in overall good agreement with those obtained through experimental data because this model detected the effect of interaction between both drugs (i.e., TCH and IMC) and the carrier (i.e., PLA: PCL nanofibres) on the release rate.

4.8. Mass Loss of Fibre Mats

The biodegradability of PLA:PCL fibre mats in terms of mass loss is displayed in Figure 11. After the degradation period from the first 3 to 72 h in PBS, it is well noted that the weights of both PLA:PCL/TCH and PLA:PCL/IMC fibre mats have very trivial variations. The difference of mass has been identified between them to just commence after 168 h. It is worth mentioning that mass loss tends to increase over the degradation time up to 336 h by reaching the levels of 2.9% and 2.5% for PLA:PCL/TCH and PLA:PCL/IMC fibre mats, respectively. Overall, PLA:PCL/HNT-ASP/TCH nanocomposite fibre mats consistently gain the highest mass losses in all degradation time, as exemplified by 3.44% at 336 h. The small mass loss may be ascribed to PCL within the blends, thus hindering the water penetration inside the nanofibres. Additionally, water absorption has been found to take place slowly since PCL behaves with hydrophobic and semi-crystalline characteristics.

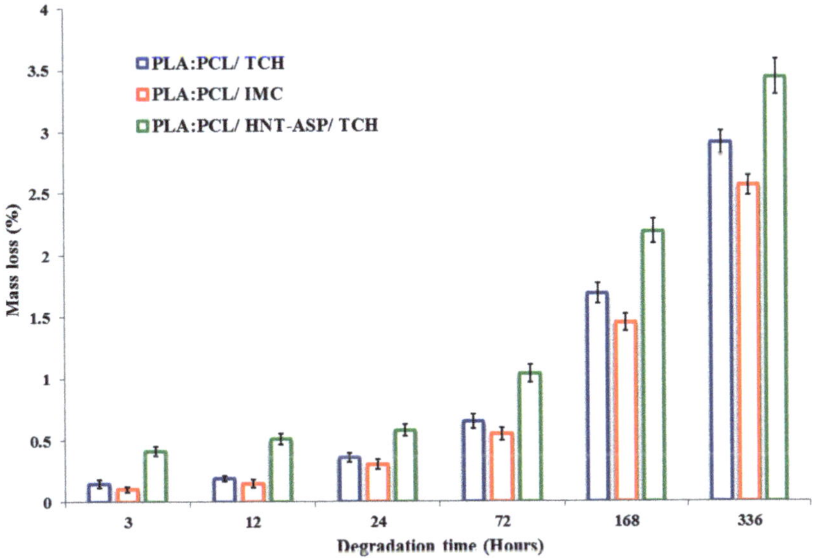

Figure 11. Mass loss of electrospun PLA:PCL fibre mats during both short and long degradation periods.

In spite of smaller fibre diameters of PLA:PCL/TCH fibre mats, their degradation seems to be relatively high as compared with those of PLA:PCL/IMC counterparts. When hydrophilic drug TCH is released from fibre mats, it may accelerate the degradation process of hydrophobic nanofibres in a more rapid manner due to its increased wettability, as opposed to that of hydrophobic IMC drug. Moreover, HNTs have been detected to improve the PBS absorption and also increase the porosity at the nanofiber surfaces for PLA:PCL/HNT-ASP/TCH nanocomposite mats. Such a phenomenon presents the strong evidence for the good incorporation of TCH drug molecules onto HNT-ASP in order to reduce the drug release rate. However, such nanocomposite fibre mats may undergo relatively high degradability.

5. Conclusions

It was found that the addition of TCH appears to increase the electrical conductivity of PLA:PCL solution, thus considerably decreasing nanofiber diameters. However, nanofibre diameters are shown to be further altered insignificantly when loaded with IMC. Loaded TCH and IMC drugs help to decrease the degree of crystallinity, the T_g and T_m of PCL within PLA:PCL fibre mats. FTIR spectra confirm the successful encapsulation of IMC and TCH into PLA:PCL fibre mats and corresponding nanocomposites. Thermal degradation of PLA may be delayed with the addition of these two drugs, as well as HNT-ASP/TCH in nanocomposite fibre mats despite the little variation to PCL. When hydrophilic drug TCH is loaded into HNT-ASP and hydrophobic PLA:PCL blends, it decreases the drug release and overcomes the weak interaction between TCH and PLA:PCL blends. Such a typical characteristic is evidenced by excellent agreement achieved between Ritger–Peppas model and Zeng model and experimental data in order to clearly understand the mechanism of drug release kinetics. These models might not completely fit drug release data because they do not consider the erosion/biodegradation and dimensional alteration of the carriers. The mass loss in relation to the degradation effect signifies good TCH-embedding effect onto HNT-ASP leading to the reduction in drug release rate.

Author Contributions: Conceptualization, H.J.H. and Y.D.; methodology, H.J.H., and Y.D.; formal analysis, H.J.H. and Y.D.; resources, H.J.H. and Y.D.; data curation, H.J.H; writing—original draft preparation, H.J.H.; writing—review and editing, H.J.H., Y.D., S.J. and S.R.; project administration, H.J.H. and Y.D. All authors have read and agreed to the published version of the manuscript.

Funding: This research received no external funding.

Institutional Review Board Statement: Not applicable.

Informed Consent Statement: Not applicable.

Data Availability Statement: Data is contained within the article.

Conflicts of Interest: The authors declare no conflict of interest.

References

1. Agarwal, S.; Wendorff, J.; Greiner, A. Use of electrospinning technique for biomedical applications. *Polymer* **2008**, *49*, 5603–5621. [CrossRef]
2. He, S.-W.; Li, S.-S.; Hu, Z.-M.; Yu, J.-R.; Chen, L.; Zhu, J. Effects of three parameters on the diameter of electrospun poly (ethylene oxide) nanofibers. *J. Nanosci. Nanotechnol.* **2011**, *11*, 1052–1059. [CrossRef] [PubMed]
3. Han, J.; Chen, T.; Branford-White, C.; Zhu, L. Electrospun shikonin-loaded PCL/PTMC composite fiber mats with potential biomedical applications. *Int. J. Pharm.* **2009**, *382*, 215–221. [CrossRef] [PubMed]
4. Haroosh, H.J.; Dong, Y.; Lau, K.-T. Tetracycline hydrochloride (TCH)-loaded drug carrier based on PLA: PCL nanofibre mats: Experimental characterisation and release kinetics modelling. *J. Mater. Sci.* **2014**, *49*, 6270–6281. [CrossRef]
5. Bi, H.; Feng, T.; Li, B.; Han, Y. In Vitro and In Vivo Comparison Study of Electrospun PLA and PLA/PVA/SA Fiber Membranes for Wound Healing. *Polymers* **2020**, *12*, 839. [CrossRef]
6. Du, M.; Guo, B.; Lei, Y.; Liu, M.; Jia, D. Carboxylated butadiene-styrene rubber/halloysite nanotube nanocomposites: Interfacial interaction and performance. *Polymer* **2008**, *49*, 4871–4876. [CrossRef]
7. Haroosh, H.J.; Dong, Y.; Ingram, G.D. Synthesis, Morphological Structures, and Material Characterization of Electrospun PLA: PCL/Magnetic Nanoparticle Composites for Drug Delivery. *J. Polym. Sci. Part B Polym. Phys.* **2013**, *51*, 1607–1617. [CrossRef]
8. Haroosh, H.J.; Dong, Y.; Chaudhary, D.S.; Ingram, G.D.; Yusa, S. Electrospun PLA: PCL composites embedded with unmodified and 3-aminopropyltriethoxysilane (ASP) modified halloysite nanotubes (HNT). *Appl. Phys. A Mater. Sci. Process.* **2013**, *110*, 433–442. [CrossRef]
9. Kenawy, E.; Abdel-Hay, F.; El-Newehy, M.; Wnek, G. Processing of polymer nanofibers through electrospinning as drug delivery systems. *Mater. Chem. Phys.* **2009**, *113*, 296–302. [CrossRef]
10. Kweon, H.Y.; Yoo, M.K.; Park, I.K.; Kim, T.H.; Lee, H.C.; Lee, H.S.; Oh, J.S.; Akaike, T.; Cho, C.S. A novel degradable polycaprolactone networks for tissue engineering. *Biomaterials* **2003**, *24*, 801–808. [CrossRef]
11. Sun, M.; Downes, S. Physicochemical characterisation of novel ultra-thin biodegradable scaffolds for peripheral nerve repair. *J. Mater. Sci. Mater. Med.* **2009**, *20*, 1181–1192. [CrossRef]
12. Deng, X.L.; Sui, G.; Zhao, M.L.; Chen, G.Q.; Yang, X.P. Poly (L-lactic acid)/hydroxyapatite hybrid nanofibrous scaffolds prepared by electrospinning. *J. Biomater. Sci. Polym. Ed.* **2007**, *18*, 117–130. [CrossRef]
13. Haroosh, H.J.; Chaudhary, D.S.; Dong, Y. Electrospun PLA/PCL fibers with tubular nanoclay: Morphological and structural analysis. *J. Appl. Polym. Sci.* **2012**, *124*, 3930–3939. [CrossRef]
14. Leonés, A.; Peponi, L.; Lieblich, M.; Benavente, R.; Fiori, S. In Vitro Degradation of Plasticized PLA Electrospun Fiber Mats: Morphological, Thermal and Crystalline Evolution. *Polymers* **2020**, *12*, 2975. [CrossRef]
15. Lv, G.; He, F.; Wang, X.; Gao, F.; Zhang, G.; Wang, T.; Jiang, H.; Wu, C.; Guo, D.; Li, X. Novel nanocomposite of nano Fe3O4 and polylactide nanofibers for application in drug uptake and induction of cell death of leukemia cancer cells. *Langmuir* **2008**, *24*, 2151–2156. [CrossRef]
16. Kim, G.; Yoon, H.; Park, Y. Drug release from various thicknesses of layered mats consisting of electrospun polycaprolactone and polyethylene oxide micro/nanofibers. *Appl. Phys. A Mater. Sci. Process.* **2010**, *100*, 1197–1204. [CrossRef]
17. Rezk, A.I.; Kim, K.-S.; Kim, C.S. Poly (ε-Caprolactone)/Poly (Glycerol Sebacate) Composite Nanofibers Incorporating Hydroxyapatite Nanoparticles and Simvastatin for Bone Tissue Regeneration and Drug Delivery Applications. *Polymers* **2020**, *12*, 2667. [CrossRef] [PubMed]
18. Kupka, V.; Dvořáková, E.; Manakhov, A.; Michlíček, M.; Petruš, J.; Vojtová, L.; Zajíčková, L. Well-blended PCL/PEO electrospun nanofibers with functional properties enhanced by plasma processing. *Polymers* **2020**, *12*, 1403. [CrossRef] [PubMed]
19. Xie, Z.; Buschle Diller, G. Electrospun poly (D, L lactide) fibers for drug delivery: The influence of cosolvent and the mechanism of drug release. *J. Appl. Polym. Sci.* **2010**, *115*, 1–8. [CrossRef]
20. Zeng, L.; An, L.; Wu, X. Modeling Drug-Carrier Interaction in the Drug Release from Nanocarriers. *J. Drug Deliv.* **2011**, *2011*, 15. [CrossRef]
21. Barani, H.; Khorashadizadeh, M.; Haseloer, A.; Klein, A. Characterization and release behavior of a thiosemicarbazone from electrospun polyvinyl alcohol core-shell nanofibers. *Polymers* **2020**, *12*, 1488. [CrossRef]

22. Xu, X.; Zhong, W.; Zhou, S.; Trajtman, A.; Alfa, M. Electrospun PEG–PLA nanofibrous membrane for sustained release of hydrophilic antibiotics. *J. Appl. Polym. Sci.* **2010**, *118*, 588–595. [CrossRef]
23. Buschle-Diller, G.; Cooper, J.; Xie, Z.; Wu, Y.; Waldrup, J.; Ren, X. Release of antibiotics from electrospun bicomponent fibers. *Cellulose* **2007**, *14*, 553–562. [CrossRef]
24. Chennamaneni, S.; Zhong, B.; Lama, R.; Su, B. COX inhibitors Indomethacin and Sulindac derivatives as antiproliferative agents: Synthesis, biological evaluation, and mechanism investigation. *Eur. J. Med. Chem.* **2012**, *56*, 17–29. [CrossRef]
25. Ritger, P.L.; Peppas, N.A. A simple equation for description of solute release I. Fickian and non-fickian release from non-swellable devices in the form of slabs, spheres, cylinders or discs. *J. Control. Release* **1987**, *5*, 23–36. [CrossRef]
26. Cai, X.; Luan, Y.; Dong, Q.; Shao, W.; Li, Z.; Zhao, Z. Sustained release of 5-fluorouracil by incorporation into sodium carboxymethylcellulose sub-micron fibers. *Int. J. Pharm.* **2011**, *419*, 240–246. [CrossRef]
27. Das, R.K.; Kasoju, N.; Bora, U. Encapsulation of curcumin in alginate-chitosan-pluronic composite nanoparticles for delivery to cancer cells. *Nanomed. Nanotechnol. Biol. Med.* **2010**, *6*, 153–160. [CrossRef]
28. Chittur, K.K. FTIR/ATR for protein adsorption to biomaterial surfaces. *Biomaterials* **1998**, *19*, 357–369. [CrossRef]
29. Higuchi, T. Mechanism of sustained-action medication. Theoretical analysis of rate of release of solid drugs dispersed in solid matrices. *J. Pharm. Sci.* **1963**, *52*, 1145–1149. [CrossRef] [PubMed]
30. Dong, Y.; Ghataura, A.; Takagi, H.; Haroosh, H.J.; Nakagaito, A.N.; Lau, K.-T. Polylactic acid (PLA) biocomposites reinforced with coir fibres: Evaluation of mechanical performance and multifunctional properties. *Compos. Part A Appl. Sci. Manuf.* **2014**, *63*, 76–84. [CrossRef]
31. Andrade, J.; Pereira, C.G.; de Almeida Junior, J.C.; Viana, C.C.R.; de Oliveira Neves, L.N.; da Silva, P.H.F.; Bell, M.J.V.; dos Anjos, V.d.C. FTIR-ATR determination of protein content to evaluate whey protein concentrate adulteration. *Lwt* **2019**, *99*, 166–172. [CrossRef]
32. Kenawy, E.; Bowlin, G.; Mansfield, K.; Layman, J.; Simpson, D.; Sanders, E.; Wnek, G. Release of tetracycline hydrochloride from electrospun poly (ethylene-co-vinylacetate), poly (lactic acid), and a blend. *J. Control. Release* **2002**, *81*, 57–64. [CrossRef]
33. Mi, F.L.; Sung, H.W.; Shyu, S.S. Release of indomethacin from a novel chitosan microsphere prepared by a naturally occurring crosslinker: Examination of crosslinking and polycation–anionic drug interaction. *J. Appl. Polym. Sci.* **2001**, *81*, 1700–1711. [CrossRef]
34. Nyström, M.; Murtomaa, M.; Salonen, J. Fabrication and characterization of drug particles produced by electrospraying into reduced pressure. *J. Electrost.* **2010**, *68*, 42–48. [CrossRef]
35. Sassman, S.A.; Lee, L.S. Sorption of three tetracyclines by several soils: Assessing the role of pH and cation exchange. *Environ. Sci. Technol.* **2005**, *39*, 7452–7459. [CrossRef]
36. Boultif, A.; Louer, D. Powder pattern indexing with the dichotomy method. *J. Appl. Crystallogr.* **2004**, *37*, 724–731. [CrossRef]
37. Puppi, D.; Piras, A.; Detta, N.; Dinucci, D.; Chiellini, F. Poly (lactic-co-glycolic acid) electrospun fibrous meshes for the controlled release of retinoic acid. *Acta Biomater.* **2010**, *6*, 1258–1268. [CrossRef]
38. Li, Z.; Kolb, V.M.; Jiang, W.-T.; Hong, H. Ftir and XRD INVestigations of tetracycline intercalation in smectites. *Clays Clay Miner.* **2010**, *58*, 462–474. [CrossRef]
39. Mohanambe, L.; Vasudevan, S. Anionic clays containing anti-inflammatory drug molecules: Comparison of molecular dynamics simulation and measurements. *J. Phys. Chem. B* **2005**, *109*, 15651–15658. [CrossRef]
40. He, C.; Huang, Z.; Han, X.; Liu, L.; Zhang, H.; Chen, L. Coaxial electrospun poly (l-lactic acid) ultrafine fibers for sustained drug delivery. *J. Macromol. Sci. Part B* **2006**, *45*, 515–524. [CrossRef]
41. Natu, M.; de Sousa, H.; Gil, M. Effects of drug solubility, state and loading on controlled release in bicomponent electrospun fibers. *Int. J. Pharm.* **2010**, *397*, 50–58. [CrossRef] [PubMed]
42. Vergaro, V.; Abdullayev, E.; Lvov, Y.M.; Zeitoun, A.; Cingolani, R.; Rinaldi, R.; Leporatti, S. Cytocompatibility and uptake of halloysite clay nanotubes. *Biomacromolecules* **2010**, *11*, 820–826. [CrossRef] [PubMed]

materials

Article

Optimized Rivastigmine Nanoparticles Coated with Eudragit for Intranasal Application to Brain Delivery: Evaluation and Nasal Ciliotoxicity Studies

Mansi Bhanderi [1], Jigar Shah [1,*], Bapi Gorain [2], Anroop B. Nair [3], Shery Jacob [4], Syed Mohammed Basheeruddin Asdaq [5,*], Santosh Fattepur [6], Abdulhakeem S. Alamri [7,8], Walaa F. Alsanie [7,8], Majid Alhomrani [7,8], Sreeharsha Nagaraja [3,9] and Md. Khalid Anwer [10]

[1] Department of Pharmaceutics, Institute of Pharmacy, Nirma University, Ahmedabad 382481, India; 12mph105@nirmauni.ac.in
[2] Department of Pharmaceutical Sciences and Technology, Birla Institute of Technology Mesra, Ranchi 835215, India; bapi.gn@gmail.com
[3] Department of Pharmaceutical Sciences, College of Clinical Pharmacy, King Faisal University, Al-Ahsa 31982, Saudi Arabia; anair@kfu.edu.sa (A.B.N.); sharsha@kfu.edu.sa (S.N.)
[4] Department of Pharmaceutical Sciences, College of Pharmacy, Gulf Medical University, Ajman 4184,United Arab Emirates; sheryjacob6876@gmail.com
[5] Department of Pharmacy Practice, College of Pharmacy, AlMaarefa University, Dariyah, Riyadh 13713, Saudi Arabia
[6] School of Pharmacy, Management and Science University, Seksyen 13, Shah Alam 40100, Malaysia; dr_santosh@msu.edu.my
[7] Department of Clinical Laboratory Sciences, The Faculty of Applied Medical Sciences, Taif University, Taif 26571, Saudi Arabia; a.alamri@tu.edu.sa (A.S.A.); w.alsanie@tu.edu.sa (W.F.A.); m.alhomrani@tu.edu.sa (M.A.)
[8] Centre of Biomedical Sciences Research (CBSR), Deanship of Scientific Research, Taif University, Taif 26571, Saudi Arabia
[9] Department of Pharmaceutics, Vidya Siri College of Pharmacy, Off Sarjapura Road, Bangalore 560035, India
[10] Department of Pharmaceutics, College of Pharmacy, Prince Sattam Bin Abdulaziz University, Al-Alkharj 11942, Saudi Arabia; m.anwer@psau.edu.sa
* Correspondence: jigar.shah@nirmauni.ac.in (J.S.); sasdag@mcst.edu.sa (S.M.B.A.)

Citation: Bhanderi, M.; Shah, J.; Gorain, B.; Nair, A.B.; Jacob, S.; Asdaq, S.M.B.; Fattepur, S.; Alamri, A.S.; Alsanie, W.F.; Alhomrani, M.; et al. Optimized Rivastigmine Nanoparticles Coated with Eudragit for Intranasal Application to Brain Delivery: Evaluation and Nasal Ciliotoxicity Studies. *Materials* **2021**, *14*, 6291. https://doi.org/10.3390/ma14216291

Academic Editors: Roser Sabater i Serra and Ángel Serrano-Aroca

Received: 26 September 2021
Accepted: 20 October 2021
Published: 22 October 2021

Abstract: Rivastigmine, a reversible cholinesterase inhibitor, is frequently indicated in the management of demented conditions associated with Alzheimer disease. The major hurdle of delivering this drug through the oral route is its poor bioavailability, which prompted the development of novel delivery approaches for improved efficacy. Due to numerous beneficial properties associated with nanocarriers in the drug delivery system, rivastigmine nanoparticles were fabricated to be administer through the intranasal route. During the development of the nanoparticles, preliminary optimization of processing and formulation parameters was done by the design of an experimental approach. The drug–polymer ratio, stirrer speed, and crosslinking time were fixed as independent variables, to analyze the effect on the entrapment efficiency (% EE) and in vitro drug release of the drug. The formulation (D8) obtained from 2^3 full factorial designs was further coated using Eudragit EPO to extend the release pattern of the entrapped drug. Furthermore, the 1:1 ratio of core to polymer depicted spherical particle size of ~175 nm, % EE of 64.83%, 97.59% cumulative drug release, and higher flux (40.39 ± 3.52 μg.h/cm^2). Finally, the intranasal ciliotoxicity study on sheep nasal mucosa revealed that the exposure of developed nanoparticles was similar to the negative control group, while destruction of normal architecture was noticed in the positive control test group. Overall, from the in vitro results it could be summarized that the optimization of nanoparticles' formulation of rivastigmine for intranasal application would be retained at the application site for a prolonged duration to release the entrapped drug without producing any local toxicity at the mucosal region.

Keywords: nanoparticles; rivastigmine; intranasal application; optimization; quality by design

1. Introduction

The neurological and neurodevelopmental conditions are the main reason for disability-adjusted life years (DALYs) and the second leading cause of death globally, accounting for 9 million deaths per year. The main contributors of neurological DALYs in 2016 were stroke (42.2%), migraine (16.3%,) dementia (10.4%), meningitis (7.9%), and Epilepsy (5%). Dementia results from a number of diseases and injuries that primarily or secondarily affect the brain. Alzheimer disease (AD) is the most frequent form of dementia and may contribute to 60–70% of cases [1]. The worldwide cases of AD are increasing drastically and are likely to be 74 million by 2030 [2]. The prevalence of this disease is relatively high in Asia and Africa. As the world population is growing older, the socioeconomic consequences of AD are immense and pose a serious challenge for healthcare in modern society. The main reason for the higher costs associated with AD is due to prolonged survival with costly disease-modifying treatment. Presently, for the current 35 million patients with dementia, societal cost is more than $600 billion per year, which is nearly 1% of global Gross Domestic Product [3].

Due to progressive damage or death of neurons during the neurodegenerative disease conditions, such as Parkinson disease, AD, or Huntington disease, researchers are facing the challenge to transport the drugs to the brain at effective concentrations without producing any toxic manifestation [4,5]. Aging is the chief cause of such neurodegeneration and the increasing geriatric population in the near future will further aggravate the present situation [6,7]. Conditions of patients with AD are expressed by the loss of intellectual ability and cognition and impairment of memory due to changes in the morphology of the brain. Prediction of increasing numbers of AD patients in the coming decades has initiated scientists to explore novel deliveries to transport therapeutics directly to the brain [8]. Drug delivery through non-invasive intranasal routes has gained tremendous attention by formulation scientists in the last few decades [9]. Compared to conventional drug delivery systems, nasal drug delivery represents a non-invasive approach with additional advantages such as the rapid onset of action and reduced side effects by a more targeted drug delivery [10]. This route is considered a better alternative for delivering therapeutics to the brain via the olfactory and trigeminal pathways, circumventing the rigid blood–brain barrier that limits the access of actives to the central nervous system while administered using conventional oral or parenteral routes [11].

Rivastigmine is a US FDA-approved natural para-sympathomimetic agent that reversibly inhibits acetylcholinesterase and butyl cholinesterase to treat dementia associated with AD [12]. Such inhibition of cholinesterase increases the brain concentration of acetylcholine to facilitate recovery from memory loss and cognitive deficits due to selective loss of cholinergic neurons in the cerebral cortex, nucleus basalis, and hippocampus [13,14]. However, the oral bioavailability of rivastigmine tartrate is very low, 36%, because of extensive first-pass metabolism and its hydrophilicity [15,16]. Further, this agent is associated with severe gastrointestinal side effects, when administered orally [17]. Thus, there is an urgent requirement to solve the issues with this drug, the low bioavailability, and side effects. Considering the advantages of intranasal administration of therapeutics to directly deliver the drugs to the brain, the present study attempted to deliver this potent agent to the brain using intranasal delivery. The literature signifies that in situ Pluronic F-127 hydrogel constituting Eudragit RL-100 nanoparticles has been previously investigated for the intranasal administration of rivastigmine [18]. Ex vivo studies reported significant permeability enhancement through sheep nasal mucosa in comparison to the drug solution. However, the cumulative percentage of drug released ranged between 60–80% after 24 h and the amount of drug permeated per unit area was rather limited, ranging between 16×10^{-4} to 49×10^{-4} mg/cm^2 min after 8 h. In another study, Polysorbate 80-coated poly(n-butyl cyanoacrylate) nanoparticles significantly transported (3.82 fold) rivastigmine to the brain after intravenous administration, when compared to a free drug solution [19]. However, this approach has certain drawbacks such as systemic toxicity, patient non-compliance, and distribution of drug to non-target tissues.

On the contrary, delivering therapeutics using an intranasal route needs special attention for retention at the site with a prolonged-release profile so that the released drug could be transported to the brain [10,20]. Being a popular delivery system for controlled release characteristics, properly designed nanoparticles are known to sustain the release of encapsulated drugs over a longer period from the polymeric matrix [21]. Different polymers have been used to obtain polymer/active ingredient composite systems with rapid, controlled, or targeted delivery. Among the others, PVP [22], zein [23], or PLLA [24] have been utilized to prepare nanoparticles with controllable morphology, encapsulation, release, and improvement of bioavailability.

Further, wide varieties of available components provide the options to choose the desired ingredients favoring mucoadhesion for intranasal application. For example, chitosan, a cationic polymer from a natural origin and its thiolated derivatives have received tremendous attention because of its non-irritant, non-toxic, biocompatible, and biodegradable characteristics along with significant mucoadhesive properties [8,25,26]. Chitosan is reported to disrupt the intercellular tight junctions, hence enhancing the permeability of an epithelium [27]. Similarly, Eudragit® EPO is a cationic terpolymer from the poly(methacrylate) family specifically used as a coating polymer to overcome humidity-related instability of dosage forms and also has good mucoadhesive properties [28,29]. The combination of the repeating methacrylate units within this polymer promotes its solubility under acidic conditions, which is applicable in the design of dosage forms targeted for nasal mucosal region with a pH range typically between 5.5–6.5. Excellent mucoadhesive properties of Eudragit® EPO have been demonstrated using freshly excised sheep nasal mucosa [29]. Due to the inherent clearance mechanism that exists in the nasal cavity, exploitation of mucoadhesive agents along with mucoadhesive coating may extend the time of contact between the drug and the mucus layer, disrupt tight junctions, enhance permeation, and prolong the duration of action, respectively. Therefore, the current investigation aimed to develop mucoadhesive rivastigmine loaded in chitosan and coated with eudragit for intranasal delivery, which could circumvent the first-pass metabolism of drugs and help to achieve a sustained drug release over an extended period. To our knowledge, this is the first study of its type where the rivastigmine was loaded with chitosan and coated with eudragit EPO for the purpose of intranasal delivery. Various formulation and processing parameters were optimized to obtain chitosan nanoparticles having the desired quality. Later, the optimized formulation was coated using Eudragit, to prolong the drug release, and evaluated for in vitro release and permeation study. Finally, the ciliotoxicity study of the coated nanoparticles was done to evaluate the effect on sheep nasal mucosa.

2. Materials and Methods

2.1. Materials

Rivastigmine tartrate (purity > 98%) was procured from Cadila Healthcare, Mumbai, India. Low-molecular-weight chitosan (50,000 Da; 75–85% deacetylated) was purchased from Sigma Aldrich, Bangalore, India. Eudragit EPO was received from Evonik India, Mumbai, India. Span 80 and glutaraldehyde were obtained from SD Fine Chemicals, Mumbai. Glacial acetic acid was procured from High Purity Lab Chemicals, Mumbai, India. Light and heavy liquid paraffin were obtained from Central Drug House, New Delhi, India.

2.2. Preliminary Optimization of Process Parameters for Blank Nanoparticle Preparation

To achieve various desirable properties of the nanoparticles, several preliminary trials were done. First, blank nanoparticles were prepared to select the external phase and the stirrer position. Selection of the external phase was done by experimenting with three different phases, light liquid paraffin, heavy liquid paraffin, and the combination of light and heavy liquid paraffin at a ratio of 1:1. Similarly, to optimize the location of the stirrer, it was set at the top, middle, and bottom positions to check its effect on the formation of nanoparticles.

2.3. Optimization of Parameters to Fabricate Drug-Loaded Nanoparticles

To optimize the amount of chitosan, three different batches of nanoparticles (P1, P2, and P3) were prepared using 2%, 3%, and 4% (weight percent) chitosan, respectively. The other components were fixed to fabricate the formulations and to characterize accordingly. Once the percentage of chitosan was optimized during the preliminary process, the amount of drug (batch P4 (150 mg), P5 (100 mg), and P6 (75 mg)) was modified to check the influence of the drug on the formulation characteristics. In a similar process, the volume of span 80 (batch P7 (1%), P8 (2%), and P9 (3%)) in the formulation, volume percentage of crosslinking agent (i.e., glutaraldehyde) (batch P10 (2 mL), P11 (3 mL), and P12 (4 mL)), stirrer speed (batch P13 (1000 rpm), P14 (1500 rpm), and P15 (2000 rpm)), and crosslinking time (batch P16 (2 h), P17 (3 h), and P18 (4 h)) were varied one after another and evaluated to obtain the product having the desired quality.

2.4. Full Factorial Design of the Rivastigmine-Loaded Nanoparticles

The above preliminary trials provide us the data to consider the various process and formulation variables for the preparation of the optimization batch. Accordingly, the drug-to-polymer ratio, stirrer speed, and crosslinking time were found to significantly affect the % drug release after 8 h and entrapment efficiency (% EE) of the fabricated nanoparticles. Thus, for the optimization process using the design of experiments' statistical technique, these three variables were selected as independent variables, and % drug release after 8 h and % EE were selected as dependent variables. A 2^3 full factorial design (Design Expert® software, version 12, Stat-Ease Inc. Minneapolis, MN, USA) was applied [30] to check the effects of independent variables on dependent variables at two levels demonstrating low and high, respectively (Table 1). The design formulation batches (D1–D8) representing independent variables with coded values and their levels are depicted in Table 1.

Table 1. Independent variables with levels and coded values.

Independent Variables			
	Levels		
	+1		−1
A (Drug: Polymer)	1:5		1:2
B (Stirrer Speed (rpm))	1500		1000
C (Crosslinking Time (h))	4		2
Design Matrix with the Independent Variables and Their Coded Values			
Batches	Values of Independent Variables		
	A (Drug: Polymer)	B (Stirrer Speed (rpm))	C (Crosslinking Time (h))
D1	−1	−1	−1
D2	+1	−1	−1
D3	−1	−1	+1
D4	+1	−1	+1
D5	−1	+1	−1
D6	+1	+1	−1
D7	−1	+1	+1
D8	+1	+1	+1

2.5. Preparation of Nanoparticles

The emulsion crosslinking method was selected for the preparation of mucoadhesive chitosan nanoparticles containing rivastigmine tartrate by a method described previously with minor modifications [31]. Accurately weighed amounts of drug and chitosan were added to 10 mL of prepared acetic acid solution (2% *w/v*) placed in a glass beaker. The dispersion was initially stirred with a glass rod, followed by sonication in a bath sonicator (Trans-o-sonic, D-compact, Mumbai, India) to facilitate the dissolution. The Sonication process continued until chitosan dissolved completely and formed a transparent gel. Meanwhile, the required quantity of light liquid paraffin and Span 80 were taken in another beaker. The mixture was stirred for 10 min using a stirrer (Remi Instruments, Ahmedabad, India) at room temperature. Thereafter, the drug–polymer solution was added dropwise to the external oil phase with continuous stirring at constant rpm for 15 min.

Thereafter, glutaraldehyde solution (strength) was added dropwise to the emulsion to facilitate crosslinking and stirring continued for 3–5 h. The suspension was left to stand for 20 min to allow nanoparticles to sediment under the force of gravity. The supernatant was discarded, and the remaining portion constituting nanoparticles with small amounts of oil was separated employing a vacuum filter (NOVA instruments, Mumbai, India). The nanoparticles were then washed 4–5 times using petroleum ether to remove the traces of oil from the surface of nanoparticles and freeze dried at 60 °C for 24 h. Finally, it was air-dried for 24 h at room temperature (25 ± 1 °C) and stored in a cool and dry place.

2.6. *In Vitro Release of Rivastigmine from the Designed Batches*

The dialysis tube method was used to carry out the in vitro drug release study of the fabricated nanoparticles. Accurately weighed nanoparticles equivalent to 6 mg of drug dispersed in 2 mL of simulated nasal fluid (pH 6.4) were taken in a cellophane dialysis bag (molecular cutoff 12–14 kDa) and placed in the receptor compartment (20 mL of simulated nasal fluid). The temperature of the receiver fluid was maintained at 37 ± 0.5 °C and 100 rpm was set throughout the study [32]. A sample (3 mL) was withdrawn from the receptor compartment at various time intervals and a fresh solution of the same volume was replaced to maintain sink condition. The collected samples were analyzed at 220 nm, and Lambert–Beer's equation was used for the calculation of % of drug released at different time points. Kinetics and possible mechanisms of drug release from formulations were evaluated by fitting the data into various mathematical models, as described elsewhere [33].

2.7. *% Entrapment Efficiency and Drug Loading*

To determine the % EE of the fabricated formulation, accurately weighed (70 mg) nanoparticles were pulverized in a glass mortar. Then, 50 mg of the crushed nanoparticles were weighed and placed into a flask containing 50 mL of methanol. The flask was shaken using a magnetic stirrer (400 rpm) for 24 h and kept aside for 4 h to solubilize the entrapped drug in the nanoparticles. Filtration of the solution was done using Whatman filter paper (Grade 602 h) to separate the polymers from the supernatant. Finally, the concentration of the drug in the methanol was measured using a spectrophotometer (UV 1800, Shimadzu, Japan) [34] at 220 nm to find out the amount of drug entrapped in the nanoparticles [35].

The % EE was estimated using the following formula:

$$\%\ \text{EE} = \frac{\text{Amount of drug in nanoparticles}}{\text{Amount of drug used}} \times 100 \quad (1)$$

The percentage of drug loading was estimated using the same procedure followed for EE. Then % drug loading was calculated by using the formula mentioned below.

$$\%\ \text{Drug loading} = \frac{\text{Amount of drug in nanoparticles}}{\text{Amount of nanoparticles formed}} \times 100 \quad (2)$$

2.8. *Coating of the Optimized Nanoparticles*

Eudragit EPO is soluble in acetone; thus, the emulsion solvent evaporation technique was employed for the coating of prepared nanoparticles [36]. Different core-(nanoparticles)-to-coat ratios were selected to check their effect on drug release. Light liquid paraffin was selected as the external phase and Tween 80 was used as an emulsifier for the coating. Optimized batch D8 was selected as the core and the coating was done by the method described in the literature with minor modifications [37]. Briefly, the required quantity of polymer was dissolved in acetone and previously prepared chitosan nanoparticles were added to the polymeric solution with constant stirring. The mixture was then added dropwise to light liquid paraffin and Tween 80 solution with magnetic stirring at constant speed for 2 h to evaporate acetone completely. The hardened nanoparticles were recovered by centrifugation and washed three times in petroleum ether to remove the excess of oil. Nanoparticles were then lyophilized and stored in a cool and dry place. Different ratios of nanoparticles and the coating polymer-eudragit (1:1 (batch C1), 1:3 (batch C2), and 1:5 (batch C3)) were tested to fabricate the coated nanoparticles.

2.9. FTIR Analysis

Compatibility study was carried out using a Jasco FT-IR spectrophotometer (FP-6500, Tokyo, Japan). The IR spectrum of the pure drug (2 mg), chitosan (2 mg), and nanoparticles' formulation D8 (equivalent to 2 mg of drug) was studied by preparing potassium bromide (KBr) pellets with 98 mg KBr [38]. The prepared pellets were then scanned 50 times over the range of 4000–400 cm^{-1} wavenumber. The characteristic absorption peaks of rivastigmine tartrate and chitosan at different wavenumbers were compared with the peaks obtained in the nanoparticles' formulation.

2.10. In Vitro Release of Rivastigmine from the Coated Nanoparticles

The in vitro release of rivastigmine from the eudragit-coated nanoparticles (batches C1, C2, and C3) and the optimized uncoated nanoparticle (batch D8) were studied similarly to the procedure mentioned in Section 2.6.

2.11. In Vitro Drug Permeability Study

In vitro rivastigmine transport from the nanoparticles was studied using the freshly excised sheep nasal mucosa obtained from a local slaughterhouse (Ahmedabad, India). Nasal mucosa was sandwiched between the donor and receptor cells with the mucosal side directed towards the donor cell. The active diffusion area for drug permeation was 1.13 cm^2. The receiver fluid was filled with simulated nasal fluid (20 mL), stirred with a magnetic bar at 50 rpm, and the temperature was maintained at 37 ± 0.5 °C [39]. Different ratios (1:1, 1:3, 1:5) of coated nanoparticles equivalent to 12 mg of the drug were dispersed in 1 mL of simulated nasal fluid and placed in the donor compartment. For comparison purposes, a drug solution with the same amount of drug was used as a control. Aliquots were withdrawn at various intervals for 24 h and were replaced with an equal volume of fresh medium. Samples were filtered through a 0.2-µm Millex syringe-driven membrane unit, diluted appropriately, and assayed at λmax 220 nm. The flux and enhancement were calculated according to the formula described elsewhere [40,41].

2.12. Measurement of Particle Size, Particle Size Distribution, and Zeta Potential

For the measurement of the particle size, size distribution, polydispersity index, and zeta potential of the nanoparticles, a Zeta sizer (Nano ZS90, Malvern Instruments, Malvern, UK) was utilized. Nanoparticles were dispersed in water and the measurement was done at room temperature [42].

2.13. In Vitro Mucoadhesion Testing

This analysis was performed to evaluate the mucoadhesive property of the fabricated nanoparticles according to the method described in literature [43]. To perform this test, a freshly cut 2-cm-long piece of sheep nasal mucosa was collected and tied to a glass slide with the help of clips. Accurately weighed 50-mg nanoparticles were placed on the apical surface of the mucosa. The setup was kept in the humidity control chamber ($75 \pm 5\%$ humidity) for 10 min to allow the anionic mucosa to interact with cationic Eudragit-coated nanoparticles. Then, this slide was hung at an angle (45°) under the burette tip, which allowed the prepared phosphate buffer (pH 7.0) to flow at a rate of 2 mL/min. The number of dry nanoparticles disadhered from the setup was collected on the Whatman filter paper. The collected nanoparticles were weighed and used to calculate the % mucoadhesion using the following formula [43].

$$\% \text{ Mucoadhesion} = (\text{Amount of nanoparticles adhere to the mucosa})/(\text{Total amount of nanoparticles applied to the mucosa}) \times 100 \quad (3)$$

2.14. Scanning Electron Microscopy (SEM)

The surface morphology of optimized nanoparticles was observed under the scanning electron microscope. For SEM analysis, nanoparticles (coated and uncoated) were suspended in petroleum ether and a small amount of this suspension was put on an aluminum stud. Following dehydration and fixation, the nanoparticles were coated with gold-palladium by using the SEM coating system POLORON E5100 (sputter coater) in a neutral environment and the morphology of the formulated nanoparticles was measured using a Nova Nano SEM-450 (FEI, Hillsboro, OR, USA). The images were recorded at HT-15 kV with a high-voltage electron beam, and a probe current of 3×10^{-10} A was passed through it to scan the nanoparticles [44].

2.15. Differential Scanning Calorimetry (DSC)

The thermal nature of rivastigmine tartrate, the physical mixture, and rivastigmine-loaded nanoparticles-optimized formulation was studied using the DSC 4000 system (Perkin Elmer, Waltham, MA, USA). A sample (5 mg) of rivastigmine nanoparticles was precisely weighed and hermetically sealed in aluminum pans. Thermograms were captured by heating samples at a constant rate of 10 °C/min from 30–300 °C. An unfilled sealed pan was employed as a reference [45].

2.16. Nasal Ciliotoxicity Studies

Nasal ciliotoxicity studies of nanoparticles of rivastigmine were conducted using sheep nasal mucosa by following the ethical guidelines (VSCP/EC/11508/2020/4, dated 24 August 2020). For this study, three sheep nasal mucosal scrapings, namely, A, B, and C, with uniform thickness were collected and mounted on Franz diffusion cells [46]. 'A' was treated with 0.5 mL of phosphate buffer as the negative control, 'B' with 0.5 mL of selected nanoparticles (batch C1) of rivastigmine for 2 h as a test, and 'C' with 0.5 mL of isopropyl alcohol for 2 h as a positive control. After 2 h, the nasal mucosa was rinsed with saline fluid and subjected to histological studies using hematoxylin-eosin staining [47]. The stained slides were examined using a light microscope (under 400× magnification) and the image was captured using a camera attached with the microscope (ZEISS, Axioscope 5, Jena, Germany).

2.17. Statistical Analysis

Data were examined using one-way ANOVA, followed by Turkey's multiple comparison post-test. The statistical differences between values exhibiting $p < 0.05$ were considered significant.

3. Results and Discussion

3.1. Preliminary Optimization of Process Parameters for Blank Nanoparticle Preparation

When three different phases were taken for the selection of the external phase for the preparation of nanoparticles, a stable emulsion was formed with heavy liquid paraffin, whereas the flakes were formed with the combination of light and heavy paraffin. On the other hand, light liquid paraffin was found to be suitable for the fabrication of nanoparticles. Therefore, it was selected as an external phase for further investigations. When the stirrer was set at top positions, it showed the formation of small and irregular-sized nanoparticles. Alternatively, setting the stirrer at the bottom position resulted in the development of big and spherical nanoparticles. It was shown that when the stirrer was set at the middle position, it produced better and desirable results when compared to the other two positions. Therefore, the middle position of the stirrer was selected for further studies.

3.2. Optimization of Preliminary Batches of the Drug-Loaded Nanoparticles

Preliminary studies (with batches P1–P18) were carried out to select the percentage of chitosan, the amount of drug, the volume of span 80, the crosslinking agent, stirrer speed, and crosslinking time by varying one after another and evaluated for % EE, drug release, and mucoadhesion. It was found that the concentration of chitosan (in batches P1, P2, and P3) had a significant effect on % EE and a minor effect on drug release and mucoadhesion. As the concentration of chitosan was increased, % EE was also proportionately increased. It could be attributable to the fact that when the concentration of chitosan was increased in the formulation, viscosity of the solution correspondingly increased, and, finally, the diffusion of the drug became difficult, which prevented drug release from the particles [48]. A similar observation was noted in the drug release profile as well. Initially, faster drug release was observed at a lower concentration. Alternatively, at higher concentrations of chitosan, initially slower drug release was observed followed by faster release. Increasing the concentration of chitosan in the formulation revealed a comparable effect as that noticed in the drug release as well as drug permeability. At lower concentrations, drug permeability was higher, as drug diffusion became easier from the less dense polymeric structure. There was no significant effect found in mucoadhesion but it was shown that mucoadhesion was slightly increased with an increase in the concentration of chitosan. It was observed that % EE, drug release, and mucoadhesion was good in batch P2 (with 3% chitosan) and P3 (with 4% chitosan) but batch P2 had good syringe ability when compared to batch P3. Thus, batch P2 with 3% chitosan was selected for further studies.

On the other hand, it was observed that % EE was significantly affected by the drug-to-polymer ratio. The % EE was inversely proportional to the drug to polymer ratio in the formulation, where 150 mg (batch P4, drug: polymer, 1:2) showed 45.34% EE but 75 mg (batch P6, drug: polymer, 1:4) showed 71.45% EE. In addition, the drug-to-polymer ratio was found to have a significant effect on drug release as well as drug permeability. It was observed that drug release and permeability retarded markedly at a higher drug-to-polymer ratio (1:4). This could be explained by the fact that the slower the rate of swelling of the polymer, the more controlled the drug release from the matrix [49,50]. Based on the results demonstrating higher % EE, controlled drug release, and sustained permeability, batch P6 with 75 mg drug and 1:4 drug-to-polymer ratio was selected for further studies.

Due to the non-ionic and lipophilic nature of Span 80, it tended to form a stable coating over the spherically dispersed droplets of the emulsion. Based on the results drawn from different batches (P7–P9) of formulations, the aggregation of the nanoparticles was observed at a higher concentration of Span 80. Alternatively, including a higher percentage of Span 80 did not significantly affect % EE, drug release, and drug permeability. Therefore, the batch P8 with 2% of Span 80 was selected for further investigations to avoid aggregation between nanoparticles.

The amount of the cross-linking agent in batches P10–P12 revealed significant alteration in % EE. The higher the amount of crosslinking agent (glutaraldehyde; 4 mL), the higher % EE, which might be because of preventing the leaching of drug from the polymeric structure during the washing step of the nanoparticles [51]. It was proven since the drug release from the polymeric matrix was also retarded from nanoparticles with a higher amount of crosslinking agent. Increasing the concentration of the crosslinking agent might have increased the density of the polymer matrix, thereby retarding the release of the drug from these polymeric particles [52]. A similar effect was observed for drug permeability, where the higher amount of crosslinking agent reduced the permeability of the drug. On the other hand, mucoadhesion was found to be decreased with an increased concentration of the crosslinking agent in the formulation. This might be correlated to the decrease in the availability of free cationic groups of chitosan at higher crosslinking [53]. With increasing the amount of crosslinking agent, particle size was decreased but shrinkage of nanoparticles occurred due to higher crosslinking of particles. Therefore, 3% of glutaraldehyde (batch P11) was considered as the optimum concentration required for further studies as it was providing minimum particle size (150–200 nm), maximum % EE (64.32%), reasonable drug release (65.72% within 8 h), and strong mucoadhesion (85%).

Stirrer speed is the most critical factor in the preparation of nanoparticles. Although, in the present research, the stirring speed did not have any significant effect on % EE, the drug release was significantly affected. As the speed was increased (1000 to 2000 rpm), the particle size of the nanoparticles was found to be decreased from 350 nm (batch P13) to 200 nm (batch P15). On the contrary, the surface area of particles increased; thus, a higher drug release (88.56%) was recorded at the higher stirrer speed (2000 rpm). A similar observation was also made for an in vitro drug permeability test. Alternatively, the particle size was found to enlarge at a lower stirrer speed, which would not be suitable for nasal delivery, as reported earlier [54]. Therefore, 1500 rpm was considered ideal.

On the other hand, the crosslinking time during the fabrication of the nanoparticles had a significant effect on % EE. Improper crosslinking of the polymer matrix resulted in incomplete encapsulation of drugs in the polymer matrix. Therefore, the loaded drug was leaked during the washing period of the fabricated nanoparticles. Additionally, the release of drugs from the nanoparticles (batches P16–P18) was also influenced by the duration of crosslinking. The drug release was largely retarded with increased time of crosslinking since a higher degree of crosslinking causes particles to became denser and thereby hinder their release from the polymer matrix [55]. Further, mucoadhesion of the nanoparticles was reduced due to a higher degree of crosslinking with chitosan contributed by a prolonged period of crosslinking [56]. The particle size was also affected by the extent of crosslinking. As the time increased, particle size decreased significantly. Based on the initial optimization of the different process parameters and compositions, the final composition and parameters of the optimized formulation are depicted in Table 2.

Table 2. Formulation composition and characterization parameters of batch P17 to fabricate nanoparticles containing rivastigmine.

Components	Optimized Values
Rivastigmine	75 mg
Chitosan	3%
Span 80	2%
Glutaraldehyde	3 mL
Stirrer speed	1500 rpm
Crosslinking time	3 h
Characterization	
Parameter	Outcome
Average particle size	150 nm
% EE	67.92
% Drug loading	11.98
Aggregation	No
% Mucoadhesion	89
% Cumulative drug release (within 8 h)	82.32

3.3. *Optimization of the Formulation Parameters Using 2^3 Full Factorial Design*

3.3.1. Drug Release from Rivastigmine Nanoparticles

Rivastigmine nanoparticles' formulation was optimized using three factor, two levels statistical design. Responses of dependent variables for designed batches are shown in Table 3. The statistical data showed that the interaction of three independent variables, drug–polymer ratio (A), crosslinking time (B), and stirrer speed (C) on % drug release after 8 h from rivastigmine nanoparticles and % EE, where the model (2^3 full factorial design) was found to be significant and the significant influence of model terms A, B, and C on % rivastigmine release from the nanoparticles formulation was confirmed by the *p*-values ($p < 0.05$).

A polynomial equation was generated based on the interaction of three model terms on the % release of rivastigmine from the developed formulations (Equation (4)). The positive coefficient value (+2.08) of C in Equation (4) indicates that the increasing % of drug release could be achieved with increasing the level of stirring speed, whereas the negative coefficient of model terms A (−2.75) and B (−3.39) represented decreasing % of drug release with the increasing level of drug–polymer ratio and crosslinking time. This could be correlated to the findings of the preliminary results. The increased crosslinking time and drug–polymer ratio increased the barrier to release the entrapped drug from the polymeric nanoparticles [57].

Table 3. Responses of dependent variables for designed batches.

Batches	Actual Responses	
	Cumulative Drug Release (%) after 8 h	% EE
D1	78.77	43.46
D2	73.33	57.33
D3	72.55	49.33
D4	67.24	75.63
D5	83.77	40.22
D6	77.90	55.33
D7	76.14	45.06
D8	70.75	73.33

$$\% \text{ drug release} = 75.06 - 2.75\text{A} - 3.39\text{B} + 2.08\text{C} + 0.0762\text{AB} - 0.0637\text{AC} - 0.3087\text{BC} \quad (4)$$

Further, the similar effect of model terms A, B, and C is reflected in the 3D surface plot (Figure 1a–c), where it is presented that the increasing levels of model terms A led to decreasing % drug release, whereas an increasing level of stirring speed led to increasing % drug release from the fabricated rivastigmine nanoparticles' formulation.

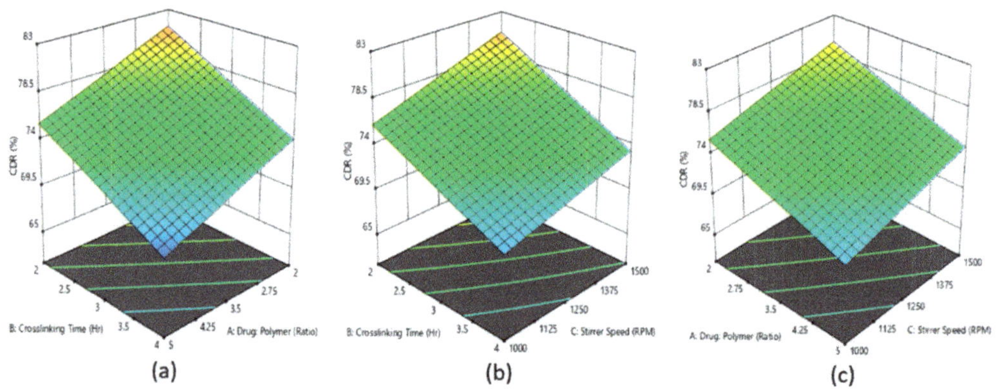

Figure 1. Effect of the interaction of (**a**) drug–polymer ratio, (**b**) crosslinking time, and (**c**) stirring speed is represented in the 3D surface plot on percentage of drug released from rivastigmine nanoparticle formulation.

3.3.2. Entrapment Efficiency of Rivastigmine Nanoparticle

The statistical data indicated that the interaction of three independent variables, drug–polymer ratio (A), crosslinking time (B), and stirrer speed (C) on % EE of the rivastigmine nanoparticles, where the model 2^3 full factorial design was found to be significant and the significant influence of model terms A, B, and C on % EE of rivastigmine in the nanoparticle formulation was confirmed by the *p*-values ($p < 0.05$).

A polynomial equation was produced based on the interaction of three model terms on % EE of the developed formulations (Equation (5)), where the positive coefficient values for all the three model terms represents the increment of % EE with increasing level of model terms.

$$\% \text{ EE} = 54.96 + 10.44\text{A} + 5.88\text{B} - 1.48\text{C} + 3.20\text{AB} + 0.4043\text{AC} - 0.1662\text{BC} \quad (5)$$

Further, a similar effect of model terms A, B, and C is reflected in the 3D surface plot (Figure 2a–c), where it is presented that the increasing levels of model terms A and B led to increasing % EE of the rivastigmine in nanoparticles.

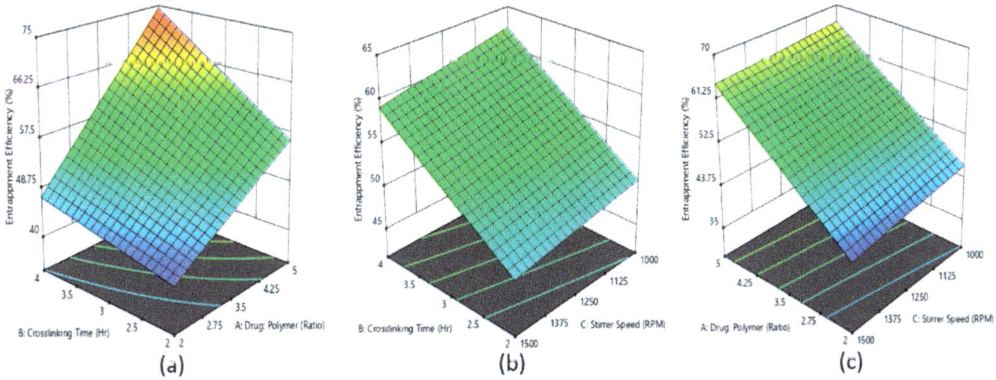

Figure 2. Effect of the interaction of (**a**) drug–polymer ratio, (**b**) crosslinking time, and (**c**) stirrer speed is represented in the 3D surface plot on percentage entrapment efficiency of rivastigmine nanoparticle formulation.

The 3D surface plot of % EE in Figure 2 signifies that the drug-to-polymer ratio had a more significant effect on % EE when compared with crosslinking time. As the drug-to-polymer ratio increased from 1:2 to 1:5, % EE increased from 40 to 57, while crosslinking time increased from 2 to 4 h and % EE increased from 40 to 48. These results are in good agreement with the % EE existing literature [51]. However, it was observed that the stirrer speed had a negative effect on % EE. Indeed, the increase in stirrer speed led to a decrease in % EE. This is probably because when the stirrer speed was increased, the particles could have broken and help the drug escape into the external phase, which resulted in lower entrapment efficiency.

3.3.3. Checkpoint Batches

To identify the design space, the overlay curve was drawn considering percent entrapment efficiency at least above 60–65% and drug release at 8 h below 75%, as shown in Figure 3. Based on the suggestions from the optimization software to validate the optimization process, two checkpoint batches (D9 and D10) were selected from the overlay plots (Figure 3a–c) and the respective analysis was done for those batches. From the result displayed in Table 4, it was observed that practical values of the checkpoint batches were very close to the values of batches that were obtained from the overlay plot. Therefore, it could be concluded that the model used for the interpretation was validated.

Table 4. Predicted and actual cumulative drug release (%) and entrapment efficiency (%) values of the checkpoint batches.

Batches	Drug: Polymer	Crosslinking Time (h)	Stirrer Speed (rpm)	Predicted Value		Actual Value	
				Cumulative Drug Release (%)	% EE	Cumulative Drug Release (%)	% EE
D9	1:5	3	1500	74.19	64.68	71.45	63.34
D10	1:4	3	1000	71.81	60.64	68.76	59.31

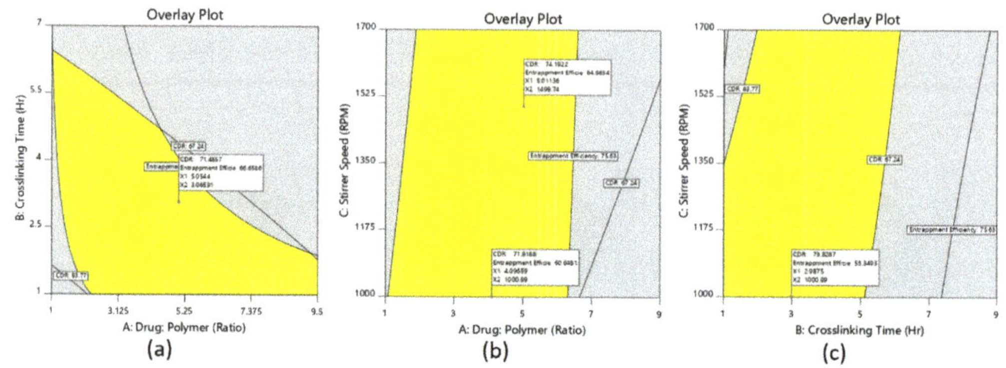

Figure 3. The overlay plots of the nanoparticle formulation, (**a**) overlay plot of the drug-to-polymer ratio and stirrer speed, (**b**) overlay plot of the drug-to-polymer ratio and crosslinking time, (**c**) overlay plot of stirrer speed and crosslinking time.

From the optimization of different data and validation of the process, the desirability was set to obtain the optimized formulation. Based on desirable characteristics for % EE, particle size, and cumulative drug release (%), the parameters of the optimized nanoparticle formulation are depicted in Table 5, which is also similar to the design batch D8.

Table 5. Optimized formulation parameters for the development of rivastigmine nanoparticles (D8) and characterization parameters.

Components	Optimized Parameters
Drug (mg)	60
Chitosan (%)	3
Span 80 (%)	2
Glutaraldehyde (mL)	3
Stirrer speed (rpm)	1500
Crosslinking time (h)	4
Characterization	
Parameter	Outcome
Average particle size	145 nm
% EE	73.33
% Drug loading	11.76
Aggregation	No
% Mucoadhesion	89
% Cumulative drug release (within 8 h)	70.75

3.4. Coating of the Optimized Nanoparticles

The drug-loaded optimized chitosan nanoparticle (D8) was then coated with Eudragit EPO to extend the release of nanoparticles to provide 1-day delivery and provide mucoadhesion. Eudragit EPO was chosen as the coating polymer for the chitosan nanoparticles because this polymer is widely used as a coating material in the pharmaceutical field and enhances the mechanical strength and restricts the dissolution rate of chitosan. In addition, it exhibits mucoadhesive properties and has also been used for nasal drug delivery [58,59]. The size of all the coated nanoparticles (batches C1–C3) was found to be <220 nm, with the lowest average size being recorded by batch C1 (~175 nm) followed by batch C2 (~190 nm) and batch C3 (~210 nm). Further, there was no major alteration in % EE from the coated nanoparticles in the formulation, which varied between 64.83% to 69.82% for the core-to-coat ratio of 1:1 to 1:5, respectively.

3.5. FTIR Analysis

Figure 4 shows the FTIR spectra of pure rivastigmine, chitosan, and drug-loaded nanoparticles (batch D8). The predominant peaks represent the main functional groups of pure rivastigmine that showed characteristic spectral peak positions at 3318 cm^{-1} representing N–H stretching and 1715 cm^{-1} representing C=O stretching (carbamate band); 1401 cm^{-1} refers to C–N stretching in tertiary amines and 954 cm^{-1} corresponds to the =C–H bending. Additionally, there are also other vibrational bands depicted due to the presence of stretching vibrational bands of C=C of the structural aromatic ring, O–H band of the tartrate, and formed N–H between the tartrate and the drug. The observed FTIR spectra of the drug are quite similar to the reported [60]. The spectra of optimized nanoparticles' formulation also showed all essential peaks of the pure drug with a minor reduction in peak intensity. Additionally, no evidence of the shift in characteristic peaks of the drug was observed. Therefore, it can be confirmed that there were no compatibility issues between the drug and other excipients used in the optimized formulation.

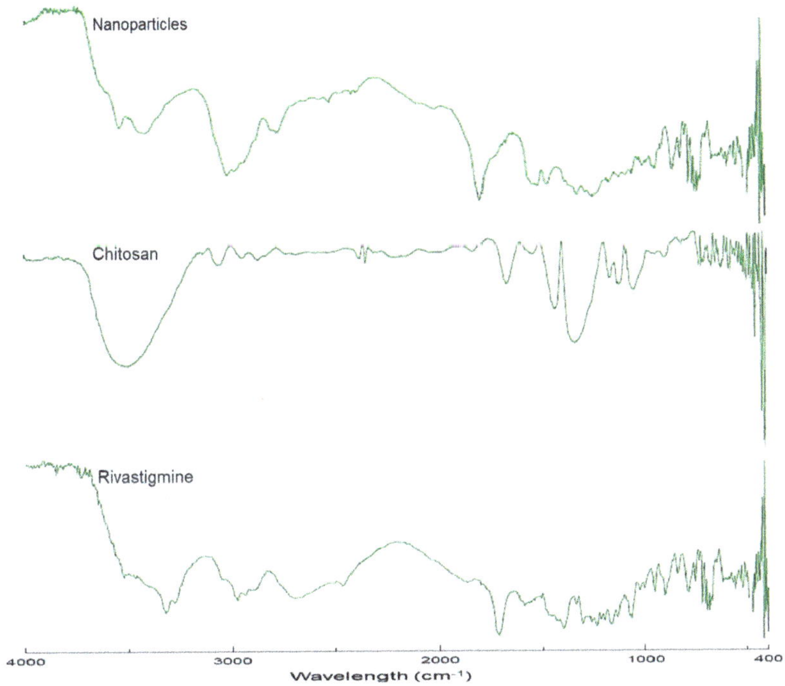

Figure 4. FTIR spectra of rivastigmine tartrate, chitosan, and rivastigmine-loaded nanoparticles (batch D8).

3.6. In Vitro Release of Rivastigmine from the Coated Nanoparticles

The cumulative release profiles of rivastigmine from the formulated batches of eudragit-coated nanoparticles (batches C1, C2, and C3) and the uncoated batch D8 are presented in Figure 5. Chitosan and its derivatives have been extensively probed as mucoadhesive agents to increase the nasal absorption of hydrophilic drugs and macromolecules due to its ability to decrease mucociliary clearance, enhancing membrane permeability [54], besides interfering with the formation of tight junctions in naso-respiratory epithelial cells [61]. However, results indicated that the cumulative percentage of drug released from rivastigmine loaded in uncoated chitosan nanoparticles using the dialysis sac method was only ~80% in 24 h [16].

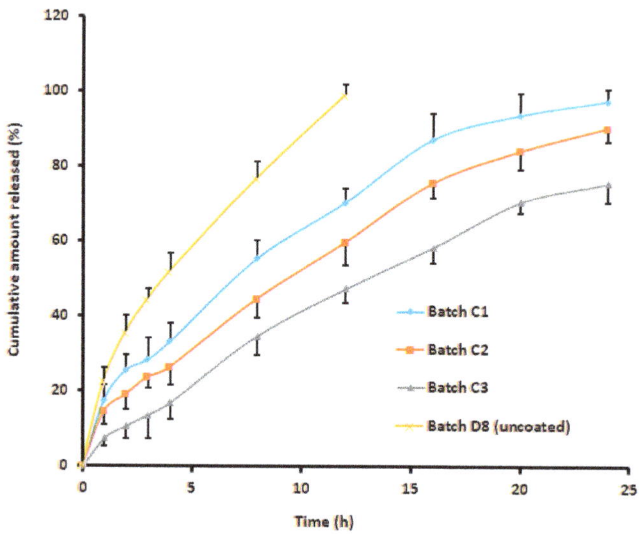

Figure 5. Comparison of cumulative percentage of rivastigmine release from batches with different core-to-coat ratios and control (uncoated). Batch C1 (nanoparticle: eudragit, 1:1), batch C2 (nanoparticle: eudragit, 1:3), and batch C3 (nanoparticle: eudragit, 1:5). The data represent an average ± SD of six trials.

Coating with eudragit slowed the release rate of the entrapped drug, which might have been due to an extra barrier created over the chitosan nanoparticles. The cumulative release of rivastigmine from the coated formulations revealed 97.59%, 90.55%, and 75.74% release from the coated formulations with the core-to-coat ratio of 1:1, 1:3, and 1:5, respectively, within the time frame of 24 h (Figure 5). Indeed, this observation could be related to the well-known fact that surface-specific dissolution rate and equilibrium solubility increase with a decrease in particle size. Release kinetics were assessed for batch C1. The goodness of fit models was selected by evaluating r^2 value, the sum of squares of residuals (SSR), and Fischer Ratio (F) to avoid error in the prediction of the release mechanism. The data indicate a higher r^2 value (0.990), least SSR value (87.92), and F value (10.99) with the Korsmeyer–Peppas model. The *n* value noticed (0.572) signifies the diffusion mechanism was anomalous transport [62]. Hence, it was concluded that the release of rivastigmine from batch C1 followed the Korsmeyer–Peppas matrix diffusion-controlled mechanism.

3.7. In Vitro Permeability of Rivastigmine from the Coated Nanoparticles

In vitro release profile does not always correlate with in vivo plasma level-time profile; hence, the suitability of a nasal drug delivery system is generally demonstrated using a permeation study, preferably using sheep nasal mucosa. The permeation of molecules across the biological membrane is generally influenced by the physicochemical properties of the permeant as well as the physiological features of the membrane [63]. The study was conducted for both the coated formulations (1:1, 1: 3, and 1:5) and drug solutions containing equivalent amounts (12 mg) of rivastigmine. Figure 6 exhibits a greater extent of rivastigmine permeation from coated nanoparticles at all ratios through the sheep mucosa in comparison to the drug solution. It was reported earlier that the cumulative percentage of rivastigmine permeated through porcine nasal mucosa from chitosan nanoparticles without any coating was only 70.1% in 24 h [16]. The statistical analysis suggested that the difference in the cumulative amount of drug diffused at 24 h between coated nanoparticles and drug solution was significant ($p < 0.0001$). The steady-state flux (40.39 ± 3.52 μg h/cm^2) noticed with 1:1 core–coat ratio of nanoparticles was more than 1:3 (36.23 ± 3.97 μg h/cm^2) and 1:5 (33.19 ± 3.64 μg h/cm^2) core–coat ratio of nanoparticles. The higher permeation of rivastigmine from the coated nanoparticles demonstrated by flux values could be directly corroborated with the release profile displayed in Figure 5. In addition, the increase in permeation of nanoparticles at 1:1 ratio could be attributed to a minimum particle size that enabled it to penetrate the nasal mucosal layer more efficiently than the larger particle size associated with 1:3 (190 nm) and 1:5 (~200 nm) coated nanoparticles. On the other hand, the lower diffusion of the pure drug solution may have been due to the polar character of rivastigmine, since, for effective transport across nasal mucosa, the drug should have preferably a lipophilic property [64]. From the results, it can be concluded that eudragit-coated chitosan nanoparticles act as an effective drug transporting and targeting carrier because of nano size and unique physicochemical properties.

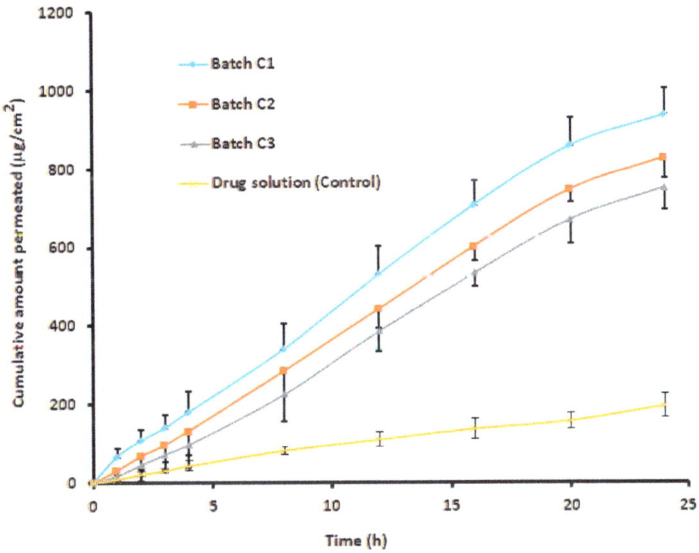

Figure 6. Comparison of rivastigmine permeation across isolated sheep mucosa from batches with different core-to-coat ratios. Batch C1 (nanoparticle: eudragit, 1:1), batch C2 (nanoparticle: eudragit, 1:3), and batch C3 (nanoparticle: eudragit, 1:5). The data represent an average ± SD of six trials.

The in vitro release profile and drug permeability analysis reported from coated formulations with a core-to-coat ratio of 1:1 (batch C1) showed the best results among the three formulations tested. Considering a gradual release and higher drug permeation, these coated nanoparticles (batch C1) with a core-to-coat ratio of 1:1 were selected as an optimized formulation for further characterization.

3.8. Particle Size, Particle Size Distribution, and Zeta Potential

The particle size and zeta potential distribution of coated and uncoated batches are shown in Table 6 and Figure 7. The designed batch D8 (uncoated) showed nanoparticles with an average particle size of 144.2 ± 28.4 nm (Table 6) and polydispersity index was 0.24 ± 0.03, while batch C1 (coated) showed nanoparticles with an average particle size of 175.4 ± 41.1 nm (Table 6) and a polydispersity index of 0.19 ± 0.03, indicating uniform and narrow size distribution of particles. The zeta potential values exhibited all positive values and greater than 18 ± 3.6 mV for coated and uncoated nanoparticles (Table 6), confirming the stability of nanoparticles and avoiding aggregation behavior [42].

Table 6. Particle size and zeta potential of rivastigmine nanoparticles.

Batches	Particle Size (nm)	Zeta Potential (mV)
D1	163.2 ± 33.1	22 ± 5.8
D2	166.8 ± 32.8	23 ± 3.1
D3	159.1 ± 29.6	19 ± 4.1
D4	162.4 ± 33.3	25 ± 2.8
D5	154.2 ± 30.7	22 ± 3.7
D6	151.9 ± 28.5	20 ± 4.1
D7	148.9 ± 29.2	20 ± 1.8
D8	144.2 ± 28.4	18 ± 3.6
C1	175.4 ± 41.1	20 ± 2.9
C2	182.3 ± 40.9	23 ± 3.5
C3	192.1 ± 43.9	18 ± 3.1

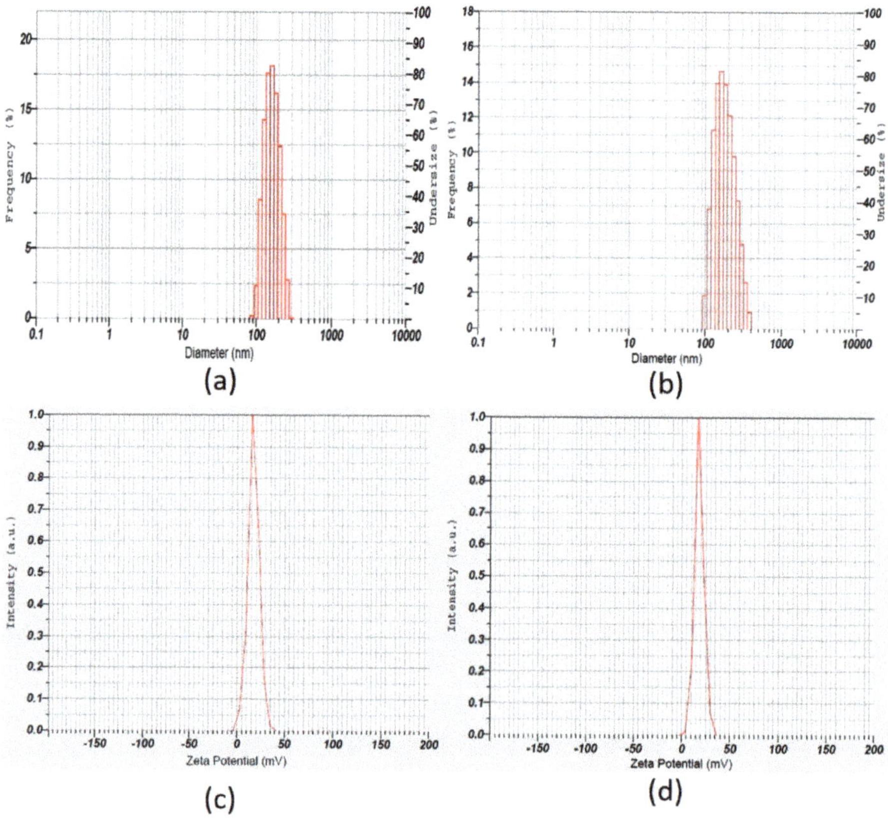

Figure 7. Upper panel: Representative size distribution curve of (**a**) uncoated (batch D8), (**b**) coated nanoparticles (batch C1). Lower panel: Representative zeta potential distribution pictures of (**c**) uncoated (batch D8), (**d**) coated nanoparticles (batch C1).

3.9. *In Vitro Mucoadhesion Testing*

To check the adhesion of coated nanoparticles to the nasal mucosa (absorption site) for a prolonged period, in vitro mucoadhesion testing was performed. The results exhibited that all coated nanoparticle batches (C1 to C3) had good mucoadhesive properties ranging from 91–93% and could suitably adhere to sheep nasal mucosa. The results proved that the selected core-to-coat ratio provides higher adhesion probably due to good interaction between the Eudragit® EPO and mucus, as reported early [29].

3.10. SEM Analysis of the Nanoparticles

The shape and surface of prepared nanoparticles were examined by SEM (Figure 8a,b). From the SEM micrographs, it was noticed that the optimized uncoated nanoparticles' (batch D8) size was ~145 nm, whereas, upon coating, the size increased slightly from ~145 nm to ~175 nm. Further analysis indicated that all the fabricated coated nanoparticles were less than 200 nm. Our findings are similar to the data reported in the literature [65]. Further, the SEM images in Figure 8 also demonstrate the spherical morphology of the formulated nanoparticles. Furthermore, the image (Figure 8b) indicates that there is no defective coating on the chitosan nanoparticles, which confirms the coating process was done properly on the chitosan nanoparticles.

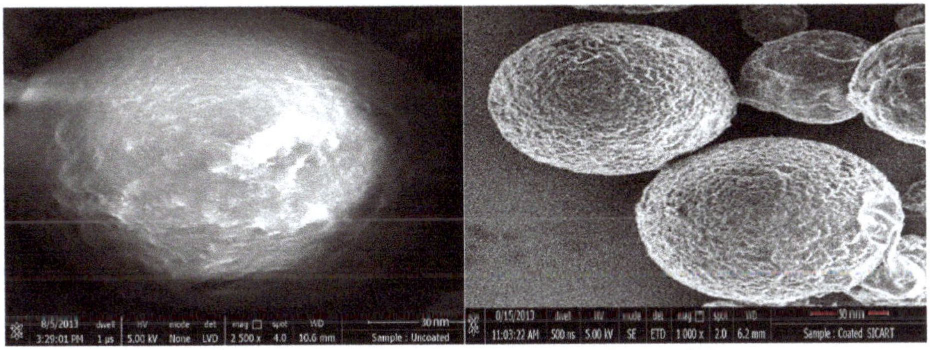

(a) (b)

Figure 8. Representative scanning electron pictures of (**a**) uncoated (batch D8), (**b**) coated nanoparticles (batch C1).

3.11. DSC Analysis of Nanoparticle Formulation

DSC thermograms of the drug, physical mixture, and nanoparticle formulation (batch C1) are shown in Figure 9. The Thermogram of rivastigmine tartrate shows a sharp endothermic peak at 126.02 °C, confirming the melting point of the drug. The physical mixture shows a sharp endothermic peak of the drug at 126.02 °C and the broad and diminished endothermic peak of chitosan and eudragit EPO in the range of 180–220 °C and 260–300 °C, respectively, which indicates that the drug and polymers were not interacting with each other. However, no characteristic peak of the drug was seen in the thermogram of nanoparticle formulation, which is probably because the drug was encapsulated in the polymeric nanoparticles, as described in the literature [16,66].

3.12. Nasal Ciliotoxicity Studies

Figure 10 illustrates the nasal toxicity studies of nanoparticles of rivastigmine (batch C1) performed using sheep nasal mucosa. It is evident from Figure 10 that the sheep nasal mucosal cells treated with phosphate buffer (Figure 10A) and nanoparticles of rivastigmine (Figure 10C) showed intact, undamaged nasal cells with intact basement membrane, suggesting no indication of toxicity of the formulation. However, mucosa treated with isopropyl alcohol (positive control) showed destruction of the normal architecture of the mucosal membrane (Figure 10B). Results of histological studies implied that the formulated rivastigmine nanoparticles are safe to apply to the nasal mucosa, where the formulation will be retained for a longer period due to significant mucoadhesive properties.

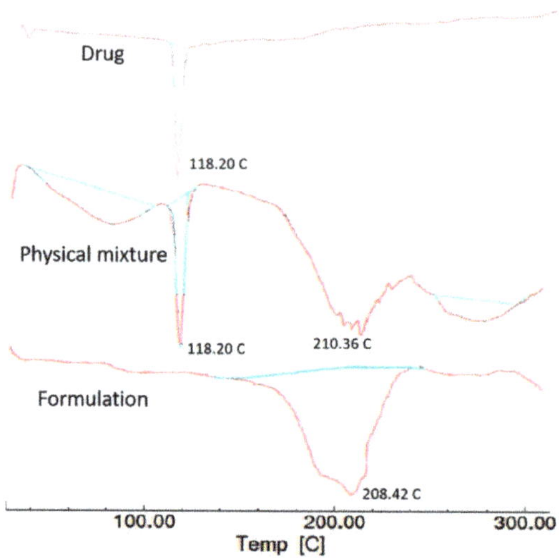

Figure 9. Differential scanning calorimetry patterns of rivastigmine tartrate, physical mixture, and optimized nanoparticle formulation (batch C1).

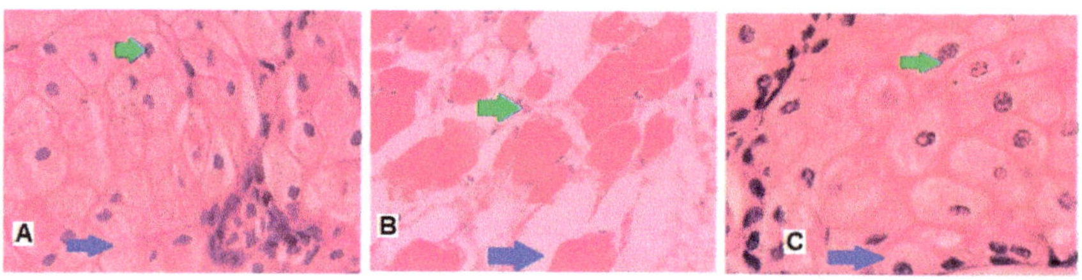

Figure 10. Light microscopic examination of sheep nasal mucosa stained using hematoxylin and eosin stain. (**A**) Negative control: nasal mucosa treated with phosphate buffer shows undamaged nasal cells with unimpaired basement membrane (blue arrow), intact glandular cells with a well-defined nucleus (green arrow). (**B**) Positive control: mucosa treated with isopropyl alcohol illustrates widespread destruction of nasal tissues including the injury to the basement membrane (blue arrow) and glandular cells (with consequent loss of nucleus) (green arrow). (**C**) Test: nasal tissue exposed to rivastigmine drug demonstrates unharmed basement membrane (blue arrow), undamaged glandular cells with distinct nuclei (green arrow).

4. Conclusions

Rivastigmine-loaded chitosan nanoparticles were successfully optimized using pharmaceutical quality design. The preliminary optimization process disclosed the effect of formulation variables, viz. different concentrations of polymer, drug, surfactant, crosslinking agent, and processing variables, such as stirring speed and crosslinking time on the quality of the product. Based on the collected data, the levels of independent variables, i.e., drug–polymer ratio, stirrer speed, and crosslinking time, were set to obtain an optimized batch with spherical nanoparticles with no aggregation, good mucoadhesive property, in vitro release profile, and in vitro permeation. The core-to-polymer ratio (1:1) was fixed to fabricate the coated nanoparticles that control the release pattern of the entrapped drug for a prolonged period. The prepared coated nanoparticles showed size less than 200 nm and released the entrapped drug gradually for 24 h (97.59%) from the 64.83% drug entrapped nanoparticles. Further, in vitro drug permeation through the sheep nasal mucosa revealed greater rivastigmine permeation. Finally, the application of the coated nanoparticles resulted in no toxicity to the nasal cilia of the experimental sheep nasal mucosa. Therefore, developed rivastigmine-nanoparticles could be used as an alternate delivery system to overcome the poor bioavailability issues of the drug. However, this novel drug delivery system should be evaluated further to establish the safety and efficacy in the in vivo experimental models before advancing towards the clinical stage.

Author Contributions: Formal Analysis, M.B., J.S., B.G., A.B.N., S.J., S.M.B.A., S.F., A.S.A., W.F.A. and M.A.; Investigation, M.B., J.S., B.G., A.B.N., S.J., S.M.B.A., S.F., A.S.A., W.F.A., M.A., S.N. and M.K.A.; Methodology, M.B., J.S., B.G., A.B.N., S.J., S.M.B.A., S.F., A.S.A., W.F.A., M.A., S.N. and M.K.A.; Writing—Original Draft Preparation, M.B., S.F., A.S.A., W.F.A., M.A. and M.K.A.; Writing—Review and Editing, J.S., B.G., A.B.N., S.J., S.M.B.A. and S.N. All authors have read and agreed to the published version of the manuscript.

Funding: Majid Alhomrani would like to acknowledge TURSP (2020/257).

Institutional Review Board Statement: Nasal ciliotoxicity studies of nanoparticles of rivastigmine were conducted using sheep nasal mucosa by following the ethical guidelines (approval number: VSCP/EC/11508/2020/4, dated 24 July 2020).

Informed Consent Statement: Not applicable.

Data Availability Statement: All the relevant data are included in the manuscript.

Acknowledgments: The authors are thankful to Nirma University, India, and AlMaarefa University, Riyadh, for providing support to do this research.

Conflicts of Interest: The authors declare no conflict of interest.

References

1. Feigin, V.L.; Nichols, E.; Alam, T.; Bannick, M.S.; Beghi, E.; Blake, N.; Culpepper, W.J.; Dorsey, E.R.; Elbaz, A.; Ellenbogen, R.G. Global, regional, and national burden of neurological disorders, 1990–2016: A systematic analysis for the Global Burden of Disease Study 2016. *Lancet Neurol.* **2019**, *18*, 459–480. [CrossRef]
2. Singh, S.; Dhanawat, M.; Gupta, S.; Kumar, D.; Kakkar, S.; Nair, A.; Verma, I.; Sharma, P. Naturally inspired pyrimidines analogues for alzheimer's disease. *Curr. Neuropharmacol.* **2021**, *19*, 136–151. [CrossRef]
3. Maresova, P.; Mohelská, H.; Dolejs, J.; Kuca, K. Socio-economic Aspects of Alzheimer's Disease. *Curr. Alzheimer Res.* **2015**, *12*, 903–911. [CrossRef]
4. Hussain, R.; Zubair, H.; Pursell, S.; Shahab, M. Neurodegenerative diseases: Regenerative mechanisms and novel therapeutic approaches. *Brain Sci.* **2018**, *8*, 177. [CrossRef] [PubMed]
5. Nair, A.B.; Jigar, S.; Vishal, C.; Hiral, S.; Snehal, P. Delivery of biomolecules to the central nervous system using a polysaccharide nanocomposite. In *Polysaccharide-Based Nano-Biocarrier in Drug Delivery*; CRC Press: Boca Raton, FL, USA, 2018; pp. 105–128.
6. Hou, Y.; Dan, X.; Babbar, M.; Wei, Y.; Hasselbalch, S.G.; Croteau, D.L.; Bohr, V.A. Ageing as a risk factor for neurodegenerative disease. *Nat. Rev. Neurol.* **2019**, *15*, 565–581. [CrossRef] [PubMed]
7. Gorain, B.; Rajeswary, D.C.; Pandey, M.; Kesharwani, P.; Kumbhar, S.A.; Choudhury, H. Nose to brain delivery of nanocarriers towards attenuation of demented condition. *Curr. Pharm. Des.* **2020**, *26*, 2233–2246. [CrossRef]
8. Chin, L.Y.; Tan, J.Y.P.; Choudhury, H.; Pandey, M.; Sisinthy, S.P.; Gorain, B. Development and optimization of chitosan coated nanoemulgel of telmisartan for intranasal delivery: A comparative study. *J. Drug Deliv. Sci. Technol.* **2021**, *62*, 102341. [CrossRef]
9. Islam, S.U.; Shehzad, A.; Ahmed, M.B.; Lee, Y.S. Intranasal Delivery of Nanoformulations: A potential way of treatment for neurological disorders. *Molecules* **2020**, *25*, 1929. [CrossRef]
10. Keller, L.-A.; Merkel, O.; Popp, A. Intranasal drug delivery: Opportunities and toxicologic challenges during drug development. *Drug Deliv. Transl. Res.* **2021**, 1–23. [CrossRef]
11. Chatterjee, B.; Gorain, B.; Mohananaidu, K.; Sengupta, P.; Mandal, U.K.; Choudhury, H. Targeted drug delivery to the brain via intranasal nanoemulsion: Available proof of concept and existing challenges. *Int. J. Pharm.* **2019**, *565*, 258–268. [CrossRef]
12. Birks, J.S.; Evans, J.G. Rivastigmine for Alzheimer's disease. *Cochrane Database Syst. Rev.* **2015**, *10*, CD001191. [CrossRef]
13. Farlow, M.R.; Lilly, M.L. Rivastigmine: An open-label, observational study of safety and effectiveness in treating patients with Alzheimer's disease for up to 5 years. *BMC Geriatr.* **2005**, *5*, 3. [CrossRef] [PubMed]
14. Vyas, S.; Kothari, S.; Kachhwaha, S. Nootropic medicinal plants: Therapeutic alternatives for Alzheimer's disease. *J. Herb. Med.* **2019**, *17*, 100291. [CrossRef]
15. Shah, B.M.; Misra, M.; Shishoo, C.J.; Padh, H. Nose to brain microemulsion-based drug delivery system of rivastigmine: Formulation andex-vivocharacterization. *Drug Deliv.* **2013**, *22*, 918–930. [CrossRef] [PubMed]
16. Fazil, M.; Shadab; Haque, S.; Kumar, M.; Baboota, S.; Sahni, J.K.; Ali, J. Development and evaluation of rivastigmine loaded chitosan nanoparticles for brain targeting. *Eur. J. Pharm. Sci. Off. J. Eur. Fed. Pharm. Sci.* **2012**, *47*, 6–15. [CrossRef]
17. Khoury, R.; Rajamanickam, J.; Grossberg, G.T. An update on the safety of current therapies for Alzheimer's disease: Focus on rivastigmine. *Ther. Adv. Drug Saf.* **2018**, *9*, 171–178. [CrossRef] [PubMed]

18. Salatin, S.; Barar, J.; Barzegar-Jalali, M.; Adibkia, K.; Alami-Milani, M.; Jelvehgari, M. Formulation and Evaluation of Eudragit RL-100 Nanoparticles Loaded In-Situ Forming Gel for Intranasal Delivery of Rivastigmine. *Adv. Pharm. Bull.* **2020**, *10*, 20–29. [CrossRef]
19. Wilson, B.; Samanta, M.K.; Santhi, K.; Kumar, K.P.S.; Paramakrishnan, N.; Suresh, B. Poly(n-butylcyanoacrylate) nanoparticles coated with polysorbate 80 for the targeted delivery of rivastigmine into the brain to treat Alzheimer's disease. *Brain Res.* **2008**, *1200*, 159–168. [CrossRef]
20. Kamboj, S.; Bala, S.; Nair, A.B. Solid lipid nanoparticles: An effective lipid based technology for poorly water soluble drugs. *Int. J. Pharm. Sci. Rev. Res.* **2010**, *5*, 78–90.
21. Bourganis, V.; Kammona, O.; Alexopoulos, A.; Kiparissides, C. Recent advances in carrier mediated nose-to-brain delivery of pharmaceutics. *Eur. J. Pharm. Biopharm. Off. J. Arb. Pharm. Verfahr. e.V* **2018**, *128*, 337–362. [CrossRef]
22. Prosapio, V.; De Marco, I.; Reverchon, E. PVP/corticosteroid microspheres produced by supercritical antisolvent coprecipita-tion. *Chem. Eng. J.* **2016**, *292*, 264–275. [CrossRef]
23. Nunes, R.; Baião, A.; Monteiro, D.; Das Neves, J.; Sarmento, B. Zein nanoparticles as low-cost, safe, and effective carriers to improve the oral bioavailability of resveratrol. *Drug Deliv. Transl. Res.* **2020**, *10*, 826–837. [CrossRef] [PubMed]
24. Brzeziński, M.; Kost, B.; Wedepohl, S.; Socka, M.; Biela, T.; Calderón, M. Stereocomplexed PLA microspheres: Control over morphology, drug encapsulation and anticancer activity. *Colloids Surf. B Biointerfaces* **2019**, *184*, 110544. [CrossRef] [PubMed]
25. Akrawi, S.H.; Gorain, B.; Nair, A.B.; Choudhury, H.; Pandey, M.; Shah, J.N.; Venugopala, K.N. Development and optimization of naringenin-loaded chitosan-coated nanoemulsion for topical therapy in wound healing. *Pharmaceutics* **2020**, *12*, 893. [CrossRef] [PubMed]
26. Sreeharsha, N.; Rajpoot, K.; Tekade, M.; Kalyane, D.; Nair, A.B.; Venugopala, K.N.; Tekade, R.K. Development of metronidazole loaded chitosan nanoparticles using QbD approach—A novel and potential antibacterial formulation. *Pharmaceutics* **2020**, *12*, 920. [CrossRef]
27. Smith, J.; Wood, E.; Dornish, M. Effect of Chitosan on epithelial cell tight junctions. *Pharm. Res.* **2004**, *21*, 43–49. [CrossRef]
28. Linares, V.; Yarce, C.J.; Echeverri, J.D.; Galeano, E.; Salamanca, C.H. Relationship between degree of polymeric ionisation and hydrolytic degradation of Eudragit® E polymers under extreme acid conditions. *Polymers* **2019**, *11*, 1010. [CrossRef]
29. Porfiryeva, N.; Nasibullin, S.F.; Abdullina, S.G.; Tukhbatullina, I.K.; Moustafine, R.I.; Khutoryanskiy, V.V. Acrylated Eudragit® E PO as a novel polymeric excipient with enhanced mucoadhesive properties for application in nasal drug delivery. *Int. J. Pharm.* **2019**, *562*, 241–248. [CrossRef]
30. Nair, A.B.; Shah, J.; Aljaeid, B.M.; Al-Dhubiab, B.E.; Jacob, S.; Nair; Shah; Dhubiab, A. Gellan gum-based hydrogel for the transdermal delivery of nebivolol: Optimization and evaluation. *Polymers* **2019**, *11*, 1699. [CrossRef]
31. Souto, E.B.; Da Da Ana, R.; Souto, S.B.; Zielińska, A.; Marques, C.; Andrade, L.N.; Horbańczuk, O.K.; Atanasov, A.G.; Lucarini, M.; Durazzo, A.; et al. In vitro characterization, modelling, and antioxidant properties of polyphenon-60 from green tea in Eudragit S100-2 Chitosan microspheres. *Nutrients* **2020**, *12*, 967. [CrossRef]
32. Rukmangathen, R.; Yallamalli, I.M.; Yalavarthi, P.R. Formulation and biopharmaceutical evaluation of risperidone-loaded chitosan nanoparticles for intranasal delivery. *Drug Dev. Ind. Pharm.* **2019**, *45*, 1342–1350. [CrossRef] [PubMed]
33. Nair, A.; Gupta, R.; Vasanti, S. In vitro controlled release of alfuzosin hydrochloride using HPMC-based matrix tablets and its comparison with marketed product. *Pharm. Dev. Technol.* **2007**, *12*, 621–625. [CrossRef] [PubMed]
34. Shah, H.; Nair, A.B.; Shah, J.; Jacob, S.; Bharadia, P.; Haroun, M. Proniosomal vesicles as an effective strategy to optimize naproxen transdermal delivery. *J. Drug Deliv. Sci. Technol.* **2021**, *63*, 102479. [CrossRef]
35. Nair, A.; E Al-Dhubiab, B.; Shah, J.; Attimarad, M. Poly(lactic acid-coglycolic acid) nanospheres improved the oral delivery of candesartan cilexetil. *Indian J. Pharm. Educ. Res.* **2017**, *51*, 571–579. [CrossRef]
36. Subudhi, M.B.; Jain, A.; Jain, A.; Hurkat, P.; Shilpi, S.; Gulbake, A.; Jain, S.K. Eudragit S100 coated citrus pectin nanoparticles for colon targeting of 5-Fluorouracil. *Materials* **2015**, *8*, 832–849. [CrossRef]
37. Salatin, S.; Barar, J.; Barzegar-Jalali, M.; Adibkia, K.; Kiafar, F.; Jelvehgari, M. Development of a nanoprecipitation method for the entrapment of a very water soluble drug into Eudragit RL nanoparticles. *Res. Pharm. Sci.* **2017**, *12*, 1–14. [CrossRef]
38. Nair, A.B.; Al-Dhubiab, B.E.; Shah, J.; Jacob, S.; Saraiya, V.; Attimarad, M.; SreeHarsha, N.; Akrawi, S.H.; Shehata, T.M. Mucoadhesive buccal film of almotriptan improved therapeutic delivery in rabbit model. *Saudi Pharm. J. SPJ Off. Publ. Saudi Pharm. Soc.* **2020**, *28*, 201–209. [CrossRef]
39. Sharma, D.; Sharma, R.K.; Sharma, N.; Gabrani, R.; Sharma, S.K.; Ali, J.; Dang, S. Nose-To-Brain Delivery of PLGA-Diazepam Nanoparticles. *AAPS PharmSciTech* **2015**, *16*, 1108–1121. [CrossRef]
40. Nair, A.; Vyas, H.; Shah, J.; Kumar, A. Effect of permeation enhancers on the iontophoretic transport of metoprolol tartrate and the drug retention in skin. *Drug Deliv.* **2011**, *18*, 19–25. [CrossRef]
41. Anroop, B.; Ghosh, B.; Parcha, V.; Kumar, A.; Khanam, J. Synthesis and comparative skin permeability of atenolol and propranolol esters. *J. Drug Deliv. Sci. Technol.* **2005**, *15*, 187–190. [CrossRef]
42. Shah, J.; Nair, A.B.; Jacob, S.; Patel, R.K.; Shah, H.; Shehata, T.M.; Morsy, M.A. Nanoemulsion based vehicle for effective ocular delivery of moxifloxacin using experimental design and pharmacokinetic study in rabbits. *Pharmaceutics* **2019**, *11*, 230. [CrossRef]
43. Belgamwar, V.S.; Patel, H.S.; Joshi, A.S.; Agrawal, A.; Surana, S.J.; Tekade, A.R. Design and development of nasal mucoadhesive microspheres containing tramadol HCl for CNS targeting. *Drug Deliv.* **2011**, *18*, 353–360. [CrossRef]
44. Jacob, S.; Nair, A.; Aldhubiab, B. Preparation and evaluation of niosome gel containing acyclovir for enhanced dermal deposition. *J. Liposome Res.* **2017**, *27*, 283–292. [CrossRef]
45. Chaudhary, S.; Nair, A.B.; Shah, J.; Gorain, B.; Jacob, S.; Shah, H.; Patel, V. Enhanced solubility and bioavailability of dolutegravir by solid dispersion method: In vitro and in vivo evaluation—A potential approach for HIV therapy. *AAPS PharmSciTech* **2021**, *22*, 127. [CrossRef]

46. Kumbhar, S.A.; Kokare, C.R.; Shrivastava, B.; Gorain, B.; Choudhury, H. Antipsychotic potential and safety profile of TPGS-based mucoadhesive aripiprazole nanoemulsion: Development and optimization for nose-to-brain delivery. *J. Pharm. Sci.* **2021**, *110*, 1761–1778. [CrossRef]
47. Morsy, M.A.; Abdel-Latif, R.G.; Nair, A.B.; Venugopala, K.N.; Ahmed, A.F.; Elsewedy, H.S.; Shehata, T.M. Preparation and evaluation of atorvastatin-loaded nanoemulgel on wound-healing efficacy. *Pharmaceutics* **2019**, *11*, 609. [CrossRef] [PubMed]
48. El Hady, W.E.A.; Mohamed, E.A.; Soliman, O.A.E.-A.; EL Sabbagh, H.M. In vitro–in vivo evaluation of chitosan-PLGA nanoparticles for potentiated gastric retention and anti-ulcer activity of diosmin. *Int. J. Nanomed.* **2019**, *14*, 7191–7213. [CrossRef] [PubMed]
49. Mohammed, M.A.; Syeda, J.T.M.; Wasan, K.M.; Wasan, E.K. An overview of chitosan nanoparticles and its application in non-parenteral drug delivery. *Pharmaceutics* **2017**, *9*, 53. [CrossRef] [PubMed]
50. Jhaveri, J.; Raichura, Z.; Khan, T.; Momin, M.; Omri, A. Chitosan Nanoparticles-Insight into Properties, Functionalization and Applications in Drug Delivery and Theranostics. *Molecules* **2021**, *26*, 272. [CrossRef] [PubMed]
51. Sedyakina, N.; Kuskov, A.; Velonia, K.; Feldman, N.; Lutsenko, S.; Avramenko, G. Modulation of Entrapment Efficiency and In Vitro Release Properties of BSA-Loaded Chitosan Microparticles Cross-Linked with Citric Acid as a Potential Protein–Drug Delivery System. *Materials* **2020**, *13*, 1989. [CrossRef]
52. Abbas, A.K.; Alhamdany, A.T. Floating microspheres of enalapril maleate as a developed controlled release dosage form: Investigation of the effect of an ionotropic gelation technique. *Turk. J. Pharm. Sci.* **2020**, *17*, 159–171. [CrossRef] [PubMed]
53. Malik, N.S.; Ahmad, M.; Minhas, M.U.; Tulain, R.; Barkat, K.; Khalid, I.; Khalid, Q. Chitosan/Xanthan gum based hydrogels as potential carrier for an antiviral drug: Fabrication, characterization, and safety evaluation. *Front. Chem.* **2020**, *8*, 50. [CrossRef]
54. Gänger, S.; Schindowski, K. Tailoring Formulations for intranasal nose-to-brain delivery: A review on architecture, physico-chemical characteristics and mucociliary clearance of the nasal olfactory mucosa. *Pharmaceutics* **2018**, *10*, 116. [CrossRef] [PubMed]
55. Markl, D.; Zeitler, J.A. A review of disintegration mechanisms and measurement techniques. *Pharm. Res.* **2017**, *34*, 890–917. [CrossRef] [PubMed]
56. Szekalska, M.; Sosnowska, K.; Zakrzeska, A.; Kasacka, I.; Lewandowska, A.; Winnicka, K. The influence of chitosan cross-linking on the properties of alginate microparticles with metformin hydrochloride—In vitro and in vivo evaluation. *Molecules* **2017**, *22*, 182. [CrossRef]
57. Boni, F.I.; Prezotti, F.G.; Cury, B. Gellan gum microspheres crosslinked with trivalent ion: Effect of polymer and crosslinker concentrations on drug release and mucoadhesive properties. *Drug Dev. Ind. Pharm.* **2016**, *42*, 1283–1290. [CrossRef]
58. Porfiryeva, N.N.; Semina, I.I.; Salakhov, I.A.; Moustafine, R.I.; Khutoryanskiy, V.V. Mucoadhesive and mucus-penetrating interpolyelectrolyte complexes for nose-to-brain drug delivery. *Nanomed. Nanotechnol. Biol. Med.* **2021**, *37*, 102432. [CrossRef]
59. Pawłowski, Ł.; Bartmański, M.; Strugała, G.; Mielewczyk-Gryń, A.; Jażdżewska, M.; Zieliński, A. Electrophoretic deposition and characterization of Chitosan/Eudragit E 100 coatings on titanium substrate. *Coatings* **2020**, *10*, 607. [CrossRef]
60. Salatin, S.; Barar, J.; Barzegar-Jalali, M.; Adibkia, K.; Jelvehgari, M. Thermosensitive in situ nanocomposite of rivastigmine hydrogen tartrate as an intranasal delivery system: Development, characterization, ex vivo permeation and cellular studies. *Colloids Surf. B Biointerfaces* **2017**, *159*, 629–638. [CrossRef]
61. Huang, T.W.; Wei, C.-K.; Su, H.-W.; Fang, K.-M. Chitosan promotes aquaporin formation and inhibits mucociliary differentiation of nasal epithelial cells through increased TGF-β1 production. *J. Tissue Eng. Regen. Med.* **2017**, *11*, 3567–3575. [CrossRef]
62. Nair, A.B.; Shah, J.; Jacob, S.; Al-Dhubiab, B.E.; Sreeharsha, N.; Morsy, M.A.; Gupta, S.; Attimarad, M.; Shinu, P.; Venugopala, K.N. Experimental design, formulation and in vivo evaluation of a novel topical in situ gel system to treat ocular infections. *PLoS ONE* **2021**, *16*, e0248857. [CrossRef] [PubMed]
63. Anroop, B.; Ghosh, B.; Parcha, V.; Khanam, J. Transdermal delivery of atenolol: Effect of prodrugs and iontophoresis. *Curr. Drug Deliv.* **2009**, *6*, 280–290. [CrossRef] [PubMed]
64. Richter, T.; Keipert, S. In vitro permeation studies comparing bovine nasal mucosa, porcine cornea and artificial membrane: Androstenedione in microemulsions and their components. *Eur. J. Pharm. Biopharm. Off. J. Arb. Pharm. Verfahr. e.V* **2004**, *58*, 137–143. [CrossRef] [PubMed]
65. Pereira, G.G.; Santos Oliveira, R.; Albernaz, M.S.; Canema, D.; Weismüller, G.; Barros, E.B.; Magalhães, L.; Lima Ribeiro, M.H.M.; Pohlmann, A.; Guterres, S.S. Microparticles of Aloe vera/vitamin E/chitosan: Microscopic, a nuclear imaging and an in vivo test analysis for burn treatment. *Eur. J. Pharm. Biopharm. Off. J. Arb. Pharm. Verfahr. e.V* **2014**, *86*, 292–300. [CrossRef]
66. Joshi, S.A.; Chavhan, S.S.; Sawant, K.K. Rivastigmine-loaded PLGA and PBCA nanoparticles: Preparation, optimization, characterization, in vitro and pharmacodynamic studies. *Eur. J. Pharm. Biopharm. Off. J. Arb. Pharm. Verfahr. e.V* **2010**, *76*, 189–199. [CrossRef] [PubMed]

materials

MDPI

Article

Temperature-Responsive Polysaccharide Microparticles Containing Nanoparticles: Release of Multiple Cationic/Anionic Compounds

Takumi Sato and Yoshihiko Murakami *

Department of Organic and Polymer Materials Chemistry, Tokyo University of Agriculture and Technology, Naka-cho, Koganei-shi 2-24-16, Tokyo 184-8588, Japan; takumi.sato.2021@gmail.com
* Correspondence: muray@cc.tuat.ac.jp; Tel.: +81-42-388-7387

Abstract: Most drug carriers used in pulmonary administration are microparticles with diameters over 1 μm. Only a few examples involving nanoparticles have been reported because such small particles are readily exhaled. Consequently, the development of microparticles capable of encapsulating nanoparticles and a wide range of compounds for pulmonary drug-delivery applications is an important objective. In this study, we investigated the development of polysaccharide microparticles containing nanoparticles for the temperature-responsive and two-step release of inclusions. The prepared microparticles containing nanoparticles can release two differently charged compounds in a stepwise manner. The particles have two different drug release pathways: one is the release of nanoparticle inclusions from the nanoparticles and the other is the release of microparticle inclusions during microparticle collapse. The nanoparticles can be efficiently delivered deep into the lungs and a wide range of compounds are released in a charge-independent manner, owing to the suitable roughness of the microparticle surface. These polysaccharide microparticles containing nanoparticles are expected to be used as temperature-responsive drug carriers, not only for pulmonary administration but also for various administration routes, including transpulmonary, intramuscular, and transdermal routes, that can release multiple drugs in a controlled manner.

Keywords: polysaccharide; carrageenan; nanoparticle; pulmonary drug delivery

Citation: Sato, T.; Murakami, Y. Temperature-Responsive Polysaccharide Microparticles Containing Nanoparticles: Release of Multiple Cationic/Anionic Compounds. *Materials* **2022**, *15*, 4717. https://doi.org/10.3390/ma15134717

Academic Editors: Ángel Serrano-Aroca and Roser Sabater i Serra

Received: 5 June 2022
Accepted: 4 July 2022
Published: 5 July 2022

1. Introduction

A variety of nanoparticles have been developed, including inorganic nanoparticles formed from silica and gold and polysaccharide nanoparticles formed from alginate and chitosan [1–4], as drug carriers in various drug-dosage forms, including oral and intravenous administration [5–7]. Among these, cationic nanoparticles have been reported to exhibit antimicrobial activities and membrane permeabilities, which can be used as anticancer agents [8,9]. Biologically-derived polymers, such as chitosan and poly-L-lysine, are often used as cationic polymers to form cationic nanoparticles [10–12]. Chitosan is a linear polymer prepared by deacetylating natural chitin, which is obtained from crustaceans, such as shrimps and crabs. Poly-L-lysine is a polymer produced by the fermentation of lysine by *Streptomyces albulus* [12]. Nanoparticles formed from chitosan or poly-L-lysine have been reported on several occasions [12–15]. For example, mixing chitosan or poly-L-lysine with tripolyphosphoric acid (TPP) has been reported to form nanoparticles through electrostatic interactions between the amino groups of the cationic polymer and phosphate groups of TPP [16–19]; such nanoparticles are easily formed by ultrasonication in the absence of surfactants.

Pulmonary administration [20,21], in which drugs are inhaled and absorbed into the body through the lungs, is a medication method that has several advantages, including simplicity of administration [22], excellent immediate efficacy [23], and efficient treatment of lung diseases [24,25]. However, current pulmonary administration uses drug carriers that are mainly microparticles with diameters over 1 μm; only a few examples have used nanoparticles because nanoparticles less than 1 μm in diameter are too small for pulmonary administration. This is because they are expelled from the body through exhalation [26].

We previously developed temperature-responsive carrageenan microparticles that are efficiently delivered to the lungs [27]. These particles, which are readily prepared by the sol–gel transition of carrageenan, can rapidly release their inclusions by collapsing in response to temperature. Carrageenan particles are formed from water-in-oil (w/o) emulsions and readily encapsulate the compounds dissolved or dispersed in the carrageenan solution. However, as carrageenan particles are negatively charged, encapsulating negatively charged compounds is difficult owing to electrostatic repulsion. Therefore, developing carrageenan microparticles capable of encapsulating a wide range of single or multiple compounds is important.

In the present study, we developed temperature-responsive microparticles by complexing cationic nanoparticles with anionic microparticles. Dispersing these cationic nanoparticles in an aqueous carrageenan solution during microparticle preparation facilitates their complexation with anionic microparticles through electrostatic interactions. Furthermore, these cationic nanoparticles can contain anionic compounds, also through electrostatic interactions. Therefore, the previously developed carrageenan microparticles [27] can contain positively charged compounds, whereas the microparticles developed in this study facilitate the concurrent containment of negatively charged compounds through complexation with cationic nanoparticles. The carrageenan microparticles designed based on this idea should exhibit two-step release behavior; that is, the nanoparticles and compounds dispersed inside the microparticles are released in a temperature-responsive manner, followed by the release of another compound from within the nanoparticles (Figure 1). In the present study, the nanoparticles were first prepared using chitosan or poly-L-lysine, after which we evaluated the release behavior of the negatively charged compounds from these nanoparticles. Using the sol–gel transition of carrageenan, we then investigated the complexation of the nanoparticles in the microparticles. In addition, we prepared carrageenan microparticles containing both positively charged compounds and the aforementioned nanoparticles and evaluated the release behavior of the two compounds from the microparticles. To the best of our knowledge, the release of a wide range of compounds by combining polysaccharide microparticles and biopolymer nanoparticles based on the proposed technology that controls both the sol–gel transition and emulsion formation has not been reported.

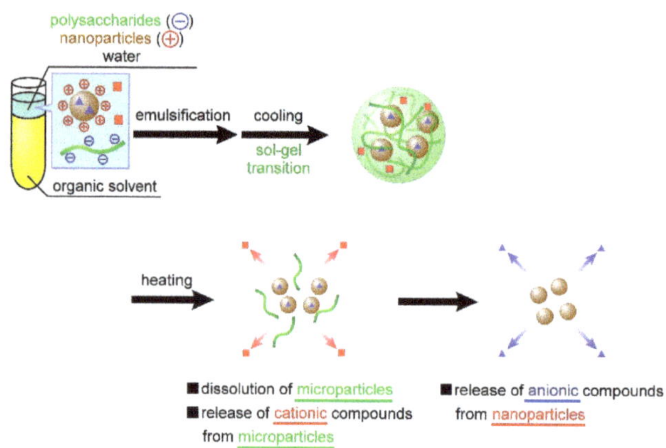

Figure 1. Temperature-responsive nanoparticle-containing or nanoparticle-decorated polysaccharide microparticles for the release of multiple compounds. The nanoparticles and compounds (red squares) dispersed inside the microparticles are released in a temperature-responsive manner, followed by the release of another compound (blue squares) from within the nanoparticles.

2. Materials and Methods

2.1. Materials

κ-Carrageenan (κ-CRG), potassium chloride, methylene blue trihydrate (Mb), pentasodium triphosphate (TPP), acetic acid, hydrochloric acid, sodium hydroxide solution, and toluene were purchased from Wako Pure Chemical Industries (Osaka, Japan). ι-Carrageenan (ι-CRG) and poly-L-lysine (405 kDa, PLL) were purchased from Sigma-Aldrich (St. Louis, MO, USA). Sodium 2-naphthalenesulfonate (Ns) was purchased from Tokyo Chemical Industry Co., Ltd. (Tokyo, Japan). Chitosan (100 kDa, CS) was obtained from Dainichiseika Color & Chemicals Mfg. Co., Ltd. (Tokyo, Japan). Poly(ethylene glycol)-*b*-poly(ε-caprolactone) block copolymer (PEG-*b*-PCL) was synthesized as a polymeric surfactant according to a previously reported method [28,29] with slight modifications (M_ns of the PEG and PCL units were 3500 and 4300, respectively; M_w/M_n ratios of PEG and PEG-*b*-PCL were 1.09 and 1.35, respectively). All other reagents were of analytical grade and used without further purification. The chemical structures of κ-CRG, ι-CRG, CS, PLL, TPP, Mb, and Ns are shown in Figure S1.

2.2. Preparation of CS(Ns) and PLL(Ns) Nanoparticles

The CS(Ns) and PLL(Ns) nanoparticles were prepared according to previously reported methods [16,18,19,30–32]. Here, *A*(*m*) nanoparticles refer to nanoparticles *A* that contain *m* inside them. If a compound is contained within the nanoparticles, then *m* is the name of the compound; otherwise, *m* is simply written as "-" if the nanoparticle has no compound. CS(Ns) nanoparticles were prepared by dropping an aqueous solution (5 mL, pH 5) of TPP (4.2 mg) into an acetate buffer solution (10 mL, pH 5) containing CS (20 mg) and Ns (20 mg) under sonication (20 kHz, 5 min) with an ultrasonic homogenizer (UH-50, SMT Co., Ltd., Tokyo, Japan). The solution was ultrasonicated for another 10 min, then stirred at 300 rpm for 30 min and centrifuged at 9200 rpm for 20 min to produce CS(Ns) nanoparticles. In contrast, PLL(Ns) nanoparticles were prepared by dropping an aqueous solution (5 mL, pH 4) containing TPP (0.92 mg) into an aqueous solution (10 mL, pH 4) of PLL (10 mg) and Ns (20 mg) under sonication (20 kHz, 5 min) with an ultrasonic homogenizer. The pH was then adjusted to 7 with NaOH, and then ultrasonicated for another 10 min. The PLL(Ns) nanoparticles were subsequently obtained following the same procedure used for the CS(Ns) nanoparticles.

2.3. Preparation of the CRG(CS(Ns), Mb) and CRG(PLL(Ns), Mb) Microparticles

The CRG(CS(Ns), Mb) and CRG(PLL(Ns), Mb) microparticles were prepared according to a previously reported method [27]. Here, *A*(*m*, *n*) refers to microparticle *A* that contains nanoparticle *m* and compound *n* inside them. If the microparticle contains only one nanoparticle or compound, then only *m* is provided, whereas *m* is simply written as "-" if the microparticle contains no compound or nanoparticle. An aqueous polysaccharide solution was prepared by dissolving κ-CRG (2.5 *w*/*v*%), ι-CRG (2.5 *w*/*v*%), potassium chloride (32 mM), and Mb (7.3 mM) in Milli-Q water (70 °C, 40 mL). A mixture of κ-CRG and ι-CRG was used instead of κ-CRG or ι-CRG alone because the mixture gelates rapidly with moderate rigidity [27]. The organic solvent was prepared by dissolving PEG-*b*-PCL (0.5 mM) in toluene (10 mL). The w/o emulsion was then prepared by emulsifying (70 °C, 12,000 rpm, 5 min) the polysaccharide solution (0.35 g), organic solvent, and nanoparticle dispersion (1.6 mg/mL CS(Ns) or 0.6 mg/mL PLL(Ns) nanoparticles, 100 μL) with a high-speed homogenizer (NS-51 K and NS-10, Microtec Co., Ltd., Chiba, Japan). The emulsion was gradually cooled to 25 °C in a water bath, and the microparticles were washed three times by the repetitive addition of toluene and subsequent centrifugation to remove excess PEG-*b*-PCL. The CRG(CS(Ns), Mb) and CRG(PLL(Ns), Mb) microparticles were finally obtained after drying naturally in a perfluoroalkoxylalkan beaker. For comparison, the formation of the microparticles using only PLL (Ns) nanoparticles without CRG was also investigated. The w/o emulsion was prepared by mixing a solution of PEG-*b*-PCL (0.5 mM) in toluene (10 mL), Milli-Q water (0.35 mL), and nanoparticle dispersion (100 μL) with a homogenizer (70 °C, 12,000 rpm, 5 min). The white solid was collected by the same process used to prepare the CRG-containing microparticles.

2.4. Characterization of Nanoparticles and Microparticles

The hydrodynamic diameters of the nanoparticles and microparticles were determined by dynamic light scattering (DLS). A particle dispersion (1 mg/mL) was placed in the DLS instrument (Zetasizer Nano-ZS; Malvern Instruments, Malvern, UK) and the effects of temperature (70 °C) and homogenization (12,000 rpm, 5 min) on the nanoparticle diameter were evaluated. The particle zeta potentials were also determined using the same instrument. All the experiments were performed three times.

The nanoparticle morphologies were examined by transmission electron microscopy (TEM; JEM-1400, JEOL, Tokyo, Japan). A CS(Ns) or PLL(Ns) nanoparticle dispersion (5 μL) was applied to a grid (Microgrid Cu200, JEOL, Tokyo, Japan), and the solution was removed using a filter paper after 1 min. A drop of gadolinium acetate (2.5 *w*/*w*%, 5 μL) was applied for 1 min to negatively stain the sample. The excess staining solution was removed using a filter paper and the sample was dried for 15 min. An accelerator voltage of 120 kV was used for the TEM.

The microparticles morphologies were examined by scanning electron microscopy (SEM; VE-9800, KEYENCE, Osaka, Japan) with an accelerator voltage of 1.3 kV. The specimens were prepared by placing the microparticles on an aluminum plate and coating them with an ~10-nm-thick platinum thin film under reduced pressure using an MSP-1S ion coater (Vacuum Device, Ibaraki, Japan).

The nanoparticle inclusions in the microparticles were examined by differential scanning calorimetry (DSC). The CRG(CS(Ns), Mb) and CRG(PLL(Ns), Mb) microparticles (5 mg) were placed in a sealed aluminum pan and subjected to DSC at 10 °C/min from 0 to 500 °C using a differential scanning calorimeter (DSC-60A plus, SHIMADZU, Kyoto, Japan). To examine in detail the complexation of nanoparticles inside the microparticles, we subjected the following samples to DSC: CS(Ns) and PLL(Ns) nanoparticles, CRG(Mb) microparticles, CS, PLL, Mb, Ns, TPP, PEG-*b*-PCL, and κ,ι-CRG (i.e., a physical mixture of κ-CRG and ι-CRG).

The following quantities were determined by spectrofluorometry (FP-6500, JASCO Co., Ishikawamachi, Japan; λ_{ex} = 273 nm, λ_{em} = 324 nm for Ns, λ_{ex} = 640 nm, λ_{em} = 680 nm for Mb): (1) encapsulation ratios and efficiencies of Ns in the CS(Ns) and PLL(Ns) nanoparticles; (2) encapsulation ratios and efficiencies of Mb in the CRG(CS(Ns), Mb) and CRG(PLL(Ns), Mb) microparticles; (3) retention ratio and efficiencies of the CS (Ns) nanoparticles in the CRG(CS(Ns), Mb) microparticles; and (4) retention ratio and efficiencies of the PLL(Ns) nanoparticles in the CRG(PLL(Ns), Mb) microparticles. The nanoparticles and microparticles were placed in a dialysis membrane and immersed in phosphate-buffered saline (PBS). At regular intervals (10 min), the solution outside the dialysis membrane was collected, and its fluorescence intensity was determined by spectrofluorometry. The number of fluorescent substances (Ns or Mb) contained in each particle was determined from the fluorescence intensity at saturation. The encapsulation rate and efficiencies of Ns for the nanoparticles and Mb for the microparticles were calculated using the amounts of particles recovered, fluorescent substance used in the preparation of the particles, and fluorescent substance contained in the particles. The retention ratio and efficiencies of the nanoparticles in the microparticles were calculated based on the encapsulation ratio of Ns in the nanoparticles and weight of Ns released from the microparticles.

The encapsulation ratio and encapsulation efficiency were determined using the following equations:

Encapsulation ratio [%]:

$$\frac{\text{Weight of Ns or Mb in the nanoparticles or microparticles [mg]}}{\text{Weight of the nanoparticles or microparticles [mg]}} \times 100 \quad (1)$$

Encapsulation efficiency [%]:

$$\frac{\text{Weight of Ns or Mb in the nanoparticles or microparticles [mg]}}{\text{Weight of Ns or Mb (mg)}} \times 100 \quad (2)$$

The retention ratio and efficiency were determined using the following equations:
Retention ratio [%]:

$$\frac{\text{Weight of Ns in the microparticles [mg]}}{\text{Microparticle weight [mg]} \times (\text{encapsulation ratio of Ns in the nanoparticles [\%]})/100} \times 100 \quad (3)$$

Retention efficiency [%]:

$$\frac{\text{Weight of Ns in the microparticles [mg]}}{\text{Weight of the Ns [mg]} \times (\text{encapsulation ratio of Ns in the nanoparticles [\%]})/100} \times 100 \quad (4)$$

2.5. Nanoparticle and Microparticle Ns- and Mb-Release Profiles

The profiles depicting the release of Ns and Mb from the nanoparticles and microparticles were constructed using a dialysis method (n = 3) according to a previous report [33]. CS(Ns) or PLL(Ns) nanoparticles (1 mg), and CRG(CS(Ns), Mb) or CRG(PLL(Ns), Mb) microparticles (3 mg) were dispersed in PBS solution (1 and 3 mL, respectively). The nanoparticle (1 mL) or microparticle (3 mL) dispersion was dialyzed against PBS solution (39 and 37 mL, respectively) through a Spectra/Por6 dialysis membrane (molecular weight cut-off: 10,000; Spectrum Houston, TX, USA). Ns or Mb was released from the nanoparticles and the microparticles in a sustained manner under gentle stirring. Each sample was collected periodically (1 mL/15 min) from the exterior of the dialysis membrane and the same amount of PBS (1 mL) was added to the solution. The particle-release behavior was evaluated using three temperature patterns: (1) constant at 10 °C, (2) constant at 70 °C, and (3) ramped from 10 °C to 70 °C at 1 h after the start of experiment. The following values were determined by spectrofluorometry: (1) release ratio of Ns from the CS (Ns) or PLL (Ns) nanoparticles and (2) release ratio of Mb from the CRG(CS(Ns), Mb) or CRG(PLL(Ns), Mb) microparticles. All the releases experiments were performed three times.

3. Results and Discussion

3.1. Preparation and Characterization of the CS(Ns) and PLL(Ns) Nanoparticles

Controlling the charge states of both molecules in a solution is important when forming CS(Ns) and PLL(Ns) nanoparticles through electrostatic interactions between cationic polymers and TPP. Most of the amino groups of CS are positively charged at approximately pH 4.43 because its pK_a is approximately 6.5 [17,34], whereas 80% of its amino groups are deprotonated at pH 7.14 [17]. Meanwhile, three of the five phosphate groups of TPP are negatively charged at pH 4–5 [17]. Therefore, the CS nanoparticles were formed by electrostatic interactions in acetate buffer solution at pH 5. In contrast, because PLL has a pK_a of ~10 [35–37], most of its amino groups are positively charged, even at approximately pH 7. In addition, because four of the five phosphate groups of TPP are negatively charged at pH 7 [17], TPP exhibited stronger electrostatic interactions with PLL than with CS, resulting in the formation of nanoparticles with higher structural stability. However, strong electrostatic PLL–TPP interactions may also promote cross-linking reactions between nanoparticles to form nanoparticle aggregates. Therefore, the aqueous TPP solution was adjusted to pH 4 during particle formation to reduce electrostatic interactions between PLL and TPP and inhibit aggregate formation, the solution was shifted to pH 7 after particle formation to facilitate crosslinking within each nanoparticle.

Figure 2 shows the TEM images of the CS(Ns) and PLL(Ns) nanoparticles and their respective diameter distributions evaluated by DLS. Figure 2A,B display the TEM images before heating and agitation. The diameters of the CS(Ns) and PLL(Ns) nanoparticles are almost equal (100–200 nm). The particles appeared white when negatively stained (as observed for the PLL nanoparticle in Figure 2B); however, they appear black (as observed for CS nanoparticle in Figure 2A) owing to the particle overlap caused by aggregation. The aggregates of the CS(Ns) nanoparticles were attributed to the nanoparticle dialysis with Milli-Q water during nanoparticle preparation, i.e., the solution becomes almost neutral during dialysis, which lowers the positive charge of CS through progressive deprotonation and, consequently, finally aggregates the nanoparticles through hydrophobic interactions. In contrast, almost no aggregates were observed for the PLL(Ns) nanoparticles, which can be ascribed to the positive charge of PLL, even at an almost neutral pH, which prevents the formation of aggregates through electrostatic repulsion between nanoparticles. As obtained by the DLS, the diameters of the CS(Ns) and PLL(Ns) nanoparticles are approximately 200 and 500 nm, respectively, which are larger than those measured by TEM. This difference is attributed to the nanoparticle swelling in water, which is consistent with previous reports whereby nanoparticles formed from water-soluble polymers swelled and increased in size when dispersed in water [38]. Because the PLL(Ns) nanoparticles are more hydrophilic than the CS(Ns) nanoparticles, they swell more easily; hence, the DLS data show that the PLL(Ns) nanoparticles are larger (Figure 2D) than the CS(Ns) nanoparticles (Figure 2C).

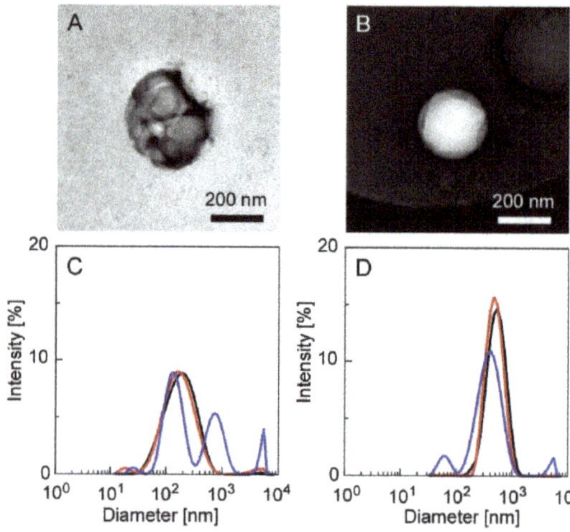

Figure 2. TEM images of the CS(Ns) and PLL(Ns) nanoparticles, and their respective diameter distributions evaluated by DLS ((**A**,**C**): CS(Ns) nanoparticles and (**B**,**D**): PLL(Ns) nanoparticles). Measurement conditions: 25 °C (black trace), 70 °C (red trace), and 70 °C after homogenization at 12,000 rpm for 5 min (blue trace).

To evaluate the effects of temperature and homogenization on the particle state during nanoparticle/microparticle complexation, we measured the particle diameters after heating (70 °C) and homogenization (12,000 rpm, 5 min), the results of which are shown in Figure 2C,D. The diameters of the CS(Ns) nanoparticles were relatively constant under heating; however, the diameters increased under heating and homogenization. Meanwhile, the average particle diameter of the PLL(Ns) nanoparticles did not vary significantly upon heating or homogenization. The different results suggest the higher structural stability of the PLL-TPP nanoparticles than the CS-TPP nanoparticles, which can be attributed to several factors. This includes the high electrostatic interaction between PLL and TPP. When the cross-linking reaction inside the PLL particles was carried out at pH 7, the four phosphate groups of TPP are in their ionized state. Meanwhile, when the cross-linking reaction inside the CS nanoparticles was carried out at pH 5, the three phosphate groups are in the ionized state. Another factor is the large molecular weight of PLL, which is four times that of CS, resulting in the intense entanglement of the PLL chains. The thermal stability of nanoparticles has been discussed to date [39]. Except for inorganic nanoparticles with high thermal conductivity, heat can affect the structural stability of nanoparticles.

In the absence of Ns encapsulation, the CS(-) and PLL(-) nanoparticles had average diameters of 150 and 300 nm, respectively (Figure S2). However, as shown in Figure 2, the diameters of the CS(-) nanoparticles increased from 150 nm to 200 nm, whereas those of the PLL(-) nanoparticles increased from 300 nm to 500 nm during the encapsulation of Ns. Nanoparticle size has been reported to depend on the amount of TPP added. In addition, electrostatic interactions between the cationic polymers and TPP decrease as the amount of added TPP decreases, resulting in an increase in the nanoparticle diameter [40]. Hence, in the present experimental system, the diameters of the obtained nanoparticles increased owing to the stronger electrostatic interactions between the positively charged polymers and Ns and concurrent weaker electrostatic interactions with TPP from the encapsulation of negatively charged Ns. In other words, we suggest that the nanoparticle size can easily be adjusted by changing the amount of inclusion or TPP added.

3.2. Ns-Release Profiles of the CS(Ns) and PLL(Ns) Nanoparticles

Figure 3 shows Ns-release profiles of the CS(Ns) and PLL(Ns) nanoparticles, which reveals that the Ns-release behavior does not significantly vary based on the type of cationic polymer. The number of fluorescent substances (Ns or Mb) contained in each particle was determined from the fluorescence intensity of the solution at saturation because it was difficult to completely collapse the particles. The nanoparticles were formed by the electrostatic interaction of cationic polymers and anionic TPPs. Since the increase in temperature increases the mobility of each molecule, the structural stability of the particles was decreased, thus facilitating the release of the inclusions. After the number of Ns released reached saturation, increasing the temperature did not further release Ns, suggesting that there was no residual Ns retained in the polycation by intermolecular interactions (i.e., the saturated amount of Ns released is the amount of Ns that was encapsulated in the particles). Both the CS(Ns) and PLL(Ns) nanoparticles released almost 100% of their inclusions within 1 h, demonstrating that these nanoparticles rapidly released Ns. Furthermore, the CS(Ns) nanoparticle dialysis membrane dispersion became cloudy with further increases in time, whereas the analogous dispersion remained transparent for the PLL(Ns) nanoparticles.

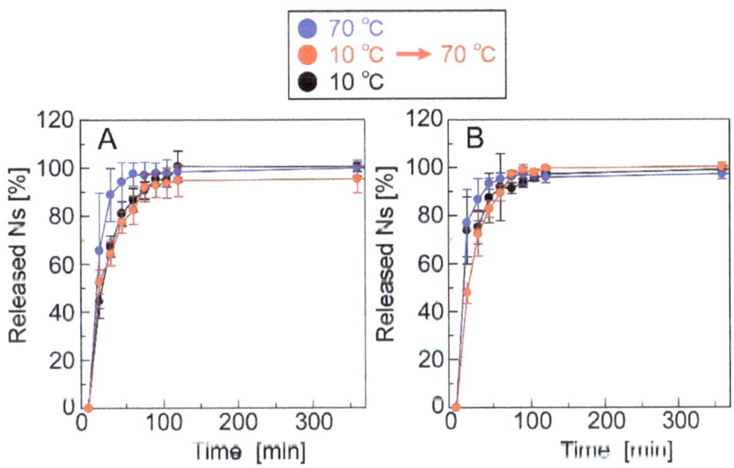

Figure 3. Ns-release profiles of (**A**) CS(Ns) and (**B**) PLL(Ns) nanoparticles.

We calculated the Ns-nanoparticle encapsulation ratios and efficiencies of CS(Ns) and PLL(Ns) (Table S1). The encapsulation ratios of CS(Ns) and PLL(Ns) were 19.6 ± 4.8% and ad 42.8 ± 5.9%, respectively. The results show that the PLL(Ns) nanoparticles encapsulated more Ns than the CS(Ns) nanoparticles, which is attributed to the different cationic properties of CS and PLL. In particular, the pH of the CS(Ns) nanoparticle dispersion is close to neutral owing to the dialysis during the preparation of the nanoparticles, and the amino groups of CS are partially deprotonated. Therefore, there was a weak interaction between CS and the negatively charged Ns, which resulted in a lower internalization ratio. The encapsulation efficiency of 33.3 ± 8.1% and 37.2 ± 5.1% for the CS(Ns) and PLL(Ns) nanoparticles, respectively, did not vary based on the type of cationic polymer used to form the nanoparticles.

3.3. Surface Morphologies of the CRG(CS(Ns), Mb) and CRG(PLL(Ns), Mb) Microparticles

Figure 4 shows the SEM images of the white solid formed from the PLL(Ns) nanoparticles in the absence of CRG, CRG(CS(Ns), Mb) microparticles, and CRG(PLL(Ns), Mb) microparticles. No spherical structures were obtained by cooling the w/o emulsion formed from water with dispersed PLL(Ns) nanoparticles only (Figure 4A). However, microparticles were formed when the w/o emulsion prepared using a nanoparticle dispersion in a CRG solution was cooled (Figure 4B,C). We confirmed that these particles were formed by the sol–gel transition of CRG by cooling the w/o emulsion in the presence of CRG only. We measured the zeta potential of the particles under the conditions where the particles did not aggregate because the surfaces of the particles significantly affect their zeta potential [41]. The zeta potentials of the CS(Ns) and PLL(Ns) nanoparticles were determined to be 38.5 ± 1.1 and 44.9 ± 0.7 mV, respectively, whereas those of the CRG(Mb), CRG(CS(Ns), Mb), and CRG(PLL(Ns), Mb) microparticles were −38.1 ± 1.5, −34.8 ± 2.3, and −34.5 ± 1.6 mV, respectively. These values are more positive than the zeta potential of the CRG(-) microparticles (−44.1 ± 0.78 mV) reported in a previous paper [27], suggesting that the positively charged nanoparticles and Mb affects the surface properties of the microparticles. Particles with rough surfaces can be delivered more easily to deep lung sites, such as the alveoli, compared to smooth particles [42]. The rough surface of the particles causes the boundary layer on the upstream side of the particle to change from laminar to turbulent. The turbulent boundary layer can remain attached to the particle surface much longer than a laminar boundary with less eddies and, hence, creates a narrower low-pressure wake with a reduced pressure drag. The reduction in pressure drag causes the particle to travel further [43]. Thus, the rough microparticles obtained in this study are expected to be useful drug carriers for pulmonary drug-delivery applications.

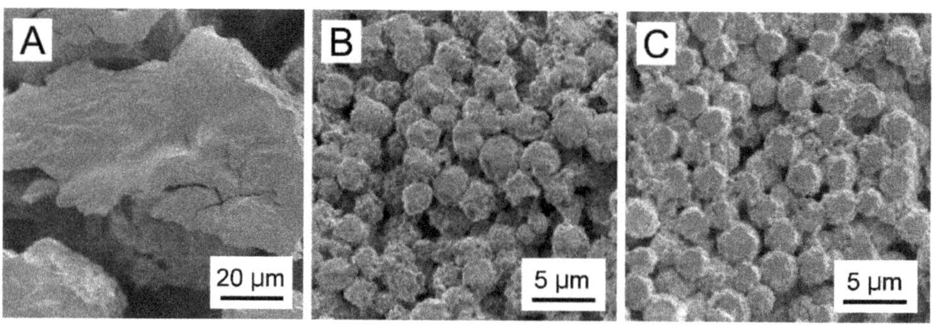

Figure 4. SEM images of (**A**) the white solid formed from PLL(Ns) nanoparticles devoid of CRG, (**B**) CRG(CS(Ns), Mb) microparticles, and (**C**) CRG(PLL(Ns), Mb) microparticles.

3.4. Nanoparticle Inclusions in the Microparticles

Figure 5 shows the DLS data for the microparticle dispersions before and after heating (70 °C), a PEG-b-PCL dispersion, an aqueous κ,ι-CRG (i.e., a physical mixture of κ-CRG and ι-CRG) solution, and a nanoparticle dispersion. The DLS data revealed that the microparticles have significantly different diameters before and after heating. The CRG(CS(Ns),

Mb) microparticles (Figure 5A) exhibited a unimodal peak at 2 μm (orange trace), whereas three peaks at 100 nm, 600 nm, and 9 μm were observed after heating (black trace). Three situations are depicted in Figure 5A: (1) PEG-b-PCL was not completely dissolved when dispersed in water and formed structures with diameters of 100–200 nm (blue trace); (2) the CS(Ns) nanoparticles formed aggregates with diameters of approximately 600 nm when heated and homogenized (green trace), as shown in Figure 2; and (3) the CRG precipitated and formed structures of several micrometers in diameter after the particles were collapsed by heat because of the insolubility of CRG in water at 25 °C (red trace). Thus, the multiple peaks, which were not observed in the pre-heated CRG(CS(Ns), Mb) microparticles, were obtained after heating, owing to the PEG-b-PCL, CS(Ns) nanoparticles (and their aggregates) and CRG. Hence, despite the lack of homogenization after heating, the CS(Ns) nanoparticles aggregated owing to the presence of anionic polysaccharides. These results strongly suggest that CS(Ns) nanoparticles are complexed inside the CRG(CS(Ns), Mb) microparticles because a peak corresponding to the CS(Ns) nanoparticles was observed for the heated CRG(CS(Ns), Mb) microparticles.

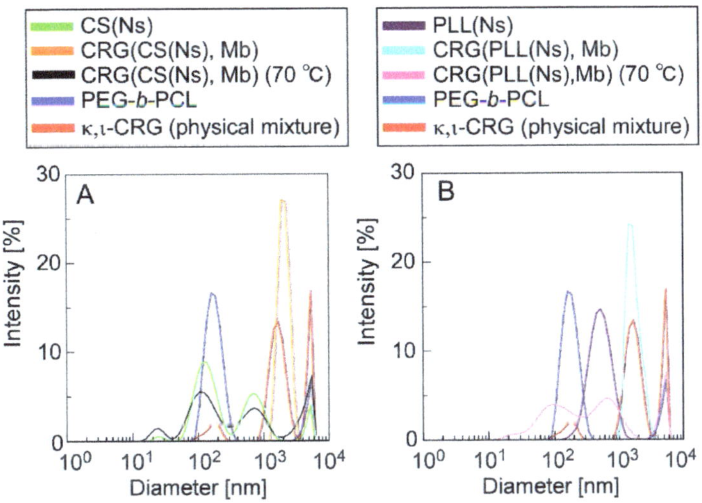

Figure 5. DLS-determined diameter distributions of (**A**) CRG(CS(Ns), Mb) and (**B**) CRG(PLL(Ns), Mb) microparticles. DLS data for CS (Ns) and PLL(Ns) nanoparticles, PEG-*b*-PCL, and a κ,ι-CRG (physical mixture of κ-CRG and ι-CRG) are also shown for comparison.

In comparison, the PLL(CS(Ns), Mb) microparticles (Figure 5B) exhibited a unimodal peak at approximately 2 μm (light blue trace), and three peaks at 100 nm, 500 nm, and 9 μm (pink trace), similar to the CRG(CS(Ns), Mb) microparticles. The peaks near 100 nm and above 9 μm can be ascribed from structures formed by the PEG-b-PCL and CRG precipitates, whereas the peak at approximately 500 nm agreed well with the particle size distribution of the PLL(Ns) nanoparticles (purple trace). These results strongly suggest that the PLL(Ns) nanoparticles were complexed inside the CRG(PLL(Ns), Mb) microparticles.

Figure 6 shows the thermal behavior of the particles, particle forming agents, and inclusions evaluated by DSC. Figure 6 reveals that the CRG(CS(Ns), Mb), CRG(PLL(Ns), Mb), and CRG(Mb) microparticles exhibit endothermic peaks at ~55 °C, and exothermic peaks at 220 or 240 °C. Furthermore, the CRG(CS(Ns), Mb) and CRG(PLL(Ns), Mb) microparticles show minor endothermic peaks at approximately 400 °C. In contrast, the CS(Ns) nanoparticles exhibit an exothermic peak at 250 °C, whereas the PLL(Ns) nanoparticles show a minor endothermic peak at approximately 300 °C. Figure 6 also shows that PEG-b-PCL is associated with the endothermic peaks at 55 and 350 °C. In addition, κ,ι-CRG and CS exhibit exothermic peaks at 200 and 300 °C, respectively, whereas PLL exhibits an endothermic peak at ~300 °C. Mb has exothermic peaks at 200 and 300 °C. Meanwhile, Ns has an endothermic peak at 400 °C, and TPP is associated with the endothermic peaks at 90 and 120 °C.

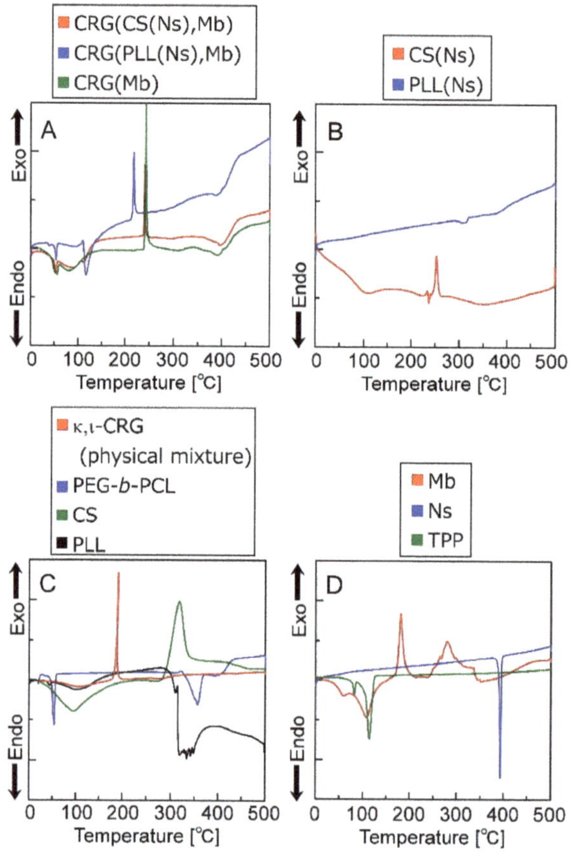

Figure 6. DSC thermograms (0–500 °C) of (**A**) microparticles, (**B**) nanoparticles, (**C**) polymers that form the particles and (**D**) low-molecular-weight molecules that form the microparticles and were encapsulated in the particles.

The CS(Ns) and PLL(Ns) nanoparticles do not exhibit endothermic peaks derived from Ns (~400 °C), suggesting that Ns is uniformly dispersed inside the nanoparticles; however, the microparticles composited with these nanoparticles exhibit a minor endothermic peak at approximately 400 °C, which suggests that a small number of Ns was released from the nanoparticles, which formed local microcrystals during microparticle preparation.

The exothermic peak corresponding to the CRG particles is shifted to a higher temperature than that of the physical mixture of CRGs, indicating that CRG is more thermally stabilized by particle formation, which is ascribed to the nanoparticle internalization and the Mb inside the CRG microparticles. Pure CS has been reported to have an exothermic peak at approximately 300 °C, whereas pure PLL has an endothermic peak at approximately 300 °C [44]. The exothermic peak reportedly shifts to a higher temperature when CS and PLL are mixed with CRG owing to the electrostatic interactions between the cationic polymers and CRG [45,46]. Furthermore, Mb exhibits an exothermic peak at approximately 280 °C and is more thermally stable than CRG. Therefore, both nanoparticle complexation and electrostatic interactions between Mb and CRG can reduce the free volume of CRG and limit polymer-chain mobility, thereby improving the thermal stability of CRG. Furthermore, an endothermic peak associated with the evaporation of the water retained by each material was observed at approximately 100 °C. Therefore, these results show that microparticles several micrometers in diameter can be prepared by the emulsion method while controlling the sol–gel transition of CRG.

3.5. Ns- and Mb-Release Profiles of the CRG(CS(Ns), Mb) and CRG(PLL(Ns), Mb) Microparticles

Figure 7 shows the release profiles of Ns and Mb from the CRG(CS(Ns), Mb) and CRG(PLL(Ns), Mb) microparticles. Approximately 80% of Ns was released by the CRG(CS(Ns) Mb) microparticles at 60 min, whereas the CRG(PLL(Ns), Mb) microparticles released almost 100% at 120 min; both release profiles were found to be temperature-independent (Figure 7A,B). The nanoparticle complexation within the microparticles slightly reduced the Ns-release ratio because the nanoparticles released almost 100% of the Ns in 60 min (Figure 3). These results suggest that the inclusions are rapidly released from the nanoparticles, even near body temperature. In contrast, the release of Mb dispersed within the microparticles was temperature-responsive (Figure 7C,D).

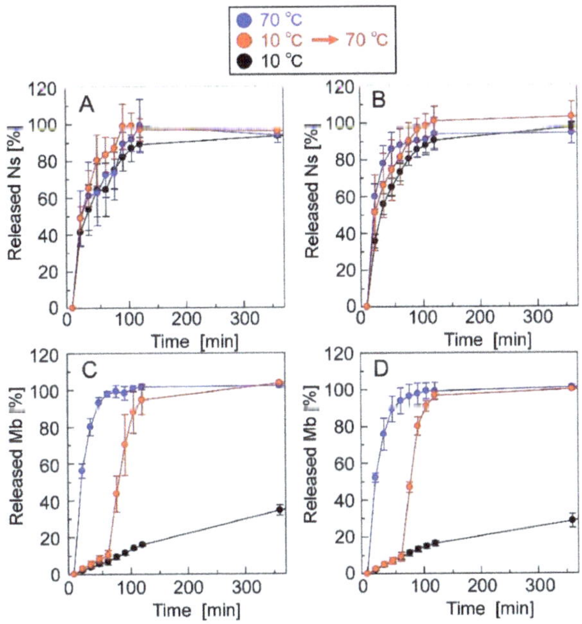

Figure 7. Ns- and Mb-release profiles of the microparticles. The microparticles used were (**A**,**C**) CRG(CS(Ns), Mb) and (**B**,**D**) CRG(PLL(Ns), Mb) microparticles.

We determined the retention ratios and efficiencies of the CS(Ns) and PLL(Ns) nanoparticles within the CRG(CS(Ns), Mb) and CRG(PLL(Ns), Mb) microparticles (Table S2). In addition, we analyzed the encapsulation ratios and efficiencies of Mb within the microparticles (Table S3). The type of the encapsulated nanoparticles has minimal effect on the retention ratios and efficiencies of the nanoparticles or compounds within the microparticles. The nanoparticles complexed more efficiently with the negatively charged microparticles, despite the positive charge of nanoparticles and Mb, which is attributed to the polymer-chain entanglement and electrostatic interactions during microparticle complexation. In other words, the interactions between the cationic nanoparticles and anionic CRG and the entanglement of the CRS chains from the sol–gel transition inhibited the nanoparticle release from the microparticles, resulting in a highly efficient nanoparticle/microparticle complexation.

At the beginning of our research, we designed microparticles containing nanoparticles to achieve a two-step release behavior. It was assumed that Mb retained in the microparticles would be released quickly, followed by the release of Ns retained in the nanoparticles, as shown in Figure 1. In this study, a uniform dispersion of the nanoparticles inside the microparticles was assumed. However, contrary to the initial assumption, we observed the reverse order of Mb and Ns release; that is, Ns retained in the nanoparticles were released quickly, followed by the release of Mb retained in the microparticles. This difference can be attributed to the non-uniformity of the nanoparticles inside the microparticles, their presence near the surface of the microparticles, or their exposure to the solvent. Several factors can cause our above assumption concerning the coexisting structures of the nanoparticles and microparticles, namely (1) crystallinity of PEG-b-PCL, (2) surface morphologies of the particle, and (3) release properties of the particles. Although Ns was hardly released from the particles during their preparation by dialysis, Ns was rapidly released in the PBS solution (Figure 3), suggesting that the state of the nanoparticle-forming compounds (polycation and Ns) was greatly affected by the salts (Na^+ or Cl^-) in the PBS solution. This is mainly ascribed to the electrostatic interaction of the amino group (NH_3^+) in the polycationic molecule and sulfate group (SO_3^- of Ns with the Na^+ and Cl^- ions in the PBS solution, thereby reducing the electrostatic interaction between polycation and Ns, resulting in the rapid release of Ns from the particles. In contrast, the release of Mb retained in the microparticles was suppressed even in the PBS solution (Figure 7). As PEG-b-PCL crystallizes at approximately 20 °C [47,48], the release of Mb was suppressed at 10 °C because of the crystallization of PEG-b-PCL oriented on the surface of the microparticles, which inhibited solvent flow into the particles, thereby releasing Mb after the collapse of the particles as the temperature increased. As Ns was rapidly released from the nanoparticles without the protective effect of the PEG-b-PLC crystallization, the nanoparticles are likely located near the particle surface, or a part of the particle is exposed to the solvent. In fact, the diameter of the nanoparticles and microparticles make it difficult to uniformly disperse the nanoparticles inside the microparticles. Considering the uneven surface of the microparticles observed in Figure 4, a part of the nanoparticles surface is exposed to the solvent owing to the slower release rate of Ns from the nanoparticles in the microparticles (Figure 7) than that of the nanoparticles alone (Figure 3). Therefore, the results obtained in the present study revealed that the prepared polysaccharide microparticles containing nanoparticles can release two different charged compounds in a stepwise manner (Figure 8) with the observed release behavior different from that assumed at the beginning of this study (Figure 1).

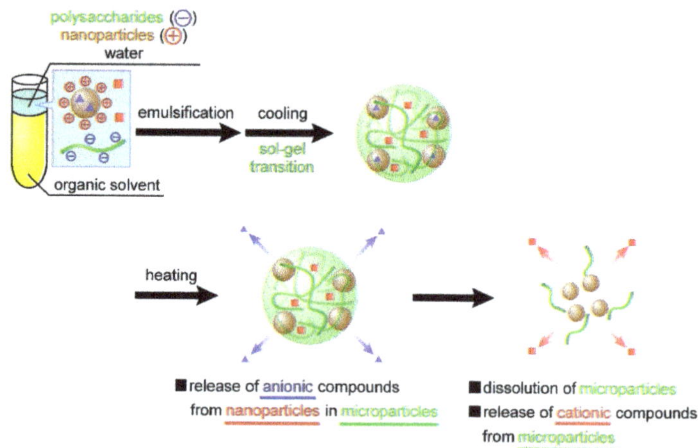

Figure 8. Release behavior of temperature-responsive nanoparticle-containing polysaccharide microparticles based on the experimental results of this study.

4. Conclusions

In this study, we developed polysaccharide microparticles containing nanoparticles for the temperature-responsive and two-step release of inclusions. The CS nanoparticles, PLL nanoparticles, and CRG microparticles separately complexed with both nanoparticles were prepared. The inclusion ability of the nanoparticles in the microparticles, and the effect of the type of cationic polymer on the nanoparticle and microparticle release behavior were evaluated. Nanoparticle/microparticle complexation was confirmed by DSL and DSC. The nanoparticles released their inclusions rapidly regardless of the type of cationic polymer and temperature. Meanwhile, the inclusions within the nanoparticles on the microparticle surfaces were released first, after which the microparticle inclusions were released through microparticle collapse, which revealed the ability of the prepared microparticles to release two differently charged compounds in two steps.

The major challenge for future research involves including the nanoparticles uniformly within the microparticles. The process presented in the paper is expected to result in the formation of nanoparticles even with the use of various other polymers, such as polyethyleneimine (pK_a: 7.0) and polyarginine (pK_a: 12.0). The hydrophilicity of the nanoparticles in the solution can be adjusted significantly by judiciously choosing a polymer with the required pK_a. Increasing the hydrophilicity of the nanoparticles or increasing the amount of nanoparticles and surfactants is expected to disperse the nanoparticles into the emulsion. In addition, further reduction of the nanoparticle size is another major approach for the uniform dispersion inside the microparticles. The nanoparticles are more likely to be dispersed inside the microparticles by further reducing the size of the nanoparticles by varying the ultrasound irradiation time and intensity, type of polycation, and concentration of polycation and TPP (electrostatic interaction). These approaches should promote the two-step release behavior of the microparticles in which the microparticles collapse in response to temperature, followed by the nanoparticle release and the subsequent release of their inclusions.

The particles developed in this study can release a variety of compounds independently of their electric charge, and thus have potential applications in drug delivery systems through various routes of administration, such as transpulmonary, intramuscular, and transdermal routes. In other words, these particles are expected to be used as temperature-responsive drug carriers and for the controlled release of multiple drugs. Although various further optimization studies need to be performed in order to fabricate particles that show temperature responsiveness at practical temperatures (around 37 °C), we believe that the fundamental results obtained in this paper (i.e., the ambitious challenge of two-step release using a combination of nanoparticles and microparticles) will make a significant contribution to research in the field of drug delivery systems.

Supplementary Materials: The following supporting information can be downloaded at: https://www.mdpi.com/article/10.3390/ma15134717/s1, Figure S1: Chemical structures of the polymers (CRGs, CS, and PLL) and compound (TPP) used to form particles, and their inclusions (Mb and Ns); Figure S2: DLS-determined diameter distributions of CS(-) and PLL(-) nanoparticles; Table S1: Encapsulation ratios and encapsulation efficiencies for Ns in CS(Ns) and PLL(Ns) nanoparticles; Table S2: Retention ratios and retention efficiencies for CS(Ns) and PLL(Ns) nanoparticles inside CRG(CS(Ns), Mb) and CRG(PLL(Ns), Mb) microparticles; Table S3: Encapsulation ratios and encapsulation efficiencies of the Mb inside CRG(CS(Ns), Mb) and CRG(PLL(Ns), Mb) microparticles.

Author Contributions: Conceptualization, Y.M.; methodology, T.S. and Y.M.; investigation, T.S. and Y.M.; writing—original draft preparation, T.S.; writing—review and editing, Y.M.; project administration, Y.M.; funding acquisition, Y.M. All authors have read and agreed to the published version of the manuscript.

Funding: This research was funded by Grants-in-Aid for Scientific Research B (JSPS, grant number: 20H04531) and Grant-in-Aid for Challenging Exploratory Research (JSPS, grant number: 21K19892).

Conflicts of Interest: The authors declare no conflict of interest.

References

1. Kanakubo, Y.; Ito, F.; Murakami, Y. Novel one-pot facile technique for preparing nanoparticles modified with hydrophilic polymers on the surface via block polymer-assisted emulsification/evaporation process. *Colloids Surf. B Biointerfaces* **2010**, *78*, 85–91. [CrossRef] [PubMed]
2. Fan, J.; Cheng, Y.; Sun, M. Functionalized Gold Nanoparticles: Synthesis, Properties and Biomedical Applications. *Chem. Rec.* **2020**, *20*, 1474–1504. [CrossRef] [PubMed]
3. Ikem, V.O.; Menner, A.; Bismarck, A. High Internal Phase Emulsions Stabilized Solely by Functionalized Silica Particles. *Angew. Chem. Int. Ed.* **2008**, *47*, 8277–8279. [CrossRef] [PubMed]
4. Zohri, M.; Arefian, E.; Javar, H.A.; Gazori, T.; Aghaee-Bakhtiari, S.H.; Taheri, M.; Fatahi, Y.; Azadi, A.; Khoshayand, M.R.; Ghahremani, M.H. Potential of chitosan/alginate nanoparticles as a non-viral vector for gene delivery: Formulation and optimization using D-optimal design. *Mater. Sci. Eng. C* **2021**, *128*, 112262. [CrossRef] [PubMed]
5. Zhang, Y.; Huang, Y.; Li, S. Polymeric Micelles: Nanocarriers for Cancer-Targeted Drug Delivery. *AAPS PharmSciTech* **2014**, *15*, 862–871. [CrossRef] [PubMed]
6. Tam, Y.Y.C.; Chen, S.; Cullis, P.R. Advances in Lipid Nanoparticles for siRNA Delivery. *Pharmaceutics* **2013**, *5*, 498–507. [CrossRef]
7. Cheng, H.; Cui, Z.; Guo, S.; Zhang, X.; Huo, Y.; Mao, S. Mucoadhesive versus mucopenetrating nanoparticles for oral delivery of insulin. *Acta Biomater.* **2021**, *135*, 506–519. [CrossRef]
8. Yim, H.; Park, S.-J.; Bae, Y.H.; Na, K. Biodegradable cationic nanoparticles loaded with an anticancer drug for deep penetration of heterogeneous tumours. *Biomaterials* **2013**, *34*, 7674–7682. [CrossRef]
9. Bilensoy, E. Cationic nanoparticles for cancer therapy. *Expert Opin. Drug Deliv.* **2010**, *7*, 795–809. [CrossRef]
10. Samal, S.K.; Dash, M.; Van Vlierberghe, S.; Kaplan, D.L.; Chiellini, E.; van Blitterswijk, C.; Moroni, L.; Dubruel, P. Cationic polymers and their therapeutic potential. *Chem. Soc. Rev.* **2012**, *41*, 7147–7194. [CrossRef]
11. Patil, N.A.; Kandasubramanian, B. Functionalized polylysine biomaterials for advanced medical applications: A review. *Eur. Polym. J.* **2021**, *146*, 110248. [CrossRef]
12. Yuan, J.; Guo, L.; Wang, S.; Liu, D.; Qin, X.; Zheng, L.; Tian, C.; Han, X.; Chen, R.; Yin, R. Preparation of self-assembled nanoparticles of ε-polylysine-sodium alginate: A sustained-release carrier for antigen delivery. *Colloids Surf. B Biointerfaces* **2018**, *171*, 406–412. [CrossRef] [PubMed]
13. Divya, K.; Jisha, M.S. Chitosan nanoparticles preparation and applications. *Environ. Chem. Lett.* **2018**, *16*, 101–112. [CrossRef]
14. Zargar, V.; Asghari, M.; Dashti, A. A Review on Chitin and Chitosan Polymers: Structure, Chemistry, Solubility, Derivatives, and Applications. *ChemBioEng Rev.* **2015**, *2*, 204–226. [CrossRef]
15. Jin, H.; Yu, Y.; Chrisler, W.B.; Xiong, Y.; Hu, D.; Lei, C. Delivery of MicroRNA-10b with Polylysine Nanoparticles for Inhibition of Breast Cancer Cell Wound Healing. *Breast Cancer Basic Clin. Res.* **2011**, *6*, 9–19. [CrossRef]
16. Fàbregas, A.; Miñarro, M.; García-Montoya, E.; Pérez-Lozano, P.; Carrillo, C.; Sarrate, R.; Sánchez, N.; Ticó, J.R.; Suñé-Negre, J.M. Impact of physical parameters on particle size and reaction yield when using the ionic gelation method to obtain cationic polymeric chitosan-tripolyphosphate nanoparticles. *Int. J. Pharm.* **2013**, *446*, 199–204. [CrossRef]
17. Mazancová, P.; Némethová, V.; Treľová, D.; Kleščíková, L.; Lacík, I.; Rázga, F. Dissociation of chitosan/tripolyphosphate complexes into separate components upon pH elevation. *Carbohydr. Polym.* **2018**, *192*, 104–110. [CrossRef]
18. Antoniou, J.; Liu, F.; Majeed, H.; Qi, J.; Yokoyama, W.; Zhong, F. Physicochemical and morphological properties of size-controlled chitosan–tripolyphosphate nanoparticles. *Colloids Surf. A Physicochem. Eng. Asp.* **2015**, *465*, 137–146. [CrossRef]
19. Agazzi, M.L.; Herrera, S.E.; Cortez, M.L.; Marmisollé, W.A.; Azzaroni, O. Self-assembled peptide dendrigraft supraparticles with potential application in pH/enzyme-triggered multistage drug release. *Colloids Surf. B Biointerfaces* **2020**, *190*, 110895. [CrossRef]
20. Takeuchi, I.; Tetsuka, Y.; Nii, T.; Shinogase, M.; Makino, K. Inhalable nanocomposite particles using amino acids with improved drug content and humidity resistance. *Colloids Surf. A Physicochem. Eng. Asp.* **2017**, *529*, 387–393. [CrossRef]
21. Al-Qadi, S.; Taboada, P.; Remuñán-López, C. Micro/nanostructured inhalable formulation based on polysaccharides: Effect of a thermoprotectant on powder properties and protein integrity. *Int. J. Pharm.* **2018**, *551*, 23–33. [CrossRef] [PubMed]
22. Loira-Pastoriza, C.; Todoroff, J.; Vanbever, R. Delivery strategies for sustained drug release in the lungs. *Adv. Drug Deliv. Rev.* **2014**, *75*, 81–91. [CrossRef] [PubMed]
23. Okuda, T.; Okamoto, H. Present Situation and Future Progress of Inhaled Lung Cancer Therapy: Necessity of Inhaled Formulations with Drug Delivery Functions. *Chem. Pharm. Bull.* **2020**, *68*, 589–602. [CrossRef]
24. Wall, D.A. Pulmonary Absorption of Peptides and Proteins. *Drug Deliv.* **1995**, *2*, 1–20. [CrossRef]
25. Pilcer, G.; Amighi, K. Formulation strategy and use of excipients in pulmonary drug delivery. *Int. J. Pharm.* **2010**, *392*, 1–19. [CrossRef] [PubMed]
26. Patton, J.S.; Byron, P.R. Inhaling medicines: Delivering drugs to the body through the lungs. *Nat. Rev. Drug Discov.* **2007**, *6*, 67–74. [CrossRef] [PubMed]
27. Sato, T.; Takami, T.; Saito, H.; Murakami, Y. Thermosensitive polysaccharide particles for pulmonary drug delivery. *Colloids Surf. A Physicochem. Eng. Asp.* **2019**, *580*, 123720. [CrossRef]
28. Anzai, R.; Takami, T.; Uchida, Y.; Murakami, Y. Poly(ε-caprolactone) (PCL) hybrid sheets containing polymeric micelles: Effects of inner structures on the material properties of the sheets. *Mater. Sci. Eng. C* **2017**, *72*, 325–331. [CrossRef]
29. Anzai, R.; Murakami, Y. Poly(ε-caprolactone) (PCL)-polymeric micelle hybrid sheets for the incorporation and release of hydrophilic proteins. *Colloids Surf. B Biointerfaces* **2015**, *127*, 292–299. [CrossRef]

30. Anandhakumar, S.; Krishnamoorthy, G.; Ramkumar, K.; Raichur, A. Preparation of collagen peptide functionalized chitosan nanoparticles by ionic gelation method: An effective carrier system for encapsulation and release of doxorubicin for cancer drug delivery. *Mater. Sci. Eng. C* **2017**, *70*, 378–385. [CrossRef]
31. Dyer, A.M.; Hinchcliffe, M.; Watts, P.; Castile, J.; Jabbal-Gill, I.; Nankervis, R.; Smith, A.; Illum, L. Nasal Delivery of Insulin Using Novel Chitosan Based Formulations: A Comparative Study in Two Animal Models Between Simple Chitosan Formulations and Chitosan Nanoparticles. *Pharm. Res.* **2002**, *19*, 998–1008. [CrossRef] [PubMed]
32. Morris, G.A.; Castile, J.; Smith, A.; Adams, G.G.; Harding, S.E. The effect of prolonged storage at different temperatures on the particle size distribution of tripolyphosphate (TPP)—Chitosan nanoparticles. *Carbohydr. Polym.* **2011**, *84*, 1430–1434. [CrossRef]
33. Ito, T.; Takami, T.; Uchida, Y.; Murakami, Y. Chitosan gel sheet containing drug carriers with controllable drug-release properties. *Colloids Surf. B Biointerfaces* **2018**, *163*, 257–265. [CrossRef]
34. Wang, Q.Z.; Chen, X.G.; Liu, N.; Wang, S.X.; Liu, C.S.; Meng, X.H. Protonation constants of chitosan with different molecular weight and degree of deacetylation. *Carbohydr. Polym.* **2006**, *65*, 194–201. [CrossRef]
35. Eckenrode, H.M.; Dai, H.-L. Nonlinear Optical Probe of Biopolymer Adsorption on Colloidal Particle Surface: Poly-l-lysine on Polystyrene Sulfate Microspheres. *Langmuir* **2004**, *20*, 9202–9209. [CrossRef]
36. Dos, A.; Schimming, V.; Tosoni, S.; Limbach, H. Acid-Base Interactions and Secondary Structures of Poly-L-Lysine Probed by ^{15}N and ^{13C} Solid State NMR and Ab initio Model Calculations. *J. Phys. Chem. B* **2008**, *112*, 15604–15615. [CrossRef] [PubMed]
37. Volodkin, D.; Ball, V.; Schaaf, P.; Voegel, J.-C.; Mohwald, H. Complexation of phosphocholine liposomes with polylysine. Stabilization by surface coverage versus aggregation. *Biochim. Biophys. Acta Biomembr.* **2007**, *1768*, 280–290. [CrossRef]
38. Bajpai, A.K.; Mahobia, S.; Bajpai, J. An In-Vitro Investigation of Swelling Controlled Delivery of Insulin from Egg Albumin Nanocarriers. *Iran. J. Pharm. Res.* **2016**, *15*, 695–711. [CrossRef]
39. Zhu, L.; Awais, M.; Javed, H.M.A.; Mustafa, M.S.; Tlili, I.; Khan, S.U.; Shadloo, M.S. Photo-catalytic pretreatment of biomass for anaerobic digestion using visible light and Nickle oxide (NiOx) nanoparticles prepared by sol gel method. *Renew. Energy* **2020**, *154*, 128–135. [CrossRef]
40. Pan, C.; Qian, J.; Zhao, C.; Yang, H.; Zhao, X.; Guo, H. Study on the relationship between crosslinking degree and properties of TPP crosslinked chitosan nanoparticles. *Carbohydr. Polym.* **2020**, *241*, 116349. [CrossRef]
41. Lunardi, C.N.; Gomes, A.J.; Rocha, F.S.; De Tommaso, J.; Patience, G.S. Experimental methods in chemical engineering: Zeta potential. *Can. J. Chem. Eng.* **2020**, *99*, 627–639. [CrossRef]
42. Nishimura, S.; Takami, T.; Murakami, Y. Porous PLGA microparticles formed by "one-step" emulsification for pulmonary drug delivery: The surface morphology and the aerodynamic properties. *Colloids Surf. B Biointerfaces* **2017**, *159*, 318–326. [CrossRef] [PubMed]
43. Chowdhury, H.; Loganathan, B.; Wang, Y.; Mustary, I.; Alam, F. A Study of Dimple Characteristics on Golf Ball Drag. *Procedia Eng.* **2016**, *147*, 87–91. [CrossRef]
44. Liang, C.; Yuan, F.; Liu, F.; Wang, Y.; Gao, Y. Structure and antimicrobial mechanism of ε-polylysine-chitosan conjugates through Maillard reaction. *Int. J. Biol. Macromol.* **2014**, *70*, 427–434. [CrossRef] [PubMed]
45. Li, T.; Wen, C.; Dong, Y.; Li, D.; Liu, M.; Wang, Z.; Janaswamy, S.; Zhu, B.; Song, S. Effect of ε-polylysine addition on κ-carrageenan gel properties: Rheology, water mobility, thermal stability and microstructure. *Food Hydrocoll.* **2019**, *95*, 212–218. [CrossRef]
46. Deka, C.; Dutta, M.; Deka, D.; Jha, D.K.; Kakati, K. Study of olive oil-loaded chitosan/carrageenan coacervate and its antibacterial property. *Int. J. Pharm. Sci.* **2016**, *6*, 1524–1533.
47. Zhang, Y.; Lu, Y.; Cao, M.; Chen, P.; Yang, B.; Miao, J.; Xia, R.; Qian, J. Y-shaped copolymers of poly(ethylene glycol)-poly(ε-caprolactone) with ketal bond as the branchpoint for drug delivery. *Mater. Sci. Eng. C* **2018**, *93*, 554–564. [CrossRef]
48. Huang, Y.; Li, L.; Li, G. An enzyme-catalysed access to amphiphilic triblock copolymer of PCL-b-PEG-b-PCL: Synthesis, characterization and self-assembly properties. *Des. Monomers Polym.* **2015**, *18*, 799–806. [CrossRef]

materials

MDPI

Review

The Role of Mesh Implants in Surgical Treatment of Parastomal Hernia

Karolina Turlakiewicz [1,2,*], Michał Puchalski [1], Izabella Krucińska [1] and Witold Sujka [2]

[1] Institute of Material Science of Textiles and Polymer Composites, Lodz University of Technology, Żeromskiego 116, 90-924 Lodz, Poland; michal.puchalski@p.lodz.pl (M.P.); izabella.krucinska@p.lodz.pl (I.K.)
[2] Tricomed S.A., Świętojańska 5/9, 93-493 Lodz, Poland; witold.sujka@tzmo-global.com
* Correspondence: karolina.turlakiewicz@dokt.p.lodz.pl

Abstract: A parastomal hernia is a common complication following stoma surgery. Due to the large number of hernial relapses and other complications, such as infections, adhesion to the intestines, or the formation of adhesions, the treatment of hernias is still a surgical challenge. The current standard for the preventive and causal treatment of parastomal hernias is to perform a procedure with the use of a mesh implant. Researchers are currently focusing on the analysis of many relevant options, including the type of mesh (synthetic, composite, or biological), the available surgical techniques (Sugarbaker's, "keyhole", or "sandwich"), the surgical approach used (open or laparoscopic), and the implant position (onlay, sublay, or intraperitoneal onlay mesh). Current surface modification methods and combinations of different materials are actively explored areas for the creation of biocompatible mesh implants with different properties on the visceral and parietal peritoneal side. It has been shown that placing the implant in the sublay and intraperitoneal onlay mesh positions and the use of a specially developed implant with a 3D structure are associated with a lower frequency of recurrences. It has been shown that the prophylactic use of a mesh during stoma formation significantly reduces the incidence of parastomal hernias and is becoming a standard method in medical practice.

Keywords: parastomal hernia; surgical mesh; hernia repair; prevention; biocompatibility

Citation: Turlakiewicz, K.; Puchalski, M.; Krucińska, I.; Sujka, W. The Role of Mesh Implants in Surgical Treatment of Parastomal Hernia. *Materials* **2021**, *14*, 1062. https://doi.org/10.3390/ma14051062

Academic Editors: Roser Sabater i Serra and Ángel Serrano-Aroca

Received: 18 January 2021
Accepted: 18 February 2021
Published: 24 February 2021

1. Introduction

Hernias are one of the most common diseases treated by surgery and occur in both women and men, regardless of age. The formation of a hernia is the result of intra-abdominal pressure that is greater than the strength of the connective tissues [1]. A typical hernia consists of a gate and a hernial sac and its contents, e.g., an intestinal loop or a fragment of the network.

The appearance of hernias is facilitated by factors that weaken the transverse fascia or increase the pressure on it, disturbances in collagen metabolism, and defects in anatomical structures [2].

Additionally, some diseases promote the development of hernias, including prostate hypertrophy, malnutrition, constipation, diabetes, hypoproteinemia, or previous surgical procedures that weakened the strength of the connective tissue [3].

Post-operative infection and recurrence are the primary problems associated with hernia repair [4]. The formation of peritoneal adhesions between the abdominal viscera and the mesh is another cause for concern and is the most crucial parameter for parastomal hernia repair. Mesh applications also remain a concern due to potential mesh-associated complications [5]. The selection of an appropriate mesh type and composition of biomaterials affects the success post-implantation and the ability to avoid complications related to specific surgeries [6]. The ideal mesh must permit repair of the fascial defect and incorporation into the surrounding body tissue while providing little adhesion, minimal immune reactions, and suitable tensile strength. The mesh must allow tissue to grow, peritoneal regeneration, and the regular healing process [7].

The high prevalence of parastomal hernia and unsatisfactory clinical results make this ailment challenging to treat. The prophylactic use of a prosthetic mesh is one of the most increasingly used approaches and may reduce the frequency of the occurrence of parastomal hernia [8].

The aim of this work is to outline the problems related to parastomal hernia, review the key properties of surgical meshes available on the market, present surgical techniques currently used to treat parastomal hernia, and characterize the related complications.

2. Parastomal Hernia

Stoma is a surgically created connection between an organ's lumen and the body's surface, i.e., the skin. Stomas are defined depending on the organ involved: colostomy—colon fistula; ileostomy—small intestine fistula; urostomy—urinary tract fistula [9].

The most common long-term complication of a stoma is a parastomal hernia, occurring in approximately 50% of patients with an isolated stoma [10,11]. It is estimated that this is an inevitable complication, and the risk of hernia increases by several percent every year after stoma surgery [12]. According to Carne et al., the incidence of parastomal hernia is assessed differently depending on the type of stoma: with loop ileostomy, hernia occurs in about 0–6.2% of cases; with loop colostomy, hernia occurs in about 0–30.8% of cases; with ileostomy, hernia occurs in approximately 1.8–28.3% of cases, and with colostomy, hernia occurs in 4–48.1% of cases [13].

There are several classifications of parastomal hernia, but so far, the most widely applied is the classification created by a team led by Professor Marek Szczepkowski MD, PhD, named the Bielanski Hospital Classification (BHC) (Table 1) [14]. Based on this classification, the European Hernia Society (EHS) classification was subsequently developed.

Table 1. Bielanski Hospital Classification (BHC) Classification.

	BHC Classification
Type I	Small, isolated parastomal hernia
Type II	Parastomal hernia with coexisting hernia in the midline scar (no significant abdominal deformation)
Type III	Isolated large parastomal hernia
Type IV	Parastomal hernia with coexisting hernia in a scar after a midline incision (significant abdominal deformity)

According to the clinical indications and types of hernia, a parastomal hernia should be treated as a special case of incisional hernia. The development of a hernia is the result of alternations within the structure of the connective tissue, including a problem with collagen synthesis [15–17].

3. Treatment

In Poland, according to National Health Fund (NFZ) data, about 40,000 patients live with a stoma, and each year, approximately 7000 stoma operations are performed, at least 2000 of which will be a parastomal hernia that is eligible for surgical repair [18].

Polish summaries published by the National Health Fund within the "Statistics of the NFZ JGP" programs were analyzed. The listed stoma procedures in which a mesh implant can be applied are presented in Table 2.

The above treatments from groups F22, F31 (A), and F32 account for 10.3% of all procedures, leading to, on average, 4323 hospitalizations per year. The mean value of hospitalization in these groups is EUR 2436. The estimated total value of hospitalizations in these groups is approximately EUR 10.5 million [19].

Repair with the use of a mesh implant (mesh graft) is currently the standard procedure used in the treatment of parastomal hernias. The previously recommended simple suturing of the fascial defect and translocation of the stoma have now been almost completely withdrawn due to the high percentage of hernial recurrences (33–76%) [20] and local infection at a level of 12% [21–23].

Table 2. Stoma procedures in which a mesh implant can be applied.

Surgical Procedure	Group
Loop ileostomy	F22
Other ileostomy	F22
Permanent ileostomy—other	F22
Anterior rectal resection with the formation of a colostomy	F31 (A)
Permanent colostomy	F32
Colostomy—other	F32
Loop colostomy	F32

The method for treating a hernia depends on the extent of damage of the fascia muscular elements of the abdominal wall, the presence of an infection focus, the possibility of using appropriate prosthetic materials, the patient's condition, and the preferences and qualifications of the doctor [24].

The use of a synthetic mesh significantly reduces the recurrence rate of parastomal hernias, but the rate of local complications still requires attention and ranges from 7% to 18% [23].

3.1. Surgical Access

Open and laparoscopic surgery is possible for all types of hernia. The decision regarding the type of surgery is made based on the surgeon's preferences and the patient's medical history and comfort. Laparoscopic surgery is associated with less pain, shorter hospitalization time, and a low probability of intraoperative complications; on the other hand, it requires general anesthesia. People with massive intraperitoneal adhesions, past inflammatory processes of the abdominal cavity, or radiotherapy will not benefit from the laparoscopic technique. For open surgery, the procedure is easier to perform and can be performed under local anesthesia but is characterized by a longer hospitalization time and a greater number of complications [25].

A study by Halabi and colleagues showed that only 10.4% of patients in the study underwent laparoscopic parastomal hernia surgery. The authors concluded that this result may be related to the high number of adhesions in the case of stoma hernias or clear clinical evidence for the management of this type of hernia [26].

3.2. Implant Position

The promising results in using mesh implants for other hernia types [27–30] have encouraged the implantation of meshes also in the treatment of parastomal hernia and for prophylactic purposes.

Studies conducted on various methods of mesh arrangements—sublay, onlay, and intraperitoneal onlay mesh (IPOM)—during parastomal hernia surgery confirmed the minimization of intraoperative and postoperative complications after implantation of a mesh device in comparison with conventional suture repair [31,32].

The simplest method for the use of a synthetic implant is to place the prosthesis on the fascia in the "onlay" position. This procedure involves first preparing the hernial sac and its contents and then reducing or removing the sac after opening. Then, after suturing the hernial gates, an appropriately selected mesh size is placed on the fascia [24]. The theoretical advantage of this technique is that patients do not need to undergo extensive dissection of the abdominal wall to create planes in which the mesh can be inserted, resulting in a shorter recovery time. The disadvantage is that the pressure in the abdominal cavity can displace the mesh, which increases the risk of recurrence, reported in up to 18.6% of cases [23]. Another disadvantage of the onlay method is the increased risk of infection because the selected implant is located near the contaminated stomal opening.

An alternative method is the "sublay" technique, which involves placing the mesh in the preperitoneal layer and fixing it with a single non-absorbable suture. The sublay method is characterized by a lower risk of infection and greater stabilization of the mesh via intra-abdominal pressure [24]. While sublay repair prevents the mesh from interacting with the abdominal organs, the intraperitoneal position (IPOM) poses an increased risk of intestinal erosion and the formation of adhesions. During IPOM repair, care should be taken to maximize tissue adhesion between the mesh and the abdominal wall to minimize seroma formation.

Hansson et al. systematically reviewed surgical techniques for parastomal hernia repair, including a total of 35 studies, which found onlay to have the highest recurrence rate and IPOM to have the lowest [23].

3.3. *Operational Technique*

In the case of parastomal hernia, the applied surgical technique has a key influence on the number of recurrences [23].

Among the techniques for the repair of parastomal hernia with the use of a mesh, the most common in the literature are the Sugarbaker technique (or the modified Sugarbaker technique using a laparoscopic approach), the keyhole technique, and the sandwich technique.

In the modified Sugarbaker technique, intraperitoneal adhesions and hernial gates are prepared after generation of the pneumoperitoneum. After dissecting the adhesions, the intestine is moved to the lateral edge of the hernial gate to create a tunnel between the parietal peritoneum and the mesh. After inserting the mesh into the peritoneal cavity, the hernial gates are covered with a margin of at least 6–7 cm in all directions or, if necessary, across all coexisting hernias. The mesh is fixed to the parietal peritoneum using tackers or sutures at intervals of 3–4 cm [33].

Open access and laparoscopic access are also used for so-called keyhole surgery. This technique uses a round or oval mesh cut from the medial side with an opening for the stoma. This mesh is inserted intraperitoneally. After being fixed, the mesh is sutured, which leads to its closure around the stoma [34,35].

The third technique is the "sandwich" technique, which is a combination of the Sugarbaker and "keyhole" techniques and uses two meshes that are placed intraperitoneally. The first implant, which is the incised mesh or the hole-type mesh, is placed around the stoma sling to cover the orifice of the hernia using the "keyhole technique". The second flat mesh implanted by the Sugarbaker technique forms a plane and, with its medial edge, covers the median wound after laparotomy, making it possible to supply the hernia through the postoperative scar in parallel.

The relapse rate for the Sugarbaker technique is 11.6%, that of the keyhole technique is 34.6% [34], and that of the sandwich technique is 2.1% [36].

The latest research describes experiments that use a dedicated synthetic composite mesh with a 3D structure (Dynamesh—IPST) for the prevention of parastomal hernia. This observational study included 88 patients divided into two cohort groups of patients: a mesh prophylaxis group (43 patients) and a non-mesh prophylaxis group (45 patients). The implant was placed in the IPOM position. During the procedure, the intestinal loop was pulled through a central funnel with 2–3 cm diameter. Then, the funnel was oriented towards the abdomen and made to fit snugly around the gut to prevent stomal loss as well as hernia recurrence. The mesh was fixed with tackers, anchors, or surgical sutures. The results of the study showed that parastomal hernia occurrence after 12 months was 11% in the mesh prophylaxis group (MP) and 54% in the non-mesh prophylaxis group (NMP). There were no significant differences in long-term complications (bowel occlusion (2—MP; 0—NMP), stenosis (2—MP; 1—NMP), or prolapse (1—MP; 2—NMP)) [37].

The aim of the study by G. Kohler et al. was to assess the occurrence of postoperative complications and the possibility of parastomal hernia in patients undergoing stoma surgery with simultaneous prophylactic placement of a 3D hernia mesh. The retrospective analysis was based on the collected data of 80 patients. A parastomal hernia developed in three patients (3.75%). No mesh-related complications were reported. In seven patients, there were complications related to the emerged stoma (infections (3), seroma (2), stenosis (1), and stomal retraction (1)). According to the authors, using the prophylactic implantation of a specially selected 3D mesh implant via the IPOM technique, stoma formation is safe, efficient, and relatively easy to perform. In contrast to flat meshes, by using meshes featuring chimneys at the boundary areas of the stoma, the stoma can be well protected against the internal organs. At the same time, the vertical funnel-shaped portion of the mesh provides protection against stoma loss [38].

Another publication compared the methods for treating a parastomal hernia, focusing on the complications that arose and the number of relapses (Table 3). The analysis performed included 135 patients, and eight different surgical techniques were used. Laparoscopic operations accounted for 46.7% (63/135). In 44 cases, the hernia recurred (32.6%), while in 24 patients (17.8%), perioperative complications occurred; 12 of these patients underwent re-operation. Only in the case of hernia repair using an implant with a 3D structure featuring both open and laparoscopic access were no hernia recurrences noted [39].

Table 3. List of complications depending on the method used [39].

Surgical Technique	Complication Occurrence (Re-Operation)	Hernia Recurrence (Re-Operation)	Number of Performed Treatments
Tension method	4 (3)	10 (6)	25
Onlay	5 (4)	12 (6)	22
Sublay	3 (1)	9 (8)	20
Laparoscopic method: keyhole	4 (2)	10 (6)	22
Laparoscopic method: Sugarbaker	1 (1)	2 (2)	4
Laparoscopic method: "sandwich"	4 (1)	1 (0)	21
Laparoscopic method with 3D implant	2 (0)	0	16
Laparotomy method with 3D implant	1 (0)	0	5

A review carried out by Francis J DeAsis et al. focused on evaluating the efficacy and safety of laparoscopic approaches for parastomal hernia repair. A total of 469 patients were deemed eligible for the present review. Three different surgical techniques were described. Most studies used an expanded polytetrafluoroethylene (ePTFE) mesh. The primary outcome analyzed was the recurrence of parastomal hernia. The recurrence rate was 10.2% for the modified laparoscopic Sugarbaker approach, whereas the recurrence rate was 27.9% for the keyhole approach. For the sandwich technique, there was one recurrence out of 47 repairs. Secondary outcomes (referring to the overall cohort) were mesh infection (1.7%), surgical site infection (3.8%), obstruction requiring reoperation (1.7%), and other complications, such as ileus, pneumonia, and urinary tract infections (16.6%). There were no intraoperative mortalities and six mortalities during the postoperative course [40].

4. Mesh Implants

In addition to the variety of surgical techniques used in the treatment of parastomal hernias, a number of materials are currently available for the replacement of fascia muscular defects in the abdominal wall [24].

Regardless of the type of material used, the material should meet a number of basic parameters that affect the body's immune response to the implant and also reduce fibrosis [41,42].

In the repair of abdominal hernias, prostheses are characterized by their durability and ease of use. The construction of mesh devices used in the surgical treatment of parastomal hernia should allow for safe intraperitoneal implantation via open and laparoscopic methods [24].

4.1. *Required Parameters and Properties*

4.1.1. Resistance and Elasticity

The tension on the abdominal wall can be calculated according to Laplace's law: tension = (diameter × pressure)/(4 × wall thickness). The maximum intra-abdominal pressure in healthy adults is generated by coughing and jumping and is approximately 170 mmHg.

Surgical meshes used to repair large hernias should withstand at least 180 mmHg before they break (tensile strength approx. 16 N/cm) [41,42].

Notably, the abdominal wall has twice the flexibility in the longitudinal direction than in the transverse direction [43,44].

The overall strength of the implant depends on the material used. Hernia recurrence increases when the mesh stretches more or less than the abdominal wall. Moreover, to meet the elasticity requirements, a hernial mesh should be elastic in all directions and have an anisotropic structure [45].

Saberski and Novitsky evaluated the anisotropy of six commercial meshes: Prolite, Pareietex, Ultrapro, Trelex (made of polypropylene), Dualmesh ePTFE, and INFINIT knitted polytetrafluoroethylene (PTFE). All implants, except for Dualmesh, feature significant anisotropic properties [46].

Due to the anisotropic properties of most of the designed products, there is a need to describe the two-direction strength properties of the meshes on the packaging, as well as the lowest strength value.

4.1.2. Pore Size

Porosity is a major indicator of tissue response. The geometry and sizes of pores define the ability of mesh ingrowth. These parameters can be controlled during the manufacturing process. Mesh pores are classified into five different groups: micro-pores (size less than 0.1 mm), small pores (0.1–0.6 mm), medium pores (1 mm), large pores (1–2 mm), and very large pores (>2 mm) [47].

The pores of meshes must be larger than 75 μm to allow the infiltration/penetration of macrophages, fibroblasts, blood vessels, and collagen. Surgical meshes with larger pores allow for the faster growth of soft tissues and are also more flexible due to the lack of formation of so-called granuloma bridging [30,41,48].

The above-mentioned bridges illustrate the process by which individual granulomas blend together and cover the entire mesh. As a result, the mesh stiffens the implant and reduces its flexibility. This phenomenon occurs in meshes with pores smaller than 800 μm [41].

Studies have shown that macroporous meshes have a positive effect on vascularization [30] and also reduce the risk of infection [49]. Meshes with larger pore sizes featuring reduced polypropylene mass lead to a decreased inflammatory reaction in the human abdominal wall [50].

Orenstein et al. reported that large pore sizes (pores larger than 1.5 mm) in meshes improved patients' quality of life after a surgical procedure with the mesh [51].

According to the review carried out by A.S.W. Jacombs, the dual concepts of effective porosity and biofilm may be crucial in mesh-related morbidity and should be investigated further. Developing new mesh implants to maintain effective porosity and reduce biofilm formation may help reduce mesh-related complications [52].

4.1.3. Surface Mass

The weight of the surgical mesh depends both on the weight of the biomaterial itself and the amount of its use (pore size). Heavy surgical meshes are made out of materials with a higher linear mass and have both small pores and high tearing strength. On the other hand, ultralight surgical meshes are made from thinner filaments and have large pores (>1 mm). Lightweight meshes, in general, reduce adverse effects, including chronic pain, fibrosis, adhesion and inflammatory response, and foreign body sensation. These meshes are also characterized by greater flexibility in comparison with heavy-weight meshes [53–56]. Despite their lower strength parameters, ultralight meshes are able to withstand pressures above the maximum intra-abdominal pressure of 170 mmHg.

Based on their weights, the meshes are classified into four categories: ultralight meshes (<35 g/m^2), light meshes (≥35–70 g/m^2), standard meshes (≥70–140 g/m^2), and heavy meshes (>140 g/m^2) [44,57,58].

4.1.4. Implant Contraction

Shrinkage of the mesh occurs due to the reduction in scar tissue around it. Scar tissue contracts to approximately 60% of the original wound surface, and smaller pores cause greater contraction [41].

It is assumed that light meshes, which reduce the amount of scar tissue formation, will have a lower degree of contraction [48]. Due to this shrinkage, it is recommended to use at least a 5-cm overlap around the defect [59].

Research carried out by Masayuki Endo et al. showed a reduction in the apparent surface area between implantation and the second day, indicating that most mesh deformation occurs prior to tissue ingrowth. In conclusion, the surface area of implanted meshes (Marlex, DynaMesh, Ultrapro) decreased by 21% by day 90 [60].

4.1.5. Biocompatibility

The biomaterials currently available on the market are physically and chemically inert. They are generally stable, non-toxic, and do not trigger an immune system response. Despite this, these materials are not biologically inert, and their presence in the patient's body causes a reaction to the foreign body (inflammation, fibrosis, calcification, thrombosis, and the formation of granulomas). Mesh implantation infections are always of concern because they are difficult to deal with without removing the mesh. The risk of infection increases in an infected surgical field, such as parastomal hernia surgery.

4.1.6. Adhesion to the Intestines

As the medical community began to locate the surgical mesh, concerns arose about the adhesion of this device. The effect on the adhesion of the implant to the intestine is determined by the structure and surface of the fibers and the size of the pores. Meshes with a high surface mass cause intense fibrotic processes, yielding strong adhesion to the abdominal wall [61].

On the other hand, microporous ePTFE meshes do not allow for tissue hypertrophy. Therefore, the risk of adhesion is lower, as the mesh is unable to adhere to the abdominal wall [62].

The presented examples show the difficulties associated with the production of a surgical mesh that would have both adhesive and anti-adhesive properties. Hence, manufacturers of surgical meshes have made attempts to design a composite device that allows the mesh to grow into the abdominal cavity without the formation of adhesions on the peritoneal side through the use of surface modification techniques and methods of knitting fabrics by combining several synthetic materials.

4.1.7. Material

For the production of surgical meshes, monofilament or multifilament yarns of various surface masses are used [63,64]. Microporous and multifilament meshes are among the devices at a higher risk of infection because macrophages and neutrophils are unable to penetrate through small pores (<10 μm). This allows bacteria (<1 μm) to survive unhindered inside the pores. Meshes with a low infection risk include those made of monofilaments with openings greater than 75 μm [65].

4.2. Implants Available on the Market

4.2.1. Synthetic

Non-resorbable synthetic surgical meshes are used for long term strengthening of damaged tissues as a result of a hernia. Mesh implants come in two different structures: knitted and woven. The conducted research showed that polypropylene (PP) is the most commonly used non-absorbable material for mesh implants [41,66–68].

Apart from PP, the materials used also include ePTFE, PTFE, and poly(ethylene terephthalate) (PET) [69,70].

First-generation implants can be divided into three categories: (1) macroporous meshes, (2) microporous meshes, (3) and macroporous meshes with multifilament and microporous components [71].

The use of non-resorbable mesh implants has many clinically proven advantages and disadvantages. On the one hand, due to the higher strength values of these meshes compared to natural tissues, these meshes lead to a reduction in flexibility, the occurrence of immune reactions in the body, and chronic pain. On the other hand, their affordable price and good overgrowth with native tissue support their worldwide use [55].

First-generation implants also include fully resorbable meshes. Unlike non-resorbable meshes, these meshes are designed to minimize inflammation and the amount of foreign material implanted. Due to progressive degradation, mechanical stability can be lost early, leading to possible hernia recurrences [65]. Table 4 presents examples of first-generation mesh implants available on the market.

Table 4. Classification of commercially available first-generation mesh implants [41].

Product	Manufacturer	Material	Filament	Surface Mass (g/m^2)	Resorbable	Pore Size (mm)
Vicryl	Ethicon	Polyglactin	Multifilament	56	Yes	0.4
Sefil	B-Baun	PGA	Multifilament	56	Yes	0.75
Goretex	Gore	e-PTFE	Multifilament	N.A.	No	0.003
Optomesh	Tricomed	PP	Monofilament	60–85	No	>1
Parietene	Covidien	PP	Monofilament	80–100	No	0.8
Prolene	Ethicon	PP	Monofilament	80–100	No	0.8
3D Max	BARD	PP	Monofilament	80–100	No	0.8
Premilene	B-Braun	PP	Monofilament	80–100	No	0.8
Polysoft	BARD	PP	Multifilament	80–100	No	0.8
Optomesh Ultralight	Tricomed	PP	Monofilament	24–35	No	>1
Prolene Light	Covidien	PP	Monofilament	36–48	No	1.0–3.6
Optilene	B-Baun	PP	Monofilament	36–48	No	1.0–3.6
Mersilene	Ethicon	PP	Monofilament	40	No	1.0–2.0
Parietex Lightweight	Medtronic	PET	Monofilament	N.A.	No	>1

4.2.2. Composite

Second-generation meshes include implants made of at least two synthetic materials with different properties on each side of the mesh (Table 5). Improvements have been made to reduce complications such as recurrent hernia, infections, and adhesions. The main advantage of composite meshes is their intraperitoneal application. The side of the mesh in contact with the peritoneum is usually smooth and microporous to prevent adherence to the intestines, while the side facing the connective tissues is rough with large pores, which positively affects tissue overgrowth [72].

Composite meshes mainly consist of a PP core covered with another synthetic or natural material. Materials used for non-adhesive coatings include, among others, titanium [73,74], chitosan [75], poly(glycolic acid) (PGA) [76], cellulose [77], and collagen [78]. Another approach for the production of composite meshes is to join filaments with different properties, such as PP, e-PTFE [79], and polyvinylidene fluoride (PVDF) [80,81].

Table 5. Classification of commercially available composite mesh implants [41].

Product	Manufacturer	Material	Filament	Surface Mass (g/m^2)	Resorbable	Pore Size
Ultrapro	Ethicon	PP/PGC-25	Monofilament	28	Partially (<140 days)	>3
Vypro, Vypro II	Ethicon	PP/polyglactin 910	Multifilament	25 and 30	Partially (42 days)	>3
Composix EX Dulex	BARD	PP/e-PTFE	Monofilament	N.A.	No	0.8
Proceed	Ethicon	PP/cellulose	Monofilament	45	Partially	N.A
TiMeshTiMesh Extralight	PFM	PP/tytan	Monofilament	16 and 35	No	>1
DynaMesh—IPST/IPOM	FEG Textiltechnik	PP/PVDF	Monofilament	60	No	1–2

Experience in the use of composite meshes for both repair and prophylactic purposes in parastomal hernia indicates a minimal risk of infection and a low risk of complications [82].

4.2.3. Biological

Biological meshes were introduced in the 1990s. Cells derived from human tissues (allograft) or animals (xenograft) are used for their production. The dermis is the most commonly used tissue due to its ability to create a larger mesh size, but prostheses with cells derived from the intestinal mucosa and pericardium are also available. Essentially, all biological meshes provide the extracellular scaffold necessary to rebuild healthy tissue, allowing mass transport by ingrowing new blood vessels and infiltrating native cells, including fibroblasts and myocytes, ultimately resulting in the deposition of a new extracellular matrix [83].

Compared to synthetic meshes, biological meshes are more biocompatible and elicit a lower inflammatory response in the body but are associated with a greater number of hernia recurrences due to their lower mechanical strength compared to synthetic meshes [84]. Although, the research carried out by Holmdahl et. al. showed comparable recurrence rates between the group of patients who received a full-thickness skin graft (8.3%) and a synthetic mesh (7.1%) [85]. Commercially available biological mesh implants are outlined in Table 6.

Table 6. Classification of commercially available biological mesh implants [86–89].

Product	Manufacturer	Material	Cross-Linking	Resistance (MPa)
CollaMend	Davol	Animal cell-free skin matrix	Yes	11
Permacol	Covidien	Animal cell-free skin matrix	Yes	39
Strattice	LifeCell	Animal cell-free skin matrix	No	18
XenMatrix	Davol	Animal cell-free skin matrix	No	14

5. Prophylactic Implantation of a Mesh Device

According to the guidelines of the European Hernia Society, the prevention of parastomal hernias in patients undergoing end colostomy surgery with prophylactic mesh implantation was satisfactory [90].

The prophylactic use of a mesh implant in permanent stoma surgery reduces the risk of a parastomal hernia by 75%. Moreover, complications occur only in individual cases, so it can be concluded that mesh implantation in this type of surgery could be routinely applied [91].

An analysis conducted by Shuanhu Wang et al. aimed at assessing the effectiveness of prophylactic mesh implantation during end colostomy. The results showed that in the case of sigmoid terminal colostomy, prophylactic mesh placement reduced the incidence of parastomal hernias and associated reoperations. There were no significant differences in stoma-related complications. Moreover, the surgical techniques of sublay and IPOM are considered to be safe and feasible, reducing the likelihood of a parastomal hernia [92].

6. Current Trends

Electrospinning and 3D printing are examples of manufacturing techniques used for the fabrication of drug-loaded devices.

The encapsulation of antimicrobial agents or drugs is one of the possible approaches that could be utilized to produce meshes with antibacterial properties. Pérez-Köhler et al. developed a new coating material known as hyaluronic acid-poly(N-isopropylacrylamide) (HApN), which forms a hydrogel that can be used as a coating for meshes only when it reaches body temperature. The authors selected two different coating formulations—one based on antibiotics (gentamicin + rifampicin) and one based on an antiseptic (chlorhexidine). The results of this study showed that HApN, when loaded with drugs, inhibited the in vitro the growth of several Gram-positive and Gram-negative bacteria [93].

The next study carried out by Nadia Qamar et al. explored the application of the fused deposition modeling in the fabrication of personalized hernial meshes with and without loading of a pharmaceutical agent (ciprofloxacin HCl). All the printed meshes (PP and polyvinyl alcohol (PVA)) showed good mechanical properties. Meshes made of PVA demonstrated a faster release of the loaded drug in comparison to the PP mesh. Moreover, in vivo testing revealed no signs of implant rejection along with a reduction in adhesion to the visceral side and faster wound healing [94].

Another solution to improve the implant properties could involve the use of metallic or diamond nanoparticles. A polypropylene–nano-diamond composite hernia mesh exhibited a significant reduction in protein absorption consistent with lower inflammatory responses; furthermore, no cytotoxicity was observed [95].

The implementation of these novel materials needs further clinical trials to determine the superiority of such materials compared to those available on the market.

7. Conclusions

The best strategy for the prevention and treatment of parastomal hernias has not yet been identified. The variety of available mesh implants—their various sizes, materials, possible spatial structures, and related surgical techniques—and the choices of mesh arrangement in relation to the layers of the abdominal wall allow surgeons to choose the best parameters depending on the preferences and needs of the patient. Choosing the right surgical mesh, however, is not the objective for a successful operation. One of the most important considerations for surgeons should be the technique used to secure the mesh in the surgical field. If the mesh is too small or under too much tension, complications with implantation will be inevitable, regardless of the material used. Despite the reduction in hernia recurrence when surgical meshes are used, it remains necessary to consider the possibility of complications such as infections, adhesions, or intestinal obstruction. Most of these disadvantages are related to the chemical and structural nature of the implant itself. The "golden mean" for the mesh is considered to be optimal integration with the abdominal wall and minimal adherence on the peritoneal side. Due to the technique's satisfactory clinical results, the prophylactic use of meshes during stoma recovery is an increasingly common approach to reduce the occurrence of parastomal hernias.

Despite the progress that has been made in the design of hernial meshes, further research is needed to understand the complex tissue–implant interactions to achieve a reduction in adhesion, infections, and immune responses, as well as better biocompatibility.

Author Contributions: Conceptualization, M.P., I.K. and W.S.; methodology, M.P. and W.S.; formal analysis, K.T.; investigation, K.T.; writing—original draft preparation, K.T.; writing—review and editing, M.P. and W.S.; supervision, M.P., W.S., and I.K. All authors have read and agreed to the published version of the manuscript.

Funding: This research received no external funding.

Institutional Review Board Statement: Not applicable.

Informed Consent Statement: Not applicable.

Data Availability Statement: The data presented in this study are openly available at [doi:10.1007/s00268-015-3187-1], [39]; at [doi:10.1308/003588410X12664192076296], [41]; at [doi.org/10.1007/s10029-013-1054-2], [86]; at [doi:10.1007/s10029-013-1070-2], [87]; at [doi:10.1007/s10029-010-0777-6], [88]; at [doi:10.1016/j.surge.2012.02.006], [89].

Acknowledgments: This article was completed while the first author was a Doctoral Candidate in the Interdisciplinary Doctoral School at the Lodz University of Technology, Poland.

Conflicts of Interest: The authors declare no conflict of interest.

References

1. Smeak, D. Abdominal Wall Reconstruction and Hernias. *Textb. Small Anim. Surg.* **2003**, *84*, 449–470.
2. Bendavid, R.; Abrahamson, J.; Arregui, M.E.; Flament, J.B.; Phillips, E.H. *Abdominal Wall Hernias: Principles and Management*; Springer: Cham, Switzerland, 2001; pp. 39–192.
3. Conze, J.; Klinge, U.; Schumpelick, V. *Surgical Treatment: Evidence-Based and Problem-Oriented W.*; Zuckschwerdt Verlag GmbH: Munich, Germany, 2001.
4. Corduas, F.; Lamprou, D.A.; Mancuso, E. Next-generation surgical meshes for drug delivery and tissue en-gineering applications: Materials, design and emerging manufacturing technologies. *Bio-Design Manuf.* **2021**, 1–33. [CrossRef]
5. Husain, S.G.; Cataldo, T.E. Late Stomal Complications. *Clin. Colon Rectal Surg.* **2008**, *21*, 31–40. [CrossRef] [PubMed]
6. Winsnes, A.; Haapamäki, M.M.; Gunnarsson, U.; Strigård, K. Surgical outcome of mesh and suture repair in primary umbilical hernia: Postoperative complications and recurrence. *Hernia* **2016**, *20*, 509–516. [CrossRef]
7. Saxena, A.K. Surgical perspectives regarding application of biomaterials for the management of large congenital dia-phragmatic hernia defects. *Pediatric Surg. Int.* **2018**, *34*, 475–489. [CrossRef]
8. Lambrecht, J.R.; Larsen, S.G.; Reiertsen, O.; Vaktskjold, A.; Julsrud, L.; Flatmark, K. Prophylactic mesh at end-colostomy con-struction reduces parastomal hernia rate: A randomized trial. *Colorectal Disease* **2015**, *17*, O191–O197. [CrossRef] [PubMed]
9. Lewandowska, I.; Kowalik, J.; Kopański, Z.; Furmanik, F. Care over patients with stoma. *J. Public Health Nurs. Med. Rescue* **2011**, *92*, 9–13.
10. Shabbir, J.; Chaudhary, B.N.; Dawson, R. A systematic review on the use of prophylactic mesh during primary stoma for-mation to prevent parastomal hernia formation. *Colorectal Dis.* **2012**, *14*, 931–936. [CrossRef]
11. Cheung, M.-T.; Chia, N.-H.; Chiu, W.-Y. Surgical treatment of parastomal hernia complicating sigmoid colostomies. *Dis. Colon Rectum* **2001**, *44*, 266–270. [CrossRef] [PubMed]
12. Israelsson, L.A. Parastomal Hernias. *Surg. Clin. N. Am.* **2008**, *88*, 113–125. [CrossRef]
13. Carne, P.W.G.; Robertson, G.M.; Frizelle, F.A. Parastomal hernia. *BJS* **2003**, *90*, 784–793. [CrossRef]
14. Szczepkowski, M. Przepuklina okołostomijna. *Nasza Troska* **2006**, *21*, 8–9.
15. Kasperk, R.; Willis, S.; Klinge, U.; Schumpelick, V. Update on incisional hernia. Parastomal hernia. *Chirurg* **2002**, *73*, 895–898. [CrossRef]
16. Klinge, U.; Conze, J.; Krones, C.J.; Schumpelick, V. Incisional Hernia: Open Techniques. *World J. Surg.* **2005**, *29*, 1066–1072. [CrossRef]
17. Koruth, S.; Chetty, Y.N. Hernias–Is it a primary defect or a systemic disorder? Role of collagen III in all hernias—A case control study. *Ann. Med. Surg.* **2017**, *19*, 37–40. [CrossRef]
18. Bazaliński, D.; Barańska, B.; Cipora, E. Stomia jelitowa nie musi być przyczyną niepełnosprawności. *Med. Rodz.* **2006**, *3*, 58–63.
19. Naradowy Fundusz Zdrowia- Statystyki. Available online: https://statystyki.nfz.gov.pl/Benefits/1a (accessed on 10 October 2020).
20. Bielecki, K. Co nowego w chirurgii ogólnej? *Post N Med.* **2016**, *XXIX*, 144–153.
21. Rieger, N.; Moore, J.; Hewett, P.; Lee, S.; Stephens, J. Parastomal hernia repair. *Color. Dis.* **2004**, *6*, 203–205. [CrossRef]
22. Riansuwan, W.; Hull, T.L.; Millán, M.M.; Hammel, J.P. Surgery of recurrent parastomal hernia: Direct repair or relocation? *Color. Dis.* **2009**, *12*, 681–686. [CrossRef] [PubMed]
23. Hansson, B.M.; Slater, N.J.; van der Velden, A.S.; Groenewoud, H.M.; Buyne, O.R.; de Hingh, I.H.; Bleichrodt, R.P. Surgical techniquesfor parastomal hernia repair: A systematic review of the literature. *Ann. Surg.* **2012**, *255*, 685–695. [CrossRef] [PubMed]
24. Janczak, D.; Litarski, A.; Merenda, M.; Rać, J.; Wieraszko, A.; Litarski, A. Zastosowanie siatek syntetycznych w leczeniu przepuklin brzusznych. *Polim. Med.* **2011**, *41*, 3.
25. Chen, D.C.; Morrison, J. State of the art: Open mesh-based inguinal hernia repair. *Hernia* **2019**, *23*, 485–492. [CrossRef] [PubMed]
26. Halabi, W.J.; Jafari, M.D.; Carmichael, J.C.; Nguyen, V.Q.; Mills, S.; Phelan, M.; Stamos, S.J.; Pigazzi, A. Laparoscopic versus open repair of parastomal hernias: An ACS-NSQIP analysis of short-term outcomes. *Surg. Endosc.* **2013**, *27*, 4067–4072. [CrossRef] [PubMed]
27. Bittner, R.; Schmedt, C.-G.; Leibl, B.J.; Schwarz, J. Early Postoperative and One Year Results of a Randomized Controlled Trial Comparing the Impact of Extralight Titanized Polypropylene Mesh and Traditional Heavyweight Polypropylene Mesh on Pain and Seroma Production in Laparoscopic Hernia Repair (TAPP). *World J. Surg.* **2011**, *35*, 1791–1797. [CrossRef]
28. Schopf, S.; von Ahnen, T.; von Ahnen, M.; Achardey, H. Chronic pain after laparoscopic transabdominal preperitioneal he-ria repair: A randomized comparison of light and extralight titanized polypropylene mesh. *World J. Surg.* **2011**, *35*, 302–310. [CrossRef]
29. Ruteglrd, M.; Gümüsçü, R.; Stylianidis, G.; Nordin, P.; Nilsson, E.; Haapamäki, M.M. Chronic pain, discomfort, quality of life and impact on sex life after open inguinal hernia mesh repair: An expertise based randomized clinical trial comparing lightweight and heavyweight mesh. *Hernia* **2018**, *22*, 411–418. [CrossRef] [PubMed]
30. Weyhe, D.; Belyaev, O.; Muller, C.; Meurer, K.; Bauer, K.-H.; Papapostolou, G.; Uhl, W. Improving Outcomes in Hernia Repair by the Use of Light Meshes—A Comparison of Different Implant Constructions Based on a Critical Appraisal of the Literature. *World J. Surg.* **2006**, *31*, 234–244. [CrossRef]
31. Berger, D. Prevention of parastomal hernias by prophylactic use of a specially designed intraperitoneal onlay mesh (Dy-namesh IPST). *Hernia* **2008**, *12*, 243–246. [CrossRef]

32. Tam, K.-W.; Wei, P.-L.; Kuo, L.-J.; Wu, C.-H. Systematic Review of the Use of a Mesh to Prevent Parastomal Hernia. *World J. Surg.* **2010**, *34*, 2723–2729. [CrossRef]
33. Asif, A.; Ruiz, M.; Yetasook, A.; Denham, W.; Linn, J.; Carbray, J.; Ujiki, M.B. Laparoscopic modified Sugarbaker technique results in superior recurrence rate. *Surg. Endosc.* **2012**, *26*, 3430–3434. [CrossRef]
34. Hansson, B.M.E.; Van Nieuwenhoven, E.J.; Bleichrodt, R.P. Promising new technique in the repair of parastomal hernia. *Surg. Endosc.* **2003**, *17*, 1789–1791. [CrossRef]
35. van Sprundel, T.C.; Gerritsen van der Hoop, A. Modified technique for parastomal hernia repair in patients with intracta-ble stoma-care problems. *Colorectal Dis.* **2005**, *7*, 445–449. [CrossRef] [PubMed]
36. Berger, D.; Bientzle, M. Polyvinylidene fluoride: A suitable mesh material for laparoscopic incisional and parastomal hernia repair! A prospective, observational study with 344 patients. *Hernia* **2009**, *13*, 167–172. [CrossRef]
37. Bertoglio, C.L.; Maspero, M.; Morini, L.; Zironda, A.; Alampi, B.D.; Mazzola, M.; Magistro, C.; Carnevali, P.; Ferrari, G. Permanent end-colostomy parastomal hernia prevention using a novel three-dimensional mesh. *Hernia* **2020**, 1–9. [CrossRef]
38. Köhler, G.; Hofmann, A.; Lechner, M.; Mayer, F.; Wundsam, H.; Emmanuel, K.; Fortelny, R.H. Prevention of parastomal hernias with 3D funnel meshes in intraperitoneal onlay position by placement during initial stoma formation. *Hernia* **2016**, *20*, 151–159. [CrossRef] [PubMed]
39. Köhler, G.; Mayer, F.; Wundsam, H.; Schrittwieser, R.; Emmanuel, K.; Lechner, M. Changes in the Surgical Management of Parastomal Hernias Over 15 Years: Results of 135 Cases. *World J. Surg.* **2015**, *39*, 2795–2804. [CrossRef] [PubMed]
40. Deasis, F.J.; Lapin, B.; Gitelis, M.E.; Ujiki, M.B. Current state of laparoscopic parastomal hernia repair: A meta-analysis. *World J. Gastroenterol.* **2015**, *21*, 8670. [CrossRef] [PubMed]
41. Brown, C.N.; Finch, J.G. Which mesh for hernia repair? *Ann. R. Coll. Surg.* **2010**, *92*, 272–278. [CrossRef] [PubMed]
42. Klinge, U.; Klosterhalfen, B.; Schumpelick, V. Foreign Body Reaction to Meshes of Used for the Repair of Abdominal Wall Hernias. *Eur. J. Surg.* **1999**, *165*, 665–673.
43. Junge, K.; Klinge, U.; Prescher, A.; Giboni, P.; Niewiera, M.; Schumpelick, V. Elasticity of the anterior abdominal wall and impact for reparation of incisional hernias using mesh implants. *Hernia* **2001**, *5*, 113–118.
44. Grundfest-Broniatowski, S. What would surgeons like from materials scientists? *Wiley Interdiscip. Rev. Nanomed. Nanobiotechnol.* **2013**, *5*, 299–319. [CrossRef]
45. Sanbhal, N.; Miao, L.; Xu, R.; Khatri, A.; Wang, L. Physical structure and mechanical properties of knitted hernia mesh materials: A review. *J. Ind. Text.* **2018**, *48*, 333–360. [CrossRef]
46. Saberski, E.R.; Orenstein, S.B.; Novitsky, Y.W. Anisotropic evaluation of synthetic surgical meshes. *Hernia* **2010**, *15*, 47–52. [CrossRef] [PubMed]
47. Lake, S.P.; Ray, S.; Zihni, A.M.; Thompson, D.M.; Gluckstein, J.; Deeken, C.R. Pore size and pore shape–but not mesh density–alter the mechanical strength of tissue ingrowth and host tissue response to synthetic mesh materials in a porcine model of ventral hernia repair. *J. Mech. Behav. Biomed. Mater.* **2015**, *42*, 186–197. [CrossRef]
48. Klosterhalfen, B.; Junge, K.; Klinge, U. The lightweight and large porous mesh concept for hernia repair. *Expert Rev. Med. Devices* **2005**, *2*, 103–117. [CrossRef]
49. Zogbi, L. The Use of Biomaterials to Treat Abdominal Hernias. In *Biomaterials Applications for Nanomedicine*, 1st ed.; Pignatello, R., Ed.; InTech: Rijeka, Croatia, 2008; Volume 18, pp. 359–382.
50. Cobb, W.S.; Burns, J.M.; Peindl, R.D.; Carbonell, A.M.; Matthews, B.D.; Kercher, K.W.; Heniford, B.T. Textile Analysis of Heavy Weight, Mid-Weight, and Light Weight Polypropylene Mesh in a Porcine Ventral Hernia Model. *J. Surg. Res.* **2006**, *136*, 1–7. [CrossRef] [PubMed]
51. Orenstein, S.B.; Saberski, E.R.; Kreutzer, N.L.; Novitsky, Y.W. Comparative Analysis of Histopathologic Effects of Synthetic Meshes Based on Material, Weight, and Pore Size in Mice. *J. Surg. Res.* **2012**, *176*, 423–429. [CrossRef]
52. Jacombs, A.S.W.; Karatassas, A.; Klosterhalfen, B.; Richter, K.; Patiniott, P.; Hensman, C. Biofilms and effective porosity of hernia mesh: Are they silent assassins? *Hernia* **2020**, *24*, 197–204. [CrossRef] [PubMed]
53. Bilsel, Y.; Abci, I. The search for ideal hernia repair; mesh materials and types. *Int. J. Surg.* **2012**, *10*, 317–321. [CrossRef] [PubMed]
54. Zhu, L.-M.; Schuster, P.; Klinge, U. Mesh implants: An overview of crucial mesh parameters. *World J. Gastrointest. Surg.* **2015**, *7*, 226–326. [CrossRef] [PubMed]
55. Kalaba, S.; Gerhard, E.; Winder, J.S.; Pauli, E.M.; Haluck, R.S.; Yang, J. Design strategies and applications of biomaterials and devices for Hernia repair. *Bioact. Mater.* **2016**, *1*, 2–17. [CrossRef]
56. Lee, S.D.; Son, T.; Lee, J.-B.; Chang, Y.S. Comparison of partially-absorbable lightweight mesh with heavyweight mesh for inguinal hernia repair: Multicenter randomized study. *Ann. Surg. Treat. Res.* **2017**, *93*, 322–330. [CrossRef]
57. Coda, A.; Lamberti, R.; Martorana, S. Classification of prosthetics used in hernia repair based on weight and biomaterial. *Hernia* **2012**, *16*, 9–20. [CrossRef] [PubMed]
58. Klinge, U.; Klosterhalfen, B. Modified classifcation of surgical meshes for hernia repair based on the analyses of 1000 ex-planted meshes. *Hernia* **2012**, *16*, 251–258. [CrossRef]
59. Moreno-Egea, A.; Campillo-Soto, Á.; Morales-Cuenca, G. Which should be the gold standard laparoscopic technique for handling Spigelian hernias? *Surg. Endosc.* **2014**, *29*, 856–862. [CrossRef]

60. Endo, M.; Feola, A.; Sindhwani, N.; Manodoro, S.; Vlacil, J.; Engels, A.C.; Deprest, J.A. Mesh contraction: In Vivo docu-mentation of changes in apparent surface area utilizing meshes visible on magnetic resonance imaging in the rabbit ab-dominal wall model. *Int. Urogynecology J.* **2014**, *25*, 737–743. [CrossRef]
61. Kristin, C.; Turza, M.D.; Charles, E.; Butler, M.D. Adhesions and Meshes: Synthetic versus Bioprosthetic, Adhesions and Meshes. *Plast. Reconstr. Surg.* **2012**, *130*, 5S-2.
62. Shrivastava, A.; Gupta, A.; Shrivastava, J.; Gupta, A. Erosion of small intestine with necrotising fasciitis of over lying abdominal wall after expanded poly-tetrafluoroethylene mesh implantation: A rare complication after laparoscopic inci-sional hernia repair. *J. Minimal Access Surg.* **2013**, *9*, 138. [CrossRef] [PubMed]
63. Draczynski, Z.; Chmielewska, M.; Bogun, M.; Sujka, W. The Analysis of Structure and Physicochemical Properties of Yarns Used for Manufacturing Hernia Meshes. *Autex Res. J.* **2019**, *19*, 119–126. [CrossRef]
64. Draczyński, Z.; Gajek, M.; Mikołajczyk, Z.; Modrzejewska, Z.; Sujka, W. Analysis of Open-work Knitwear Used for Hernia Mesh Manufacturing. *FIBRES Text. East. Eur.* **2018**, *26*, 77–86. [CrossRef]
65. Robinson, T.N.; Clarke, J.H.; Schoen, J.A.; Walsh, M.D. Major mesh-related complications following hernia repair. *Surg. Endosc.* **2005**, *19*, 1556–1560. [CrossRef]
66. Garcia-Ureña, M.A.; Ruiz, V.V.; Godoy, A.D.; Perea, J.M.B.; Gómez, L.M.M.; Hernández, F.J.C.; García, M.; Ángel, V. Differences in polypropylene shrinkage depending on mesh position in an experimental study. *Am. J. Surg.* **2007**, *193*, 538–542. [CrossRef] [PubMed]
67. Cobb, W.S.; Kercher, K.W.; Heniford, B.T. The Argument for Lightweight Polypropylene Mesh in Hernia Repair. *Surg. Innov.* **2005**, *12*, 63–69. [CrossRef]
68. Wahba, M. Evaluation of lightweight polypropylene mesh in Stoppa preperitoneal repair of bilateral inguinal hernias. *J. Am. Sci.* **2014**, *10*, 116–124.
69. Todros, S.; Pavan, P.G.; Natali, A.N. Synthetic surgical meshes used in abdominal wall surgery: Part I-materials and structural conformation. *J. Biomed. Mater. Res. Part B Appl. Biomater.* **2015**, *105*, 689–699. [CrossRef] [PubMed]
70. Kumar, A.S.; Fitzgerald, J.F. Biologic versus Synthetic Mesh Reinforcement: What are the Pros and Cons? *Clin. Colon Rectal Surg.* **2014**, *27*, 140–148. [CrossRef] [PubMed]
71. Baylón, K.; Rodríguez-Camarillo, P.; Elías-Zúñiga, A.; Díaz-Elizondo, J.A.; Gilkerson, R.; Lozano, K. Past, Present and Future of Surgical Meshes: A Review. *Membr. Basel* **2017**, *7*, 47. [CrossRef] [PubMed]
72. Sezer, U.A.; Sanko, V.; Gulmez, M.; Aru, B.; Sayman, E.; Aktekin, A.; Aker, F.V.; Demirel, G.Y.; Sezer, S. Polypropylene composite hernia mesh with anti-adhesion layer composed of polycaprolactone and oxidized regenerated cellulose. *Mater. Sci. Eng. C* **2019**, *99*, 1141–1152. [CrossRef] [PubMed]
73. Junge, K.; Rosch, R.; Klinge, U.; Saklak, M.; Klosterhalfen, B.; Peiper, C.; Schumpelick, V. Titanium coating of a polypro-pylene mesh for hernia repair: Effect on biocompatibility. *Hernia* **2005**, *9*, 115–119. [CrossRef] [PubMed]
74. Scheidbach, H.; Tannapfel, A.; Schmidt, U.; Lippert, H.; Köckerling, F. Influence of Titanium Coating on the Biocompati-bility of a Heavyweight PolypropyleneMesh. *Eur. Surg. Res.* **2004**, *36*, 313–317. [CrossRef]
75. Niekraszewicz, A.; Kucharska, M.; Wawro, D.; Struszczyk, M.H.; Kopias, K.; Rogaczewska, A. Development of aManu-facturingMethod for SurgicalMeshesModified by Chitosan. *Fibres Text. East. Eur.* **2007**, *15*, 105–109.
76. Norton, J.A.; Barie, P.S.; Bollinger, R.R.; Chang, A.E.; Lowry, S.F.; Mulvihill, S.J.; Pass, H.I.; Thompson, R.W. *Surgery*, 2nd ed.; Springer: New York, NY, USA, 2008.
77. Yelimlies, B.; Alponat, A.; Cubukçu, A.; Kuru, M.; Oz, S.; Erçin, C.; Gönüllü, N. Carboxymethylcellulose coated on vis-ceral face of polypropylene mesh prevents adhesion without impairing wound healing in incisional hernia model in rats. *Hernia* **2003**, *7*, 130–133. [CrossRef]
78. Lamber, B.; Grossi, J.V.; Manna, B.B.; Montes, J.H.; Bigolin, A.V.; Cavazzola, L.T. May polyester with collagen coating mesh decrease the rate of intraperitoneal adhesions in incisional hernia repair? *Arq. Bras. Cir. Dig.* **2013**, *26*, 13–17. [CrossRef]
79. Richard HKoehler, M.D.; Dennis Begos, M.D.; Dieter Berger, M.D.; Steve Carey, M.D.; Karl LeBlanc, M.D.; Adrian Park, M.D.; Bruce Ramshaw, M.D.; Roy Smoot, M.D.; Guy Voeller, M.D. Minimal adhesion to ePTFE Mesh After Laparoscopic Ven-tral Incisional Hernia Repair: Reoperative Findings in 65 Cases. *JSLS* **2003**, *7*, 335–340.
80. López-Borao, J.; Madrazo-González, Z.; Kreisler, E.; Biondo, S. Prevention of parastomal hernia after abdominoperineal exci-sion with a prophylactic three-dimensional funnel mesh. *Colorectal Dis.* **2019**, *21*, 1326–1334. [CrossRef]
81. Verbo, A.; Pafundi, P.; Manno, A.; Baccaro, R.; Veneziani, A.; Colli, R.; Coco, C. Polyvinylidene Fluoride Mesh (PVDF, DynaMesh®-IPOM) in The Laparoscopic Treatment of Incisional Hernia: A Prospective Comparative Trial versus Gore®ePTFE DUALMESH®Plus. *Surg. Technol. Int.* **2016**, *28*, 147–151.
82. Zhou, Z.; Bilkhu, A.; Anwar, S. The use of a composite synthetic mesh in the vicinity of bowel–For re-pair and prophylaxis of parastomal hernias. Does it increase the risk of short term infective complications? *Int. J. Surg.* **2017**, *45*, 67–71. [CrossRef] [PubMed]
83. Smart, N.J.; Bryan, N.; Hunt,, J.A.; Daniels, I.R. Porcine dermis implants in soft-tissue reconstruction: Current status. *Biol. Targets Ther.* **2014**, *8*, 83–90. [CrossRef]
84. Wang See, C.; Kim, T.; Zhu, D. Hernia Mesh and Hernia Repair: A Review. *Eng. Regen.* **2020**, *1*, 19–33.
85. Holmdahl, V.; Stark, B.; Clay, L.; Gunnarsson, U.; Strigard, K. One-year out-come after repair of giant incisional hernia using synthetic mesh or full- thickness skin graft: A randomised controlled trial. *Hernia* **2019**, *23*, 355–361. [CrossRef]

86. Alicuben, E.T.; Demeester, S.R. Onlay ventral hernia repairs using porcine non-cross-linked dermal biologic mesh. *Hernia* **2013**, *18*, 705–712. [CrossRef] [PubMed]
87. Cavallo, J.A.; Greco, S.C.; Liu, J.; Frisella, M.M.; Deeken, C.R.; Matthews, B.D. Remodeling characteristics and biomechanical properties of a crosslinked versus a non-crosslinked porcine dermis scaffolds in a porcine model of ventral hernia repair. *Hernia* **2013**, *19*, 207–218. [CrossRef]
88. Mulier, K.E.; Nguyen, A.H.; Delaney, J.P.; Marquez, S. Comparison of Permacol™ and Strattice™ for the repair of abdominal wall defects. *Hernia* **2011**, *15*, 315–319. [CrossRef] [PubMed]
89. Smart, N.J.; Marshall, M.; Daniels, I.R. Biological meshes: A review of their use in abdominal wall hernia repairs. *Surgery* **2012**, *10*, 159–171. [CrossRef]
90. Antoniou, S.A.; Agresta, F.; Alamino, J.M.G.; Berger, D.; Berrevoet, F.; Brandsma, H.; Bury, K.; Conze, J.; Cuccu-rullo, D.; Dietz, U.A.; et al. European Hernia Society guidelines on prevention and treatment of parastomal hernias. *Hernia* **2018**, *22*, 183–198. [CrossRef] [PubMed]
91. Cross, A.J.; Buchwald, P.L.; Frizelle, F.A.; Eglinton, T.W. Meta-analysis of prophylactic mesh to prevent parastomal hernia. *BJS* **2017**, *104*, 179–186. [CrossRef]
92. Wang, S.; Wang, W.; Zhu, B.; Song, G.; Jiang, C. Efficacy of Prophylactic Mesh in End-Colostomy Construction: A Systematic Review and Meta-analysis of Randomized Controlled Trials. *World J. Surg.* **2016**, *40*, 2528–2536. [CrossRef]
93. Pérez-Köhler, B.; Linardi, F.; Pascual, G.; Bellón, J.M.; Eglin, D.; Guillaume, O. Efficacy of antimicrobial agents delivered to hernia meshes using an adaptable thermo-responsive hyaluronic acid-based coating. *Hernia* **2020**, *24*, 1201–1210. [CrossRef]
94. Qamar, N.; Abbas, N.; Irfan, M.; Hussain, A.; Arshad, M.S.; Latif, S.; Mehmood, F.; Ghori, M.U. Personalized 3D printed ciprofloxacin impregnated meshes for the management of hernia. *J. Drug Deliv. Sci. Technol.* **2019**, *53*, 101164. [CrossRef]
95. Houshyar, S.; Sarker, A.; Jadhav, A.; Kumar, G.S.; Bhattacharyya, A.; Nayak, R.; Shanks, R.A.; Saha, T.; Rifai, A.; Padhye, R.; et al. Polypropylene-nanodiamond composite for hernia mesh. *Mater. Sci. Eng. C* **2020**, *111*, 110780. [CrossRef] [PubMed]

materials

MDPI

Correction

Correction: Turlakiewicz et al. The Role of Mesh Implants in Surgical Treatment of Parastomal Hernia. *Materials* 2021, *14*, 1062

Karolina Turlakiewicz [1,2,*], Michał Puchalski [1], Izabella Krucińska [1] and Witold Sujka [2]

1 Institute of Material Science of Textiles and Polymer Composites, Lodz University of Technology, Żeromskiego 116, 90-924 Lodz, Poland; michal.puchalski@p.lodz.pl (M.P.); izabella.krucinska@p.lodz.pl (I.K.)
2 Tricomed S.A., Świętojańska 5/9, 93-493 Lodz, Poland; witold.sujka@tzmo-global.com
* Correspondence: karolina.turlakiewicz@dokt.p.lodz.pl

We have recently been made aware by Dr. Holmdahl's (Umeå University—Department of Surgical and Perioperative Sciences) and the MDPI Editorial offices of some errors and omissions in Section 4.2.3. Biological of our recent paper [1]. The second paragraph of said Section 4.2.3. Biological currently reads as follows:

(Compared to synthetic meshes, biological meshes are more biocompatible and elicit a lower inflammatory response in the body but are associated with a greater number of hernia recurrences due to their lower mechanical strength compared to synthetic meshes. In a clinical trial comparing PP and biological meshes, 12% of patients experienced a recurrence of hernia after implantation of a biological mesh, but no recurrence was observed with the synthetic mesh [84]. Commercially available biological mesh implants are outlined in Table 6.)

To set straight the scientific record, we would like to make the following corrections:

(Compared to synthetic meshes, biological meshes are more biocompatible and elicit a lower inflammatory response in the body but are associated with a greater number of hernia recurrences due to their lower mechanical strength compared to synthetic meshes [84]. Although, the research carried out by Holmdahl et. al. showed comparable recurrence rates between the group of patients who received a full-thickness skin graft (8.3%) and a synthetic mesh (7.1%) [85]. Commercially available biological mesh implants are outlined in Table 6.)

Adding a new article resulted in the renumbering of the bibliography in the manuscript. Below amendments were made:

Citation: Turlakiewicz, K.; Puchalski, M.; Krucińska, I.; Sujka, W. Correction: Turlakiewicz et al. The Role of Mesh Implants in Surgical Treatment of Parastomal Hernia. *Materials* 2021, *14*, 1062. *Materials* **2021**, *14*, 3511. https://doi.org/10.3390/ma14133511

Received: 8 April 2021
Accepted: 18 June 2021
Published: 24 June 2021

Publisher's Note: MDPI stays neutral with regard to jurisdictional claims in published maps and institutional affiliations.

Table 6. Classification of commercially available biological mesh implants [86–89].

Product	Manufacturer	Material	Cross-Linking	Resistance (MPa)
CollaMend	Davol	Animal cell-free skin matrix	Yes	11
Permacol	Covidien	Animal cell-free skin matrix	Yes	39
Strattice	LifeCell	Animal cell-free skin matrix	No	18
XenMatrix	Davol	Animal cell-free skin matrix	No	14

5. Prophylactic Implantation of a Mesh Device

According to the guidelines of the European Hernia Society, the prevention of parastomal hernias in patients undergoing end colostomy surgery with prophylactic mesh implantation was satisfactory [90].

The prophylactic use of a mesh implant in permanent stoma surgery reduces the risk of a parastomal hernia by 75%. Moreover, complications occur only in individual cases, so it can be concluded that mesh implantation in this type of surgery could be routinely applied [91].

An analysis conducted by Shuanhu Wang et al. aimed at assessing the effectiveness of prophylactic mesh implantation during end colostomy. The results showed that in the case of sigmoid terminal colostomy, prophylactic mesh placement reduced the incidence of parastomal hernias and associated reoperations. There were no significant differences in stoma-related complications. Moreover, the surgical techniques of sublay and IPOM are considered to be safe and feasible, reducing the likelihood of a parastomal hernia [92].

6. Current Trends

Electrospinning and 3D printing are examples of manufacturing techniques used for the fabrication of drug-loaded devices.

The encapsulation of antimicrobial agents or drugs is one of the possible approaches that could be utilized to produce meshes with antibacterial properties. Pérez-Köhler et al. developed a new coating material known as hyaluronic acid-poly(N-isopropylacrylamide) (HApN), which forms a hydrogel that can be used as a coating for meshes only when it reaches body temperature. The authors selected two different coating formulations—one based on antibiotics (gentamicin + rifampicin) and one based on an antiseptic (chlorhexidine). The results of this study showed that HApN, when loaded with drugs, inhibited the in vitro the growth of several Gram-positive and Gram-negative bacteria [93].

The next study carried out by Nadia Qamar et al. explored the application of the fused deposition modeling in the fabrication of personalized hernial meshes with and without loading of a pharmaceutical agent (ciprofloxacin HCl). All the printed meshes (PP and polyvinyl alcohol (PVA)) showed good mechanical properties. Meshes made of PVA demonstrated a faster release of the loaded drug in comparison to the PP mesh. Moreover, in vivo testing revealed no signs of implant rejection along with a reduction in adhesion to the visceral side and faster wound healing [94].

Another solution to improve the implant properties could involve the use of metallic or diamond nanoparticles. A polypropylene–nano-diamond composite hernia mesh exhibited a significant reduction in protein absorption consistent with lower inflammatory responses; furthermore, no cytotoxicity was observed [95].

The implementation of these novel materials needs further clinical trials to determine the superiority of such materials compared to those available on the market.

Data Availability Statement: The data presented in this study are openly available at [doi:10.1007/s00268-015-3187-1], [39]; at [doi:10.1308/003588410X12664192076296], [41]; at [doi.org/10.1007/s10029-013-1054-2], [86]; at [doi:10.1007/s10029-013-1070-2], [87]; at [doi:10.1007/s10029-010-0777-6], [88]; at [doi:10.1016/j.surge.2012.02.006], [89].

References

84. Wang See, C.; Kim, T.; Zhu, D. Hernia Mesh and Hernia Repair: A Review. *Eng. Regen.* **2020**, *1*, 19–33.
85. Holmdahl, V.; Stark, B.; Clay, L.; Gunnarsson, U.; Strigard, K. One-year out-come after repair of giant incisional hernia using synthetic mesh or full- thickness skin graft: A randomised controlled trial. *Hernia* **2019**, *23*, 355–361.
86. Alicuben, E.T.; Demeester, S.R. Onlay ventral hernia repairs using porcine non-cross-linked dermal biologic mesh. *Hernia* **2013**, *18*, 705–712.
87. Cavallo, J.A.; Greco, S.C.; Liu, J.; Frisella, M.M.; Deeken, C.R.; Matthews, B.D. Remodeling characteristics and biomechanical properties of a crosslinked versus a non-crosslinked porcine dermis scaffolds in a porcine model of ventral hernia repair. *Hernia* **2013**, *19*, 207–218.
88. Mulier, K.E.; Nguyen, A.H.; Delaney, J.P.; Marquez, S. Comparison of Permacol™ and Strattice™ for the repair of abdominal wall defects. *Hernia* **2011**, *15*, 315–319.
89. Smart, N.J.; Marshall, M.; Daniels, I.R. Biological meshes: A review of their use in abdominal wall hernia repairs. *Surgery* **2012**, *10*, 159–171.

90. Antoniou, S.A.; Agresta, F.; Alamino, J.M.G.; Berger, D.; Berrevoet, F.; Brandsma, H.; Bury, K.; Conze, J.; Cuccu-rullo, D.; Dietz, U.A.; et al. European Hernia Society guidelines on prevention and treatment of parastomal hernias. *Hernia* **2018**, *22*, 183–198.
91. Cross, A.J.; Buchwald, P.L.; Frizelle, F.A.; Eglinton, T.W. Meta-analysis of prophylactic mesh to prevent parastomal hernia. *BJS* **2017**, *104*, 179–186.
92. Wang, S.; Wang, W.; Zhu, B.; Song, G.; Jiang, C. Efficacy of Prophylactic Mesh in End Colostomy Construction: A Systematic Review and Meta-analysis of Randomized Controlled Trials. *World J. Surg.* **2016**, *40*, 2528–2536.
93. Pérez-Köhler, B.; Linardi, F.; Pascual, G.; Bellón, J.M.; Eglin, D.; Guillaume, O. Efficacy of antimicrobial agents delivered to hernia meshes using an adaptable thermo-responsive hyaluronic acid-based coating. *Hernia* **2020**, *24*, 1201–1210.
94. Qamar, N.; Abbas, N.; Irfan, M.; Hussain, A.; Arshad, M.S.; Latif, S.; Mehmood, F.; Ghori, M.U. Personalized 3D printed ciprofloxacin impregnated meshes for the management of hernia. *J. Drug Deliv. Sci. Technol.* **2019**, *53*, 101164.
95. Houshyar, S.; Sarker, A.; Jadhav, A.; Kumar, G.S.; Bhattacharyya, A.; Nayak, R.; Shanks, R.A.; Saha, T.; Rifai, A.; Padhye, R.; et al. Polypropylene-nanodiamond composite for hernia mesh. *Mater. Sci. Eng. C* **2020**, *111*, 110780.

These changes have no material impact on the conclusions of our paper. We apologize for any inconvenience caused to the readers.

Reference

1. Turlakiewicz, K.; Puchalski, M.; Krucińska, I.; Sujka, W. The Role of Mesh Implants in Surgical Treatment of Parastomal Hernia. *Materials* **2021**, *14*, 1062. [CrossRef]

MDPI
St. Alban-Anlage 66
4052 Basel
Switzerland
Tel. +41 61 683 77 34
Fax +41 61 302 89 18
www.mdpi.com

Materials Editorial Office
E-mail: materials@mdpi.com
www.mdpi.com/journal/materials

www.ingramcontent.com/pod-product-compliance
Lightning Source LLC
Chambersburg PA
CBHW061555210926
51850CB00009B/2110
* 9 7 8 3 0 3 6 5 8 2 3 7 5 *